INTRODUCTION TO
MATERIALS &
PROCESSES

Delmar Publishers' Online Services

To access Delmar on the World Wide Web, point your browser to:

http:/ /www.delmar.com/delmar.html

To access through Gopher: gopher: / / gopher.delmar.com

(Delmar Online is part of "thomson.com", and Internet site with information on more than 30 publishers of International Thomson Publishing organization.)

For more inormation on our products and services:

email: info@delmar.com

Or call 800-347-7707

INTRODUCTION TO
MATERIALS &
PROCESSES

DR. JOHN R. WRIGHT

Professor and Dean
School of Technology at Central Connecticut State University
New Britain, Connecticut

and

DR. LARRY D. HELSEL

Professor and Chair
School of Technology at Eastern Illinois University
Charleston, Illinois

Delmar Publishers

I(T)P An International Thomson Publishing Company

Albany • Bonn • Boston • Cincinnati • Detroit • London • Madrid
Melbourne • Mexico City • New York • Pacific Grove • Paris
San Francisco • Singapore • Tokyo • Toronto • Washington

NOTICE TO THE READER

Delmar Staff

Publisher: Robert D. Lynch
Acquisition Editor: John Anderson
Production Manager: Larry Main
Design Coordinator: Nicole Reamer
Production Assistant: Lizabeth A. Lajeunesse

COPYRIGHT © 1996
By Delmar Publishers
a division of International Thomson Publishing Inc.

The ITP logo is a trademark under license.

Printed in the United States of America

For more information, contact:

Delmar Publishers
3 Columbia Circle, Box 15015
Albany, New York 12212-5015

International Thomson Publishing Europe
Berkshire House 168-173
High Holborn
London, WC1V 7AA
England

Thomas Nelson Australia
102 Dodds Street
South Melbourne, 3205
Victoria, Australia

Nelson Canada
1120 Birchmont Road
Scarborough, Ontario
Canada, M1K 5G4

International Thomson Editores
Campos Eliseos 385, Piso 7
Col Polanco
11560 Mexico D F Mexico

International Thomson Publishing GmbH
Konigswinterer Strasse 418
53227 Bonn
Germany

International Thomson Publishing Asia
221 Henderson Road
#05-10 Henderson Building
Singapore 0315

International Thomson Publishing—Japan
Hirakawacho Kyowa Building, 3F
2-2-1 Hirakawacho
Chiyoda-ku, Tokyo 102
Japan

1 2 3 4 5 6 7 8 9 10 XXX 01 00 99 98 97 96

Library of Congress Cataloging-in-Publication Data

Helsel, Larry David, 1949–
    Introduction to materials and processes / Larry D. Helsel, John R.
Wright.
        p.   cm.
    Includes index.
    ISBN 0–8273–5020–1
    1. Manufacturing processes. I. Wright, John (John R.)
    II. Title.
TS183H48 1996
670—dc20   94–3849
                                                     CIP

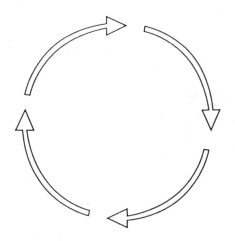

Contents

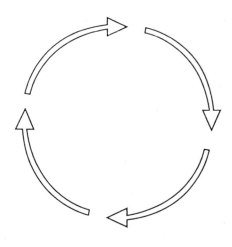

Preface

The understanding of industrial materials and how they are processed is basic to the manufacturing industry. Humankind has been reliant upon materials throughout millennia, and basic survival depended upon the ability to cut and shape basic tools and implements. Some of the early concepts of how to shape materials are still used, and many of the original materials remain the staple of our manufacturing industry.

Yet much has changed in the materials world, and new configurations of synthetic and composite materials would overwhelm the early artisan beyond his or her wildest belief. Also as exciting are the computerized processes that control the machines that bend, twist, pull, compress, and join the space-age family of materials that you and I take for granted every day.

This textbook was designed to introduce engineering and technology students to the basic concepts of materials and how they are processed. Beginning with a review of basic materials and simple machines, students are carefully led through a logical scheme to show how materials are found, developed, and processed for manufactured goods. Uniquely, this text begins with raw materials and ends up with the recycling of used products via a systems approach of studying materials. This natural ecological cycle allows students to understand the positive and negative aspects of materials processing from the design engineer to the daily consumer.

This book is also more comprehensive than the typical metallics-oriented materials textbook on the market today because it includes a major industrial materials section on the structure, properties, and characteristics of metals, polymers, ceramics, and composites. Also new to this type of textbook is the concept of identifying common machine/tool techniques and then applying them to multiple materials so that transformation processes are conceptualized into operations rather than individual pieces of equipment. The result of this approach is a more analytical understanding of industrial materials and their inherent processing commonalities and differences.

STUDENT FOCUS

This textbook is designed for engineering and technology students beginning their academic experience in manufacturing and construction technologies. Typically, materials textbooks require a comprehensive knowledge of calculus, chemistry, and physics, which restricts their use to upper-level college students. Attempts to "water down" advanced materials textbooks result in confusion and frustration by both students and instructors.

The authors, both experienced materials and processes instructors, have designed this book to effectively introduce students who have an algebra background to materials and processes at the freshman level while they are beginning to study advanced mathematics and science so that they can enjoy an early experience and be better prepared for an upper-level advanced materials analysis course. This textbook can be used with or without a laboratory component, making it an ideal introductory text for engineering or technology majors at both two-year colleges and four-year universities.

ORGANIZATION OF THE BOOK

Introduction to Materials and Processes is divided into five sections: Introduction, Industrial Materials, Designing and Developing Manufactured Goods, Material-Transformation Processes, and Materials Recovery and the Ecological/Production System. Based upon a natural ecological/production cycle, materials and processes are studied from their development as earth resources, through the production–goods producing phase, and on to their displacement as industrial waste and/or recycling.

Section 1 is introductory in nature and provides students with a historical perspective of past practices, a basic knowledge of where primary materials come from, and an introduction to the synthetics and space-age materials of the future. Section 2 is devoted to the structure, properties, and characteristics of industrial materials. Classified under metals, polymers (plastics and woods), ceramics, and composites, they are analyzed for their ability to perform as manufacturing and construction materials. The design of the products is covered in Section 3, including product-design techniques; measurement and layout procedures; and product research, development, and analysis. In Section 4, material-transformation procedures are discussed with emphasis on preprocessing materials, molding processes, casting processes, machining theory and principles, machining processes, machine cotrol, fabrication processes, material-fastening processes, assembly, and postprocessing. Section 5 concentrates on industrial waste and pollution, recycling, and materials of the future. Special emphasis is placed on the development of new products that are designed for recycling and new materials that create new opportunities in the world of manufacturing.

SPECIAL FEATURES

This textbook has several distinctive features that make it student friendly and easy to use. As an introductory text, it provides the learner with a comprehensive understanding of multiple industrial materials and the processing techonology used to convert basic resources into usable products and goods. In particular, you will find

> Easy-to-read text with illustrations and charts that explain complex concepts and technological processes
>
> Organization using an ecological/production scheme so that students can understand the total production cycle
>
> Questions and student activities at the end of each chapter
>
> A teacher's guide available for course organization
>
> Conceptual-based information to provide better understanding
>
> Algebra-based problems and formulas to facilitate use by first- and second-year college students
>
> A comprehensive approach to materials beyond metallics to include polymers, ceramics, and composites
>
> Modern machine control technology, including CAD/CAM/CIM processing techniques
>
> A design section that focuses on quality, reliability, producibility, and ecology
>
> A technical and scientific approach for explaining the properties, characteristics, and structures of materials
>
> Complete summaries included at the end of each chapter
>
> A glossary of terms with definitions
>
> A comprehensive index for easy access to information in the text

ACKNOWLEDGMENTS

The authors and publisher gratefully acknowledge the assistance of many talented reviewers who provided valuable criticism during several review stages: Brent Campbell; William J. McNelley, California State University at Chico; John A. Zaner, University of Southern Maine; Douglas Pickle, Amarillo College; Stanley B. Hopkins, New England Tech; John R. McCravy, Jr., Piedmont Technical College; W. Richard Polanin; Albert Zachwieja, Tritorn College; Michael R. Kozak, University of Northern Texas; John R. Gilsdorf; George M. Morris, Wentworth Institute of Technology; James R. Drake, Cuyahoga Community College; and Robert Simoneau, Keene State College.

The authors would also like to thank the staff at Delmar for the invaluable assistance they have provided to bring this textbook to publication, in particular, Mike McDermott for recognizing the need for a basic text in this field, Vern Anthony for picking up the project initially, and John Beck and his crew (Tom Jarmiolowski, Don Keenan, Jay Morgans, Louise Palmieri, Seth Reichgott, and David Umla) for taking the book into final editing and production.

We would also like to thank our families for their patience and support in what has been a long project for everyone.

ABOUT THE AUTHORS

Drs. Wright and Helsel were teaching colleagues at Eastern Illinois University where they were responsible for the production technology area. After several years of teaching materials and processes, they decided to collaborate on a basic materials and processes text that would be algebra-based and more comprehensive for first- and second-level college students. *Introduction to Materials and Proccesses* is a result of that effort.

Dr. John R. Wright is professor and dean of the School of Technology at Central Connecticut State University in New Britain, Connecticut. Dr. Wright has several years of industrial experience in the metals fabrication and welding area and twenty-five years of teaching experience in materials processing. He is the author of more than thirty professional publications, including journal articles, monographs, and yearbooks. He is an active member of the Society of Manufacturing Engineers, the International Technology Education Association, the National Association of Industrial Technology (NAIT), the Connecticut Manufacturing Deployment Committee, and CEO of the Institute for Industrial and Engineering Technology.

He is a graduate of Fitchburg State College, Rhode Island College, and West Virginia University. He has been recognized by ITEA, NAIT, and the Connecticut General Assembly as a leading advocate for quality education in technology and economic development through advanced manufacturing technologies to improve quality and productivity for Connecticut's small manufacturers.

Dr. Larry Helsel is professor and chair of the School of Technology at Eastern Illinois University in Charleston, Illinois. Dr. Helsel has taught manufacturing processes, materials technology, material science, and industrial systems simulations at the university level for more than fourteen years. He is a Certified Senior Industrial Technologist and a member of the National Association of Industrial Technology, Society of Manufacturing Engineers, and the American Society for Quality Control.

Dr. Helsel is a graduate of California University of Pennsylvania and The Pennsylvania State University. He has served as President of the University Division of the National Association of Industrial Technology (NAIT) and Chairman of the Executive Board of NAIT. He currently serves as Chairman of the NAIT Board of Certification.

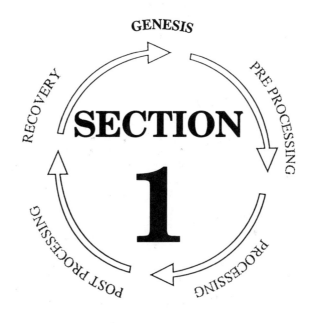

INTRODUCTION

CHAPTER 1

The World of Materials

INTRODUCTION TO BASIC MATERIALS PROCESSING

Look around you and see how many things we use each day that are made from materials. Some items are made from a single material; some are made from multiple materials. Our world is full of many different types of materials in various stages of refinement. Some materials, such as sand, coal, or wood, can be found and used in their natural forms. Others, such as steel, plastic, or glass, are made up of several natural materials that require modification (changing) before they can be used. This modification procedure is called material processing.

Material-processing techniques are used at two levels of refinement: primary and secondary. Even materials that can be found and used in their natural form usually require some type of **primary processing**. For example, sand is washed and graded; trees are cut into boards and planks; and coal is washed, cleaned, and sized before it is shipped to the customer. Materials that are combinations of natural materials, such as iron or steel, may require several different processes during the primary stage in order to get them ready for use. Other materials, such as plastics, may be by-products (additional uses) of natural materials that are intended for different uses, such as oil for energy. **Secondary processing** is the procedure of developing primary materials into goods and items used by the consumer. Very few primary processed materials can be used directly. Converting boards and planks into a house or stamping an automobile fender out of a sheet of steel is an example of secondary processing.

IN THE PAST

Many of the materials we use today have been developed during the past one hundred years. Our early ancestors did not know about materials such as plastic, silicon, or glass. They lived in a simple world of natural materials.

Early humans used materials made of stone, wood, or bone to help survive the hardships of nature. The first tool used by humans was a simple rock. The first attempt at modifying a primary material was in the form of shaping an edge on the stone. By shaping the stone, humans were relieved of the time-consuming task of searching for stones with sharp edges. They also learned which type of stone (hard or soft) to use as a spear point for hunting animals.

The next important development was to combine wood and stone into a hammer or ax. By extending the stone on a lever or handle, early humans were able to develop more hitting power than hand-held rocks permitted. They also pioneered the first fastening technique by wedging a stone between a split stick and tying it together with a vine material.

Material-processing techniques were very important for the survival of early humans. It has been theorized that the discovery of metal occurred as a result of meteor showers that deposited small amounts of metal and stone. This theory suggests early humans simply found the metals ready for use. Another theory suggests that metal was found in the fire pits of early humans as a result of the right ingredients being in the pits before the fire was started. Regardless of which theory is correct, it was not long before the use of metal (primarily copper and bronze) became very popular.

As humankind progressed through the ages, so did the knowledge and use of materials. Humans learned how to decorate and use materials (especially copper and bronze) for pleasure, and a new emphasis was placed on material goods and the artisan (craftsperson) who was able to produce them (Figure 1–1). Kings and queens commissioned (hired for pay) the artisans to craft things of beauty in honor of the gods and, of course, their own great stature. Beautiful hand-processed items were made from wood, bronze, copper, stone, and textiles. Some artisans spent an entire lifetime modifying and shaping materials into things of beauty.

The Romans expanded the knowledge of materials and were known for the use of brick and concrete in their architecture. They were famous for their road-building technique of using multiple layers of alternative natural materials.

During the seventeenth century, the use of materials began to include the development of better tools and simple machines. The most popular machine was the clock. The clockmaker was an artisan able to process materials (wood, copper, and bronze) into accurate time machines. In a sense, the clockmaker was the first mechanical engineer to learn the basic principles of machine tool fabrication. Early clockmakers had one major limitation: Their machine tools, such as the lathe, were made of wood and could cut only wood or soft metals. They needed a strong, easy-to-use material to make into machine tools (Figure 1–2).

That need was met in the early part of the eighteenth century when Abraham Darby III, at the helm of his furnace at Coalbrookdale in England, began to use coal instead of wood as a fuel to make iron. His iron was further refined by Henry Cort in 1784 with the puddling process. These two developments provided an impetus for changing the basic technological material for toolmaking from wood to metal. They also opened the door to the age of industrialization.

Machine tool development was needed to make durable tools, which in turn could manufacture products. Henry Maudslay contributed to the development of machine tools and actually ran a machine tool shop that provided training for many of the great toolmakers of his time. The most significant advancement of the time was the development of boring mills. Their development allowed the refinement of the steam engine, which provided a portable source of power for factories, mines, and transportation.

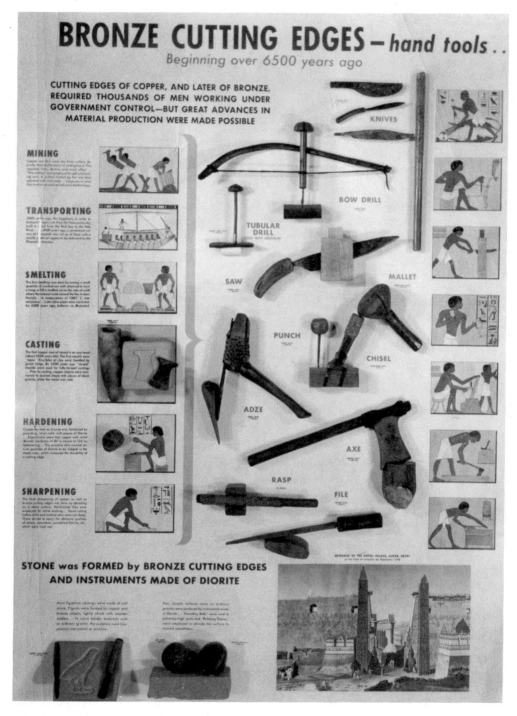

Figure 1–1 Early artisans worked with simple tools to shape soft metals such as copper and bronze *(Courtesy of DoALL Company)*

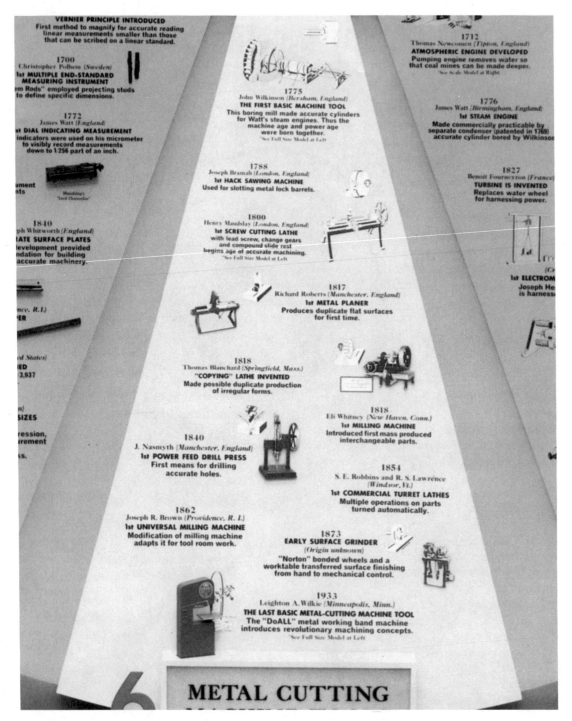

VERNIER PRINCIPLE INTRODUCED
First method to magnify for accurate reading linear measurements smaller than those that can be scribed on a linear standard.

1700
Christopher Polhem (Sweden)
1st MULTIPLE END-STANDARD MEASURING INSTRUMENT
"Rods" employed projecting studs to define specific dimensions.

1772
James Watt (England)
1st DIAL INDICATING MEASUREMENT
indicators were used on his micrometer to visibly record measurements down to 1·256 part of an inch.

1840
ph Whitworth (England)
ATE SURFACE PLATES
evelopment provided ndation for building accurate machinery.

nce, R.I.)
ER

d States)
ED
3,937

n)
SIZES

ression,
rement
s.

1712
Thomas Newcomen (Tipton, England)
ATMOSPHERIC ENGINE DEVELOPED
Pumping engine removes water so that coal mines can be made deeper.
See Scale Model at Right

1775
John Wilkinson (Bersham, England)
THE FIRST BASIC MACHINE TOOL
This boring mill made accurate cylinders for Watt's steam engines. Thus the machine age and power age were born together.
See Full Size Model at Left

1776
James Watt (Birmingham, England)
1st STEAM ENGINE
Made commercially practicable by separate condenser (patented in 1769) accurate cylinder bored by Wilkinson

1788
Joseph Bramah (London, England)
1st HACK SAWING MACHINE
Used for slotting metal lock barrels.

1800
Henry Maudslay (London, England)
1st SCREW CUTTING LATHE
with lead screw, change gears and compound slide rest begins age of accurate machining.
See Full Size Model at Left

1827
Benoit Fourneyron (France)
TURBINE IS INVENTED
Replaces water wheel for harnessing power.

1st ELECTROM
Joseph He is harness

1817
Richard Roberts (Manchester, England)
1st METAL PLANER
Produces duplicate flat surfaces for first time.

1818
Thomas Blanchard (Springfield, Mass.)
"COPYING" LATHE INVENTED
Made possible duplicate production of irregular forms.

1818
Eli Whitney (New Haven, Conn.)
1st MILLING MACHINE
Introduced first mass produced interchangeable parts.

1840
J. Nasmyth (Manchester, England)
1st POWER FEED DRILL PRESS
First means for drilling accurate holes.

1854
S. E. Robbins and R. S. Lawrence
(Windsor, Vt.)
1st COMMERCIAL TURRET LATHES
Multiple operations on parts turned automatically.

1862
Joseph R. Brown (Providence, R. I.)
1st UNIVERSAL MILLING MACHINE
Modification of milling machine adapts it for tool room work.

1873
EARLY SURFACE GRINDER
(Origin unknown)
"Norton" bonded wheels and a worktable transferred surface finishing from hand to mechanical control.

1933
Leighton A. Wilkie (Minneapolis, Minn.)
THE LAST BASIC METAL-CUTTING MACHINE TOOL
The "DoALL" metal working band machine introduces revolutionary machining concepts.
See Full Size Model at Left

METAL CUTTING

Figure 1–2 Early tools of wood were replaced by metal for strength and accuracy *(Courtesy of DoALL Company)*

Machine tools, power sources, and people were brought together by Richard Arkwright in a factory system designed to manufacture textiles and replace the homespun cottage-type industry. People began to move to the industrial centers in order to find work. As a result of this development, large urban areas began to flourish in England; and the process of industrialization was well on its way.

In America, introduction to industrialization began in Pawtucket, Rhode Island, when an Englishman named Samuel Slater brought the plans of textile technology, which he had memorized, to America. He met Moses Brown, and together they built the first American textile factory (Figure 1–3 A–D). As in England, most products were made at home in cottage-type industries. Shipbuilding was a major industry in New England, and cotton and tobacco were the staple products of the South.

Iron was first made in America at the Sagus Iron Works, which also had a rolling and slitting mill. Iron was in short supply, and nails were so hard to find that people would actually burn down old homes just to get the nails.

In 1775, the American Revolution severed all ties with England as the mother country and forced America to become self-sufficient. Invention and innovation helped make up for a shortage of skilled labor. Eli Whitney developed a system of jigs and fixtures that, when combined with machine tools, could replace many artisan-type skills. Transportation and communication technology were also very poor, causing Whitney to order his musket barrels one year in advance. A newspaper took two weeks to get from New York to Philadelphia, if it got there at all.

A rolling mill for iron was developed at the Delaware Iron Works in 1840, and iron rails were produced in 1862. It was the development of the iron rail that fueled tremendous industrial growth in manufacturing during the second half of the nineteenth century. The concept of manufacturing was also politically supported by Alexander Hamilton and Thomas Jefferson. As President, Jefferson purchased the Louisiana territory. This allowed agricultural expansion of America and increased the need for mass-produced goods. Hamilton also supported the concept of manufacturing and helped America industrialize by setting up a central banking system.

The second half of the nineteenth century was a time of discovering the scientific approach to technology. New materials such as rubber, aluminum, celluloid plastic, tungsten tool steel, and coal tars opened the door to new manufacturing possibilities. It was also a time of great expansion and increased productivity. The Bessemer and open-hearth steel-making processes replaced the slower crucible design and found a ready market for steel in the expanding railroad and shipbuilding industries (Figure 1–4).

The process of research and development became an important element in the discovery of new materials after Thomas Edison organized the first testing laboratory at Menlo Park, New Jersey, in 1878. If the nineteenth century can be viewed as the age of invention and innovation, then the twentieth century could be called the age of the industrial corporation.

The Ford Motor Company refined the idea of mass production for mass consumption. The auto industry, in general, made huge demands for more materials in larger quantities than ever before. Think about how many different types of materials are used to make an automobile. Because of the lack of technology in some areas, the auto industry began to

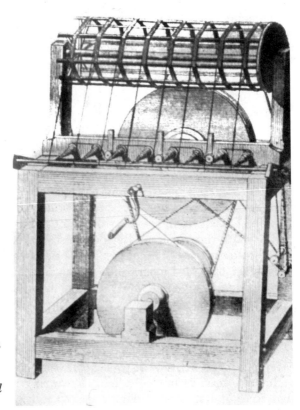

Figure 1–3 A–D   Early industrialization was focused on textiles and the factory system *(From Komacek,* Production Technology, *Copyright 1993 by Delmar Publishers Inc. Used with permission. Also courtesy of Photri, Inc.)*

Figure 1–3A

develop special processes for manufacturing their own materials. The Ford Motor Company devised a method of pouring glass onto a moving table in an endless strip. The rolling, grinding, and polishing were done on the same line.

The rubber and steel industry also began to expand in order to meet the needs of automobile manufacturing, and a number of smaller industries began to organize as part of the massive automobile submanufacturing network. Leo Baekeland developed a new and more usable type of plastic known as Bakelite. Artificial silk and acetate rayon were also by-products developed as a result of advancements made in the chemical industry. In fact, the increased use of petrochemicals (chemicals developed from oil) prior to World War II fostered the development of several new types of plastics, including the well-known polyethylenes and polystyrenes.

Research-and-development efforts were increased after World War II; and new synthetic materials such as nylons, silicones, and urethanes led to the development of an auxiliary-materials industry. Materials such as plasticizers, stabilizers, lubricants, and mold release agents were the by-products of an expanding manufacturing industry.

Figure 1–3B

THE PRESENT

Today, the primary materials of metallics, ceramics, and polymers are supported by a bounty of synthetics. The most common family members of this group are

- Acrylics
- Cellulosics
- Epoxies
- Melamines
- Nylons
- Polyesters
- Polyfluorocarbons
- Polyolefins
- Polystyrenes
- Polyurethanes
- Synthetic Rubbers
- Vinyls

Synthetics are widely used for materials all around us. We could identify many items at home or in the laboratory that are made from synthetic materials. But what about some of the uses with which we may not be familiar?

Figure 1–3C

Off the coast of New Jersey, a polypropylene material (often used to make luggage) is being used in the form of artificial seaweed to help prevent erosion of the sandy bottom. It also provides a refuge for fish and other sea animal life in that area. The housing industry also has some innovative uses of synthetic materials. New houses can now be made fireproof by applying protective layers of foamy silicone paint. We may not need nails or other types of fasteners for home construction in the future because the super-glue age of synthetic adhesives can be used for molecule-to-molecule welds. In fact, new housing may follow the design of the "House of Tomorrow" on display at Disneyland, in Anaheim, California, and be fabricated completely of plastic materials.

Fiberglass-reinforced plastic (FRP) technology provided architect Francis Xavier Dumont with the flexibility to produce unlimited shapes and colors when he designed the Taj Mahal Casino Resort in Atlantic City, New Jersey (Figure 1–5). FRP provides the flexibility and durability to hold up against the ocean's destructive elements.

Figure 1–3D

The very pliable silicone rubbers and other chemically inert polymers are being used for heart valves, eye corneas, and other human parts needing replacement because of damage or disease. Even a breast, muscle, or cartilage can be replaced by a synthetic material that feels and looks like real human tissue. It may even be possible to allow humans to breathe under water using a thin silicone film that allows oxygen to pass through it while resisting water.

The age of synthetic materials has also developed Mylar, a material that—even as thin as $1/_{1,000}$ inch—can resist the impact of an 80-mph baseball. Lexan or Merlon (polycarbonate) is a material that looks like glass but is as tough as steel, and a new rubber-like plastic called RTV 615 is transparent and can resist temperatures four times hotter than the melting point of steel.

The people who work with ceramic materials have also developed some interesting materials technology. The area of fiber optics has been refined and can now be used to re-

Figure 1–4   Steelmaking opened the door for America's industrialization *(Photo courtesy of Bethlehem Steel Corporation and Associated Photographers, Pittsburgh, PA)*

place bulky copper communications cables in metropolitan areas. New developments in silicon carbide and silicon nitride have allowed Ford Motor Company engineers to construct a ceramic turbine engine that can resist temperatures up to 2,500 degrees Fahrenheit without breaking apart. Ceramic tiles are also used on the space shuttle to protect it from high temperatures during reentry.

Even a new light bulb has been developed that replaces the tungsten filament with a piece of ceramic material that produces a brighter light with less energy.

IN THE FUTURE

The world of materials has a very promising future. Exploration for new mineral resources in developing countries, under the ocean floor, and in space could provide a bounty of materials yet unknown to humankind. Continued management of forests, oil, and natural gas

Figure 1–5 Polymers provide architects with exciting new options for structures *(Photo courtesy of Trump Taj Mahal, Atlantic City, NJ)*

with a decrease in using such resources for energy will ensure an abundant supply in the future. Recycling technology will also play an important role for conserving resources in the future and may even provide some answers for new combinations of materials not known today.

By providing known technology today, we are able to visualize some astonishing possibilities for the next twenty-five years: plastic automobiles, foam houses, self-contained homes with special silicon membranes that filter and recycle air and water, fuel-cell advancements, new solar energy collector technology, lightweight rapid transit systems for urban areas, replacement for vital parts of the human body, and huge greenhouses to grow any type of food in any part of the world.

New monomeric inks for printing dry instantly upon contact. Polymerized photography will not require the wet process commonly used by Polaroid but will completely develop picture prints in the camera in ten seconds.

Glass companies have already developed photo-ray sunglasses that adjust their tint as the wearer steps indoors or outdoors. Continued development in this area will find use in new automobile windshields and solar home construction. Because the process of photochemicals eliminates the "grain" on a transparency, it is now possible to copy a 1,245-page Bible on a single transparency less than two inches square. Astronauts and pilots will be able to carry 50,000 pages of navigational charts and technical procedures in a hand-held screening machine no larger than a small portable radio.

Some architects visualize large plastic domes covering entire cities with controlled environments, making it possible to enjoy Florida-like weather in Chicago. For years, scientists have talked about underwater cities. They may be possible now (because of silicone technology) by using the oxygen in the water itself to breathe. New "super" buildings could be constructed with new foam technology, which is light, strong, and weather resistant.

As we enter the twenty-first century, our world will continue to use metals, polymers and ceramics, and combinations of materials (composites) to help humankind deal with our complex and demanding environment. Our supermaterials, known as the "synthetics," will continue to help provide the way to new, innovative, and creative uses in the future that we have yet to imagine.

 ## REVIEW QUESTIONS

1. What are two types of processing used in the manufacturing and construction industries?
2. Explain how early humans used material processing for survival.
3. Who were the first engineers? What did they make?
4. Why was the development of machine tools important for industrialization?
5. What were the cottage industries? How did they differ from Arkwright's factory system?
6. How did the American Revolution change the technical need for America?
7. Explain how Hamilton and Jefferson helped the advancement of manufacturing.
8. What contribution did the Ford Motor Company make?
9. List and describe some of the newer materials that were used after World War II.
10. What are the synthetics? How will they affect the future in construction and manufacturing technology?

SUGGESTED ACTIVITIES

1. Set up a class field trip to a materials-processing industry.
2. Write a short research paper on one of the new materials presented in this chapter.

CHAPTER 2

Genesis of Materials

The earth's resources are finite in size but abundant in variety. Some parts of the world have large supplies of iron, while other regions may have none. Some materials can be found and used in their natural state, while others must be refined or combined before their use is practical. This chapter discusses the types of resources which are used to provide us with the basic materials of metal, wood, plastic, and ceramics.

Most of the materials that we use today are found either in a pure natural state (see Figure 2–1) or combined with other unwanted material which must be removed. In some instances, we purposely combine two or more basic materials to make one usable material. For example, a tree can be used as a structural member for a log home in its natural form. However, a steel structural member of a modern office building is made up of several natural materials and must be processed before use is possible. The process of getting basic materials ready for use is called *primary processing,* and we will explore the different techniques which are used to accomplish this task.

METALS

Metals are metallics, and there are at least eighty chemical elements which are classified as being metallic based. Of the eighty elements, industry processes about twenty with commercial success. Among those used extensively are the various irons, steels, aluminums, and a multitude of alloys. Alloys are combinations of metals put together for a special purpose. The metals family can be easily classified into two groups: ferrous and nonferrous.

Ferrous metals—primarily iron based—make up the huge family of cast irons, steels, and their alloys. The nonferrous family is much more diverse but is common in the sense that they have little or no iron as their base. Such metals as aluminum, copper, bronze, brass, magnesium, tin, lead, gold, and silver are a few of the more common nonferrous metals used by industry today.

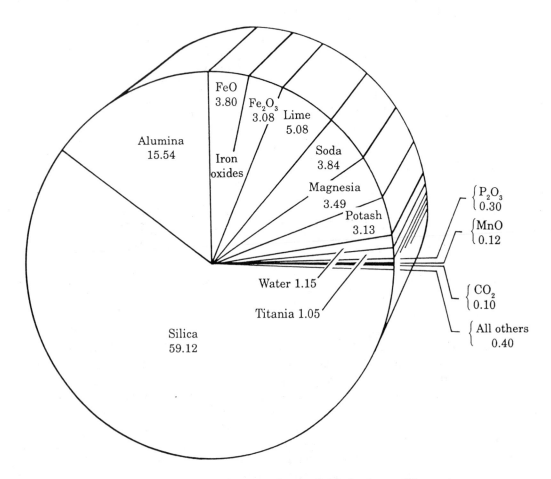

Figure 2–1   Percentage of oxides available in the earth's crust

Iron

Iron-bearing materials known as ore are found both at the surface and deep down in the earth's crust. Most of the iron that is found today is mixed with a variety of other materials. In its natural form, iron ore looks like chunks of black or red gravel. The largest centralized iron ore resource in the United States today is located in the Great Lakes region and is known as the Mesabi Range (Figure 2–2). Today, the Commonwealth of Independent States, Australia, and Brazil are the largest producers of iron ore in the world.

Not all iron ore is alike. Iron ores are classified by their chemical content. The major classifications are oxides, carbonates, silicates, and sulphides. The major iron-producing ores are hematite (Fe_2O_3), magnetite (Fe_3O_4), limonite ($HFeO_2$), and siderite ($FeCO_3$) shown in table form in Figure 2–3. The hematites and magnetites have the highest iron content.

Some ores only need to be washed and screened in order to get them ready for use. Others may need further processing such as sintering before they can be used. **Sintering** is a

Figure 2–2   Open-pit mining techniques produce iron ore from the Mesabi Range in Minnesota *(Photo courtesy of United States Steel)*

process of mixing fine iron ores with powdered coal. The materials are then ignited, air is drawn through the mixture, and the iron ore particles fuse together into a cake. The cake is then water quenched and broken into usable chunks of quality ore. Because of the abundance of low-iron-bearing ore, sintering has become a very popular process of ore beneficiation.

Extracting Processes

The extraction of iron ore is usually accomplished in one of two methods: underground mining or open-pit mining.

Underground mining is accomplished through the use of vertical or inclined shafts (drift shafts). This method is generally used in areas where the raw material (ore) is too deep to be surface mined safely and economically. Deep mining is more costly than surface min-

Class and mineralogical name	Chemical composition of pure mineral	Common designation
Oxide		
Magnetite	Fe_3O_4	Ferrous-ferric oxide
Hematite	Fe_2O_3	Ferric oxide
Ilmenite	$FeTiO_3$	Iron-titanium oxides
Limonite	$\left\{\begin{array}{l} HFeO_2 \\ FeO(OH) \end{array}\right\}$	Hydrous iron oxides
Carbonate		
Siderite	$FeCO_3$	Iron carbonate
Silicate		
Chamosite		
Stilpnomelane	Various	
Greenalite	and sometimes	Iron silicates
Minnesotaite	complex	
Grunerite		
Sulphide		
Pyrite (iron pyrites)	FeS_2	
Marcasite (white iron pyrites)	FeS_2	Iron sulphides
Pyrrhotite (magnetic iron pyrites)	FeS	

Figure 2–3   Iron ores are classified by chemical composition

ing because it requires more workers and specialized equipment to bring the material to the surface. Another problem associated with deep mining is waste. As much as 50% of the raw material is often left behind in the mine to help support the roof. Modern mining technology is reducing this problem every day, and the zero-waste mine room may soon become the state of the art.

Surface mining is often referred to as strip or open-pit mining. This type of ore recovery requires the removal of both rock and soil from the surface to expose the raw material. This process is referred to as removing the overburden. Large power shovels are used to load the ore onto trucks or railroad cars. Because of the size of the huge equipment used in surface mining, the output and recovery of iron ore is much greater than that of deep mining. This output is often double that of underground mining and has the advantage of recovering the cost of iron making. However, the cost of reclaiming the land (see Figure 2–4) due to more stringent regulations has added to the cost of this technique.

Figure 2–4   Reclaimed land restores the natural beauty of open-pit areas for ecological balance and harmony *(Photo courtesy of Colorado Mining Association—Amax, Inc.)*

Transportation

As one might imagine, the process of getting the large amount of raw materials (a finished ton of iron requires 3,350 pounds of iron ore, 570 pounds of limestone, and 1,350 pounds of coke) to the blast furnace for further refinement is unique in itself. Very large ships are used to travel across the Great Lakes to deliver the raw materials needed for the production of iron.

Because iron ore is combined with a number of impurities, it must go through a process of **beneficiation.** Beneficiation consists of (1) grinding the ore to a fine powder; (2) concentrating the iron particles and separating them from the impurities by employing magnetic separators; (3) converting 60–65% of the iron powder into pellet form (this process is known as agglomeration and increases the size of the iron pellets from ½ inch to ¾ inch in diameter); and (4) sintering, a process of baking the iron powder at 2,400 degrees Fahrenheit into pellets, which are easier to handle. Beneficiated iron ore, limestone, and coke are then combined at the blast furnace for the production of iron.

Aluminum

By a wide margin, the most abundant metal in the earth's crust is aluminum. Aluminum is one of the most popular nonferrous metals and can be found in some clays and in larger concentrations in bauxite ore. Although many clays contain about 8% aluminum, the costs for extracting it from clay are too high for commercial success. Most of our aluminum comes in the form of bauxite ore and is mined in the same manner as iron ore. Canada has large deposits of bauxite ($Al_2O_3 3H_2O$). Bauxite ore contains about 45% aluminum. The ore is crushed, dried, and screened to remove impurities. Once the ore is cleaned, it is processed into alumina (Al_2O_3), which is a pure aluminum oxide. A calcination conversion known as the Bayer process is used to separate oxides from the pure aluminum. Details of this process are discussed in Chapter 3.

Copper

Copper is produced from sulfide ores. They are mined in the same manner as iron ores. The copper ore must be refined to remove such impurities as arsenic, antimony, and sulfur. The sulfide copper ore is usually reduced by heat to an impure copper known as matte. Matte is further refined by a converter (air blasted through molten-copper matte), which produces a 99% pure copper. This type of copper is called blister copper and can be used as is or further refined by an electrolytic process. Details of the process are further discussed in Chapter 3.

Lead

Lead is a soft metal produced from a mineral called galena, which is a lead sulfide. Lead is mined in many countries, frequently in association with zinc ore. Australia, the United States, Mexico, and Canada are important producers and miners of lead. It is difficult to separate lead and zinc from each other. A flotation process which works well with the galena mineral has been developed and has revolutionized the process for making lead. Once the lead has been roasted in kilns, a process known as the Parkes method is used to purify the lead. The lead is cast into bars and graded into the following grades:

1. Acid—copper
2. Chemical
3. Corroding
4. Desilverized

Lead is perhaps best known for its heavy weight and resistance to gamma radiation. It is also used for soldering alloys, bearing alloys, ammunition, and storage batteries.

Tin

Tin is primarily used as a coating over another material (Figure 2–5). Pure tin is too soft for structural use; but because of its nontoxicity, it is widely used for can coverings and jewelry. Malaya, Indonesia, Bolivia, the Congo area, Thailand, Nigeria, China, and the Commonwealth of Independent States are the principal locations where tin ore (called cassiterite [SnO_2]) is found. Tin is often alloyed with other metals to produce pewter, soft solders, bronze, and babbitt. The most used commercial brands of tin contain a minimum of 99.8% tin.

Zinc

Zinc is also found in the mineral galena, a lead sulfide. The largest use for zinc is galvanizing another material such as steel. The life of a galvanized material depends on how thick the coating of zinc is. Heavy coatings have good corrosion resistance which can last for years. Another use for zinc is as an alloy with copper to make bronze. Zinc is found in the same countries as lead.

Magnesium

Magnesium is a lightweight metal which is used largely in aircraft, power tool, and material-handling industries. It is produced from magnesium chloride which is found in sea water and from ores called magnesite, dolomite, and brucite. Magnesium is usually alloyed

Figure 2–5 Tin is commonly used to coat containers for food.

with another material. This procedure adds strength, corrosion resistance, workability, hardenability, castability, and weldability. Most magnesium alloys are easy to weld and cast. This makes them ideal for the construction and automotive industries (Figure 2–6).

Precious Metals

The precious metals consist of gold, silver, and platinum. They are mined in the same manner as other metals and are sometimes found to be mixed with different types of base metals. In recent years, silver has been recovered from silver–lead–zinc ores and platinum has been found in copper–nickel ores.

Gold and silver are used extensively as a monetary exchange and jewelry (Figure 2–7). Industrial applications include electronic parts and tooth repair. Platinum is used in the production of gasoline and nuclear heavy water. It is also used for electrical contacts in the electronics industry.

Figure 2–6 Airplane wheels are made of magnesium because it is strong and lightweight *(Photo courtesy of the Boeing Company)*

Figure 2–7 Precious metals are used for products and as backup security for currency *(Photo courtesy of Photri, Inc.)*

NONMETALS

Nonmetallic materials are those materials which are primarily found in a variety of clays and sands. While they may also contain particles of metal, they usually do not make up the components of a metallic material. Simply stated, they are those materials which are not based on carbon chemistry. The most common of the nonmetallic materials are clay, sand, glass, cement, and abrasives.

CERAMICS

The basic raw material for most ceramic products is clay. It is an abundant material but seldom used in its natural state. Unwanted materials such as stones, roots, and grass clods must be removed in a screening process. Once the clay has been cleaned, it is classified into one of several groups. The most commonly used clays are

Kaoline	China clay	Pottery clay
Ball clay	Slip clay	Sewer pipe clay
Fire clay	Bentonite	Brick clay

Kaolins

Very pure **kaolins** usually have compositions very nearly that of kaolinite. The oxide formula for kaolinite is $Al_2O_3 2SiO_2 2H_2O$. Most kaolins are washed after they are mined because it allows the clay to be worked and shaped without rupture (splitting or breaking apart). This type of clay also retains its shape once it has been formed. Only ball clays are more plastic (shapeable) than kaolins.

Kaolins are used extensively for the manufacture of whitewares and china. Because the clay contains a small amount of iron particles, the kaolins will turn to a cream color when fired in a kiln. The purest kaolins will fire up white and are used for expensive, high-quality ceramic wares such as those shown in Figure 2–8.

Kaolins are also used for the manufacture of refractories. A refractory material is one which resists the action of high temperature and remains unmelted and undeformed (see flow chart in Figure 2–9). The most common refractories are in the form of bricks.

In the United States, Georgia is the largest producer of kaolins, with an output of 4,320,000 tons per year.

Ball Clays

The term **ball clay** designates a very plastic sedimentary clay. The name is derived from their being originally dug from the ground in blocks or balls in England. These clays are composed of kaolinite with varying amounts of impurities and organic matter. The color of ball clays is variable with the darker clays containing more iron oxide and being more plastic. Because of their high plasticity, they are used when workability and high dry strength are needed. They have a high amount of shrinkage and tend to crack upon drying. Tennessee is the largest producer of ball clays in the United States.

Fire Clays

Fire clays are used in the manufacture of brick, saggers, glass-melting pots, crucibles, and various types of refractory mortars and cements. They have a fusion point which is above 1,600 degrees Celsius and range from very plastic to clays which cannot be shaped at all. Those clays which cannot be shaped are usually referred to as the **flint clays.** Fire clays have low amounts of alkalies, calcium, magnesium, and iron oxide.

Natural fire clay refractories have given way to the newer technology of high-alumina, multibrick, and plastic nonclay with tolerance control and special properties.

The occurrence of fire clays is widespread in the United States, with a great deal of variance in content. The important deposits are located in Pennsylvania, Ohio, Maryland, and Kentucky.

Figure 2–8   Industrial products include many ceramic parts

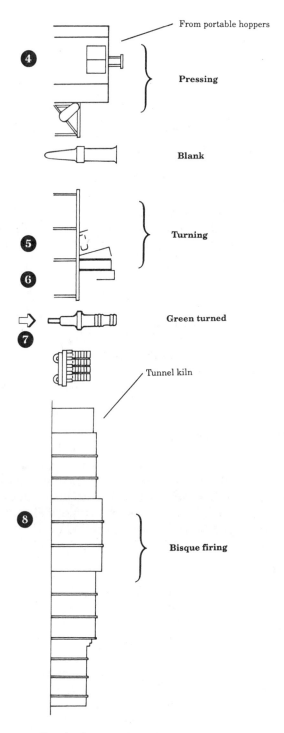

Figure 2–9 Spark plug insulator body manufacturing flow chart

China Clays

China clay is a kaolin that is used to fire white areas and fine china. Most china clay is mixed with ball clay for plasticity and other clays for color. The most famous mixture was developed by Josiah Wedgwood (1730–1795) and is still the basis for most whiteware today (Figure 2–10). His mixture included ball clay, china clay (kaolin), flint, and stone.

Slip Clays

Slip clays are very impure and have a high alkali content. When heated to 1,260 degrees Celsius, they will form into glass. The most common use for slip clay is in glazing stoneware and electrical porcelain. Important deposits of slip clay in the United States are located near the area of Albany, New York.

Bentonite

Bentonite is a very plastic and highly alkali type of clay. It is used in mixture with other clays to increase plasticity and dried strength. Its use is limited because of high shrinkage, iron oxide content, and lengthy drying time. Important deposits are found in the United States in the state of Wyoming and nearby areas.

Pottery Clays

The most common use of **pottery clays** is with the crude or less pure types of clay. While some pots are made from kaolin and ball clays, most are made from less pure clays that have good plasticity; are free of coarse material; have a low iron content; and will vitrify at less than 1,200 degrees Celsius, preferably between 90 and 100 degrees Celsius. They should be free of carbonates, sulfates, or other salts that will cause blisters or scumming when fired.

The most common types of pottery fired from this type of clay are stoneware, earthenware, garden pottery, artware, and flowerpots (Figure 2–11).

Sewer Pipe Clays

Sewer pipe clays are similar in content to that of pottery clays. They should have high dry strength, high density, and low warpage.

Brick Clays

Shales and mudstones, which are sedimentary clay deposits that have been compacted and consolidated, are used extensively for brickmaking. They may be easily ground to form a plastic extrudable material either alone or blended with varying amounts of clay (Figure 2–12). Important properties for face brick clay are plasticity, fired color, shrinkage, water absorption of the fired brick, and strength. **Brick clay** is widely abundant but is extremely good at the Triassic mudstones and shales of North Carolina.

Figure 2–10   Beautiful china is produced from high-quality kaolin *(Photo courtesy of Wedgwood USA, Inc.)*

Figure 2-11 Earthenware pottery is a popular ceramic product

Most clays are mixed together to form a variety of products. By combining the various types of clays, ceramic products can range from fine, fragile, translucent china to tough, hard, and fireproof refractories.

Cements

The term **cement** is often mistakenly used to describe a concrete structure when actually the cement is only one part of a concrete mixture. (**Concrete** is a combination of cement; water; and an aggregate such as sand, gravel, or stones.)

The most commonly used cement is called portland cement. It is a mixture or slurry of limestone; clays which contain silica, alumina, and ferric oxide; and a small amount of magnesia. The batch is heated, causing it to become molten and form a clinker. Once the clinker is cooled, it is ground into powder, which is later used to make mortar (for brick or block joints) and concrete (Figure 2-13).

Glass

Glass is made from the primary material silica. Sand is the quartz form of silica and is abundant as a raw material. However, only a high-quality sand is used for clear glass. Colored glass may use a less pure sand that contains some amounts of iron oxide. Sand for glassmaking comes from large deposits in New Jersey, Illinois, West Virginia, Pennsylvania, and Missouri.

In addition to silica, modifiers are used to reduce the high melting and working temperatures. Materials such as alkalies, alkali earths, boron oxide, lead oxide, and zinc oxide are the most common materials used to modify the heat range. To help remove gas bubbles from the glass, a small amount of materials known as melting and refining agents is added

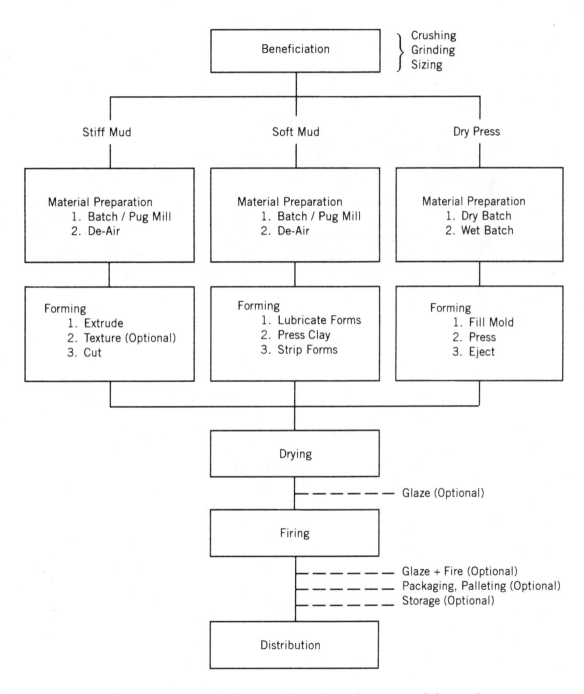

Figure 2–12 A typical vertical flow scheme for structural clay products

Figure 2–13   Concrete is an important material for commercial construction

to the glass batch (mixed raw materials). Specialized glass may also require the addition of coloring agents or opalizing agents which produce a color range or opalescent appearance.

Abrasives

An **abrasive** is a substance that is used for the removal of material by scratching or rubbing. There are many ceramic abrasives used today. They can be classified into two groups: natural and manufactured.

NATURAL ABRASIVES

The most common natural abrasives are flint, corundum, emery, garnet, sandstone, pumice, tripoli, rottenstone, and diamond.

Flint

Flint is an extremely hard form of silica. It is used for making coated abrasive (e.g., sandpaper) for use in the woodworking and leather trades.

Corundum

Abrasive-quality corundum contains 98% alumina and small amounts of silica and iron. It is the second-hardest natural material after the diamond. It is found in large rocks and is mined as an ore successfully in South Africa.

Emery

This material is an impure form of corundum and is a mixture of corundum, magnetite, and hematite (see "Metallics"). Much of our emery is mined in the United States, Greece, or Turkey. Emery is used for coated abrasives and grinding wheels, as pictured in Figure 2–14. However, newly manufactured abrasives such as fused alumina have replaced much of the use for pure emery.

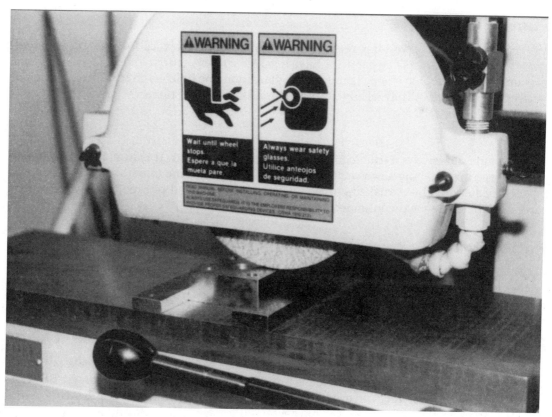

Figure 2–14 Abrasives are used to grind the most difficult materials efficiently

Garnet

The most common garnets for abrasive purposes are almandite, andradite, and shodolite. Garnet is too soft for the grinding or sanding of steel; but because it is harder than glass or quartz, it is highly valued as a coated abrasive in the woodworking trades. By "open coating" (spacing the abrasive more sparsely) the garnet, it is ideal for softwoods because it does not clog. Garnet is mined in Canada and New York.

Sandstone

This abrasive material consists of silica, which is consolidated by mineral cement such as calcite, quartz, clay, or limonite. It was used for the earliest grindstones and was quarried extensively in the past. Because of new technology and danger of sandstone grinding wheels (they broke apart easily), sandstone is not appreciably used today in industry.

Pumice

Pumice is a lava or volcanic rock. It has a frothy structure with low density which allows it to float on water. The particles have a gentle cutting action; and in powdered form, they are used for rubbing and polishing metal, glass, and fine finishes on furniture. The major sources for pumice are Italy (Sicily), Canada, and the United States.

Tripoli

This mild abrasive is a friable mineral of high silica content. It is used for scouring and cleansing scraps, rubbing down compounds, and for finishes. The mineral is mined by open-pit methods and is then crushed, milled, sieved, and classified to produce powders of the correct particle size for the intended application.

Rottenstone

Rottenstone is a soft friable rock which has been formed from a type of sandstone. True rottenstone contains 80–90% silica along with alumina and other impurities. It is used as a mild abrasive for rubbing compounds.

Diamond Abrasives

Most diamonds that are not of gem value are used industrially for abrasive purposes. The major sources for industrial diamonds are Africa, Brazil, and the Commonwealth of Independent States. Diamonds are used for heavy-duty grinding, drills, and for dressing points to true grinding wheels. They are the hardest of all known materials and expensive to use commercially. Recently, synthetic diamonds have been developed by transforming graphite into diamond through heat and pressure. The major advantage of synthetic diamonds is the control of size and shape. Today, 10–15% of industrial diamonds are synthetic.

MANUFACTURED ABRASIVES

As you may have noticed, many of the natural abrasives owe their abrasive qualities to their crystalline content of alumina. Modern fused alumina abrasives are made of a variety of types:

- Regular
- Fine crystalline
- High titania
- White

Each of these manufactured abrasives has a different grinding action. Basically, all fused alumina abrasives start off with the basic ore of bauxite. Other minerals such as calcined alumina, coke, and iron ore are combined with bauxite to produce a variety of manufactured alumina abrasives.

Silicon Carbide

Silicon carbide was discovered by E. G. Acheson, who named it carborundum. Silicon carbide exists in two crystal forms: alphia (hexagonal) and deta (cubic). The material is synthesized at high temperature by the reduction of silica by carbon and subsequent carbonization of the resulting silicon. Silicon carbide is used to grind a number of hard metals, marble, stone, glass, and ceramics.

Boron Carbide

This manufactured abrasive is harder than silicon carbide but is much more brittle. For this reason, it is not used for normal grinding operations but does have application in special processes. In some applications such as lapping tools and dies, the boron carbide powder can be used to substitute for the more expensive diamond powder.

WOOD

Of all the materials derived from plants, wood is the most common. Unlike the minerals, wood is constantly replaced by foresting-management techniques. Through careful control of felling and planting of trees (see Figure 2–15), a forest continues to produce wood indefinitely. Wood is therefore one of the most renewable material resources in the world.

A tree grows as a result of two separate processes: primary and secondary. *Primary* or opical growth is the length that the stem and branches grow. *Secondary* growth is the thickness of the stem and branches and can be identified by the number of annual rings each tree has (Figure 2–16), which is also an indicator of the tree's age.

During the process of growth, a woody tissue contains the life-giving substance for the tree and is usually the outside light-colored ring of material. The material in the center is a darker portion of wood known as the heartwood. The heartwood is not active in the growth or sap production of a tree and is regarded as dead tissue. Its dark color in many species is due to the deposition of extraneous materials such as resins, tannins, and coloring matters.

Figure 2–15   A forest is carefully managed to produce wood and maintain the basic ecological system *(Photo courtesy of USDA, Forest Products Laboratory)*

Wood is used as a raw material for lumber, lumber products, and cellulose plastic. In addition to these industrial materials, it is also used to make many other materials such as paper, oils, turpentine, and others, which will not be discussed in this text.

PLASTIC

Plastic materials are called synthetics because they are made up of a combination of several natural materials. The first plastic was used by John W. Hyatt in 1868 to make billiard balls. He combined cellulose fibers and nitric acid (nitrocellulose or gun cotton) with solid camphor. Another early plastic material, discovered by Leo H. Baekeland, was based on

Figure 2–16 The annual rings can be counted to determine the age of a tree *(Photo courtesy of USDA, Forest Products Laboratory)*

phenol-formaldehyde. Since then, major developments have occurred in the plastics industry; and the advancements have made a significant impact on the materials industry.

Plastic materials are created by chemists who use natural materials such as coal, oil, natural gas, air, water, plant fibers, and salt as their resources for combining atoms and molecules into monomers and polymers. They are called plastics because of their ability to be easily formed or molded into a variety of shapes.

Various elements can be combined to create the "intermediates" which are further combined to make the plastic resins. The principal raw material resources for plastics are coal, air, water, petroleum, natural gas, limestone, and salt.

Basic products of natural gas and petroleum also render by-products of butane, propane, ethane, methane, and naphtha, which are then refined to produce butylene, propylene, ethylene, methylene, and benzene. As these elements are combined with a variety of petrochemicals, new bonding structures are created, forming the usable plastics in the right volume.

Plastics are classified into two groups: thermoplastic or thermosetting. The basis of this classification scheme is related to the manner in which the monomer was polymerized. **Polymerization** is a chemical reaction in which molecules of a monomer are linked together to form larger molecules.

Addition Polymerization

When a monomer is polymerized and cross-links between two atoms are weak and set free by heat, pressure, or chemicals, they will link up in a more stable and relaxed position,

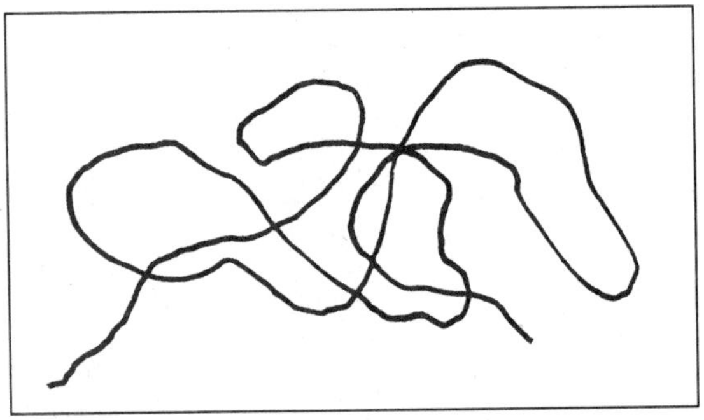

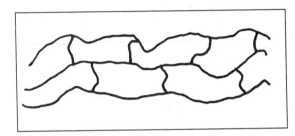

Figure 2–17 Long-chain molecules create thermoplastics

Figure 2–18 Cross-linked molecules create thermosetting materials

forming a long-chain, giant molecule. Polymers formed in this manner (Figure 2–17) are known as thermoplastic materials and have the ability to become soft when heated and hard when cooled. Because of the way they bond, they can be reheated to change shapes and re-used in manufacturing.

Condensation Polymerization

This process is similar to addition polymerization except that the monomer is changed chemically, providing bonds between atoms that connect the molecular chains. In Figure 2–18, the cross-links that connect the long chains can be seen.

The connecting of the chains by cross-linking provides rigid and strong plastics known as thermosetting materials. These plastics cannot be softened by reheating like thermoplastics and are considered as permanent after their initial shape has been processed.

SUMMARY

The genesis of materials is vast and complex. As we explore the basic raw materials that are available on the Spaceship Earth, we are reminded of the limited amount of some materials and the vast abundance of others. Yet, our resources are finite and our ability to use them wisely is important. In Chapter 3, we discuss how our basic raw materials are processed into usable products.

❓ REVIEW QUESTIONS _____

1. What are the two classifications of metals? Explain how they are different.
2. Which iron ores have the largest amount of iron content?
3. Describe the process of beneficiation and why it is used in the metal industry today.
4. List and describe two types of ore mining and how they differ from each other.
5. Which metal is the most abundant in the earth's crust?
6. Explain how tin and zinc are used with other metals.
7. What is kaolin? How is it used in industry today?
8. Identify the ingredients used to make glass. How are the gas bubbles removed?
9. Classify the abrasives into natural and manufactured. Which are used most today?
10. Why is wood considered to be a renewable resource?

SUGGESTED ACTIVITIES _____

1. Create a display of natural materials and compare them to each other.
2. Select a natural material and write a short paper on it.
3. Using a globe, identify where the natural materials are found by using colored flags. Which country has the most natural materials?
4. Identify those natural materials that can be used directly without further processing. Construct a consumer product from the material.

CHAPTER
3

Primary Processing of
Organic and Inorganic Materials

Earth and the atmosphere that surrounds it has supported life for millions of years (Figure 3–1). It has provided its inhabitants with the materials essential to sustain life (i.e., food, shelter, clothing), which humans have grown to expect. The earth, however, is a "closed system," and the materials we have grown to depend on to manufacture our essential and nonessential products are contained within this system. Until new sources of materials are discovered outside the earth and its surrounding atmosphere, the human race must rely on the plants, animals, minerals, and gases contained within our system to support material needs and wants—and life itself.

Scientists have identified 90 chemical elements from the earth's crust and its surrounding atmosphere. These chemical elements in their pure state or combined with each other provide the raw materials that are transformed into primary materials for manufacturing purposes. The raw materials extracted from the earth's crust and harvested from the forests or harnessed from the air and water are converted into thousands of materials, each with different characteristics and properties, that can be used to produce the seemingly endless variety of products available on the world market today.

The materials commonly used by manufacturing and construction industries today have been classified as inorganic and organic. **Organic materials** are the many substances that are based on carbon chemistry. Examples include natural materials (such as wood, coal, petroleum, natural gas, animal fibers, and food) and synthetics (such as plastics, solvents, adhesives, and lubricants). Opposite the organics are a variety of natural and synthetic

Figure 3–1 Earth and its atmosphere have supported life for millions of years *(Photo courtesy of NASA)*

inorganics, which are typically based on silicon and include a family of materials known as ceramics. Examples of **inorganics** include natural materials of sand, glass, brick, and concrete. The synthetic inorganic materials are some of the hardest materials available today and include materials such as aluminum oxide, silicon carbide, nitrides, borides, and others.

A commonly used and more practical classification of solid materials used in manufacturing and construction industries today includes the following:

1. *Metallic* (ferrous and nonferrous)
2. *Organic* (natural and synthetic)
3. *Ceramic* (natural and synthetic)

A relatively new classification which will be discussed in Chapter 4 is *composite*.

This chapter explores techniques used to transform the raw materials into primary materials for manufacturing. The focus of this chapter is on the conversion of natural raw materials into commercial metals, organics, and ceramics which are used in the manufacturing and construction industries to produce usable goods.

FERROUS METALS

Metal is a general term used to identify a broad classification of materials which are hard, strong, and conduct heat and electricity. Although many chemical elements classified as metals are found free in nature, as in ores which can be reduced commercially, they are seldom used in their pure state. They normally are alloyed (combined) with one another or other nonmetallics to improve their properties. Whether in their pure state or in the form of an alloy, the almost endless list of materials known as metals can be classified as *ferrous* or *nonferrous*. Ferrous metals (discussed in this section) are iron based and comprise the large and extensively used family of cast irons, carbon steel, alloy steel, and specialty steels. The various carbon steels and cast irons are approximately 95% iron. The remaining 5% consist of other chemical elements, including carbon. The percentage of carbon and its form distinguishes steel from cast iron.

From Ore to Iron

The refining of iron begins with the mining and processing of the basic raw materials which are used to form pig iron. The primary raw materials are iron oxide ore, coke, and limestone. Total worldwide mine output of iron ore in 1990 was nearly one billion tons. The United States was responsible for approximately 5% (59 million tons) of the total.

The concentrated iron oxide (60–65% pure) is shipped to the blast furnace, where it is transformed into pig iron and later refined into cast iron and steel. The basic objective in the refining process is to remove the impurities and oxygen from the ore before producing commercial iron and steel.

The blast furnace, used to refine the raw materials, is in the shape of a large cylinder which is often higher than a twelve-story building (Figure 3–2). The inside of the furnace is

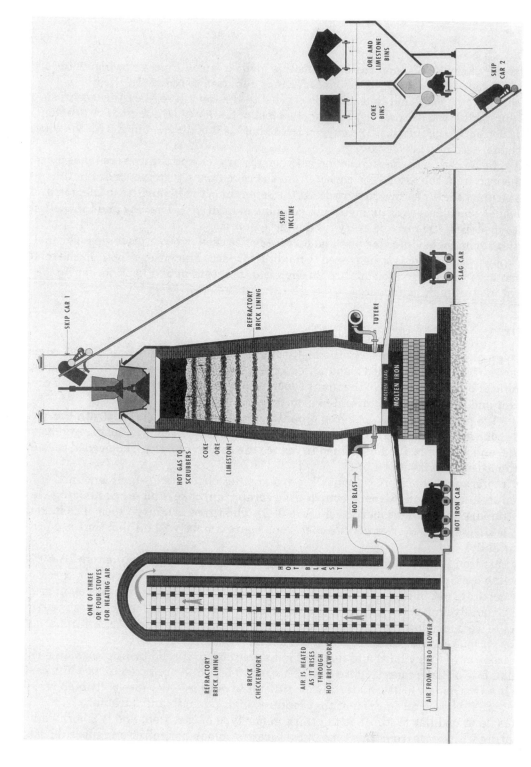

Figure 3-2 Blast furnace—iron ore is converted into pig iron by means of a series of chemical reactions that take place in the blast furnace (*Courtesy of Bethlehem Steel*)

lined with refractory bricks, and the outer shell is made of steel. Pipes carrying water are generally imbedded in the walls of the blast furnace and used to reduce the heat.

The charge (ore, coke, and limestone) is carried to the top of the blast furnace in skip cars and dumped into the furnace. A continuous blast of hot air (3,500 degrees Fahrenheit) enters the bottom of the furnace. The oxygen in the hot blast of air combines with the coke, causing combustion and heat.

During the process, a chemical reaction between the oxygen and the iron ore takes place, freeing the iron from the oxide. The carbon from the coke mixes with the oxides in the ore and is given up in the air as carbon dioxide. At the same time, the impurities in the iron and the limestone, which acts as a fluxing agent, combine to form *slag*. The slag and the molten iron are tapped from the furnace every two or three hours.

The slag is channeled into slag pits and disposed of or used in other processes. The molten iron, known as pig iron, is transported to holding furnaces known as *mixers*. In the mixers, the molten iron is mixed to equalize its chemical composition and kept hot for further refinement in steel-making furnaces.

Cast Iron

Although most pig iron is used in the molten state and further refined into steel, some of it is cast into ingots and supplied to foundries for further manufacture into various shapes. The refining of pig iron into one of the many types of cast iron involves the reduction of carbon content and the removal of unwanted elements in the iron.

During the process that converts iron ore into pig iron, the iron combines with the carbon in the coke and picks up 3.0–4.5% carbon. High carbon content makes the pig iron hard and brittle, and it must be further refined to reduce the carbon content. The needed carbon in cast iron is between 2.0 and 4.0%.

Cast iron is an alloy of iron, carbon, silicon, and other elements in small amounts. It is often produced in a cupola furnace, which uses scrap iron, coke, and other alloying elements to impart desirable properties (Figure 3–3). The three primary types of cast iron are *gray, white,* and *nodular* or *ductile* cast iron. Approximately 20 million tons are produced annually.

Gray cast iron is used in a majority of the iron castings made, approximately 70–80%. When molten cast iron solidifies, carbon in the iron may remain in the form of iron carbide, or it may separate out in the form of free graphite flakes. The flakes or free carbon in gray cast iron generally impart the following characteristics: soft, easy to machine, low strength, resistance to wear, and low thermal expansion. It is the least expensive of the cast irons and is used for various engine parts, including the engine block.

Cast iron that is very hard and nonductile has carbon in a solid or iron carbide form rather than free or flake form (Figure 3–4). This metal is usually referred to as white cast iron and is an extremely brittle material and difficult to machine. Because of its high resistance to wear, it is used on equipment that requires high strength and durability.

Malleable or nodular cast iron is the third major type of cast iron, and it is formed by heat treating white cast iron. This type of cast iron is somewhat softer and more ductile

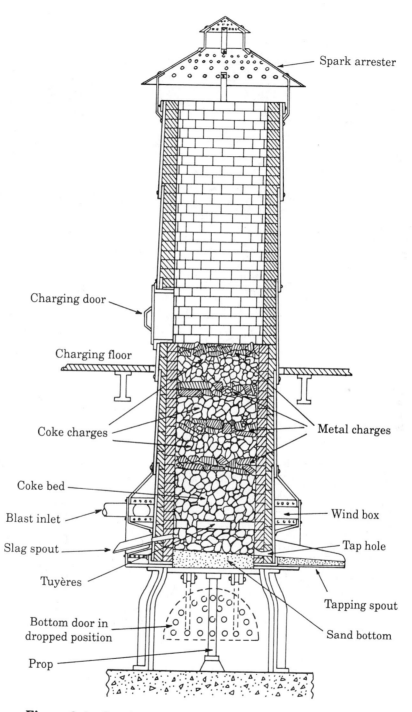

Figure 3–3 Cupola furnaces are often used to make cast iron

43

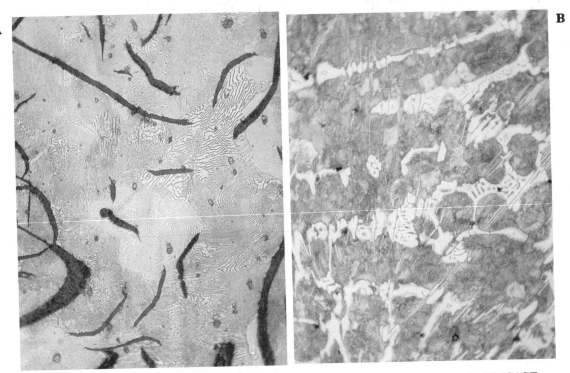

Figure 3–4   Microstructure of **(A)** gray and **(B)** white cast iron *(Photos courtesy of AFS / CAST Metals Institute)*

(easily drawn) than white cast iron but retains the toughness and strength of white cast iron. Malleable or nodular cast iron is used in the production of hardware and railroad equipment.

Steelmaking

You will recall from the discussion of refining iron oxide ore that the blast furnace process produces pig iron, which has a carbon content between 3.0 and 4.5%. In refining pig iron into steel, the carbon content as well as the silicon content must be significantly lowered. This is accomplished by one of four basic processes (see Figure 3–5):

- Bessemer
- Electric furnace
- Open-hearth
- Basic oxygen

The primary objective in all of these processes is the reduction of carbon through oxidation. The oxidation of the carbon produces carbon monoxide and ultimately carbon dioxide, which is given up into the air. In addition to reducing carbon and silicon levels, the steel-

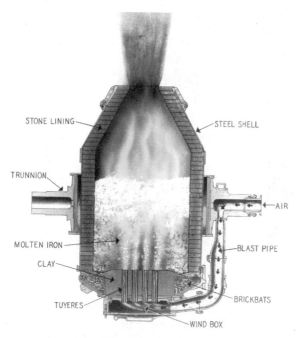

STONE LINING

STEEL SHELL

TRUNNION

AIR

MOLTEN IRON

BLAST PIPE

CLAY

TUYERES

BRICKBATS

WIND BOX

Figure 3–5A The Bessemer converter (or process) revolutionized steelmaking in the latter half of the nineteenth century and was one of the major sources of steel *(Photo courtesy of Bethlehem Steel)*

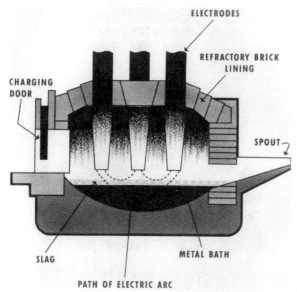

ELECTRODES

REFRACTORY BRICK LINING

CHARGING DOOR

SPOUT

SLAG

METAL BATH

PATH OF ELECTRIC ARC

Figure 3–5B The electric furnace: Because both the temperature and atmosphere can be closely controlled in an electric furnace, it is ideal for producing steel to exacting specifications. The process takes from four to twelve hours, depending on the type of steel to be produced *(Photo courtesy of Bethlehem Steel)*

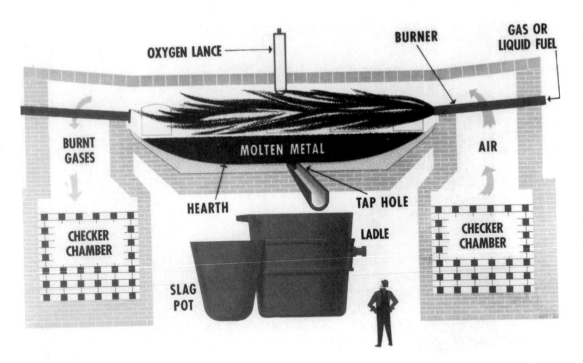

OXYGEN LANCE →

BURNER

GAS OR LIQUID FUEL

BURNT GASES

MOLTEN METAL

AIR

CHECKER CHAMBER

HEARTH

TAP HOLE

LADLE

CHECKER CHAMBER

SLAG POT

Figure 3–5C (Above) The open-hearth furnace: The name comes from the fact that the pool of molten metal covered with slag lies on the hearth of the furnace, exposed to the sweep of flames. The process takes from eight to ten hours *(Photo courtesy of Bethlehem Steel)*

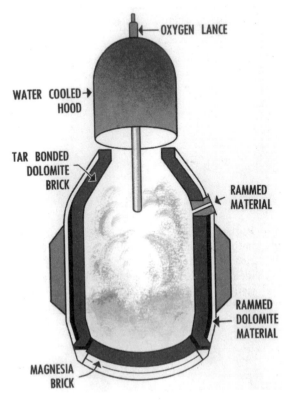

← OXYGEN LANCE

WATER COOLED → HOOD

TAR BONDED DOLOMITE BRICK

RAMMED MATERIAL

RAMMED DOLOMITE MATERIAL

MAGNESIA BRICK

Figure 3–5D The basic oxygen furnace: Designed expressly to get the best results with oxygen in steelmaking, this furnace can produce steel amazingly fast. The "broth" ingredients (iron ore, steel scrap, and molten iron) are refined into steel by blowing oxygen down from the top through a vertical lance extending to within five feet from the broth. During the blow, burnt lime and spar are added as fluxing materials *(Photo courtesy of Bethlehem Steel)*

making process involves adding alloys to achieve desired properties. All steels can be classified as carbon steel, alloy steel, or specialty steels.

Carbon steels are those whose primary elements are carbon and iron but generally have no alloying elements added. They consist of low-, medium-, and high-carbon steels. Ninety percent of products made of steel are made of carbon steel.

Alloy steels are those that achieve desired properties through the controlled addition of alloying elements (manganese, silicon, chromium, nickel, etc.). The American Iron and Steel Institute (AISI) has developed a numerical designation system for identifying alloy steels (Table 3–1). A four-digit system has been adopted to identify the major alloy and percentage of alloy in the steel. In this system, the first number represents the principal alloying element; the second number indicates the alloy content; and the final two digits represent percentage of carbon to the nearest hundredth of a percent. For example, AISI 4130 indicates chromium-molybdenum steel with 0.30% carbon.

Specialty steels is a term used to identify a group of alloy steels which have excellent corrosion-resistant and heat-resistant properties. Chromium is the principal alloy in these steels, which include stainless steel, tool steel, and other high-temperature steels. These steels are usually produced in standard mill shapes but are relatively more expensive than common carbon steel.

Standard Mill Shapes

Although some steel is used in the ingot form, most is formed into products such as plate, structural shapes, pipe, and strips before leaving the steel mill. *Forging, extruding, rolling,* and *drawing* are the most common processes used to transform metal into desired shapes. Figure 3–6 illustrates some of these processes.

Most shapes used for manufacturing or construction are produced by *rolling*. This process consists of passing the metal between rollers that revolve in opposite directions. The opening between the rollers is controlled to produce various cross-sectional dimensions (Figure 3–7). The largest percentage of mill products is produced by **hot rolling** and **cold finishing.**

Hot rolling is primarily a shaping process that can be used to produce finished mill products but is often used to produce semifinished products that require further processing. Semifinished shapes are produced by first placing ingots into a soaking pit, which heats the ingot to a uniform temperature of 2,200 degrees to 2,400 degrees Fahrenheit. Once heated to uniform temperatures, the ingots are conveyed through a series of mills which shape them into slabs, blooms, or billets.

Hot-rolled products can be further processed using a technique referred to as *cold finishing*. Cold finishing consists of cold rolling, cold reduction, and cold drawing of semifinished hot-rolled shapes. The hot-rolled products identified for cold finishing are cleaned of surface oxides (scale) by using sulfuric or hydrochloric acid followed by a water rinse.

The hot-rolled product is then rolled, reduced, or drawn at room temperature through sets of rolls that produce the desired shape. Cold finishing also changes the property of the metal and improves the surface appearance. Cold-rolled steel is normally harder and stronger than hot-rolled steel.

AISI series	Class	Major constituents
10XX	Carbon steels	Carbon steel
11XX		Resulphurized carbon steel
	Alloy steels	
13XX	Manganese	Manganese 1.75%
15XX		Manganese 1.00%
23XX	Nickel	Nickel 3.50%
25XX		Nickel 5.00%
31XX	Nickel-chromium	Nickel 1.25%, chromium 0.65% or 0.80%
33XX		Nickel 3.50%, chromium 1.55%
40XX	Molybdenum	Molybdenum 0.25%
41XX		Chromium 0.95%, molybdenum 0.20%
43XX		Nickel 1.80%, chromium 0.50% or 0.80%, molybdenum 0.25%
46XX		Nickel 1.80%, molybdenum 0.25%
48XX		Nickel 3.50%, molybdenum 0.25%
50XX	Chromium	Chromium 0.30% or 0.60%
51XX		Chromium 0.80%, 0.95%, or 1.05%
5XXXX		Carbon 1.00%, chromium 0.50%, 1.00%, or 1.45%
61XX	Chromium-vanadium	Chromium 0.80% or 0.95%, vanadium 0.10% or 0.15% min.
86XX	Multiple alloy	Nickel 0.55%, chromium 0.50%, molybdenum 0.20%
87XX		Nickel 0.55%, chromium 0.50%, molybdenum 0.25%
92XX		Manganese 0.85%, silicon 2.00%,
93XX		Nickel 3.25%, chromium 1.20%, molybdenum 0.12%
94XX		Manganese 1.00%, nickel 0.45%, chromium 0.40%, molybdenum 0.12%
97XX		Nickel 0.55%, chromium 0.17%, molybdenum 0.20%
98XX		Nickel 1.00%, chromium 0.80%, molybdenum 0.25%

Prefix	Meaning
E	Made in an electric furnace
X	Composition varies from normal limits

Suffix	Meaning
H	Steel will meet certain hardenability requirements

Other letters	Meaning
XXBXX	Steel with boron as an alloying element
XXLXX	Steel with lead additions to aid machinability

Table 3–1 Major groups in the AISI-SAE steel designation system

Figure 3–6 Forging, drawing, rolling, and extruding *(Photos courtesy of U.S. Steel and Photri, Inc.)*

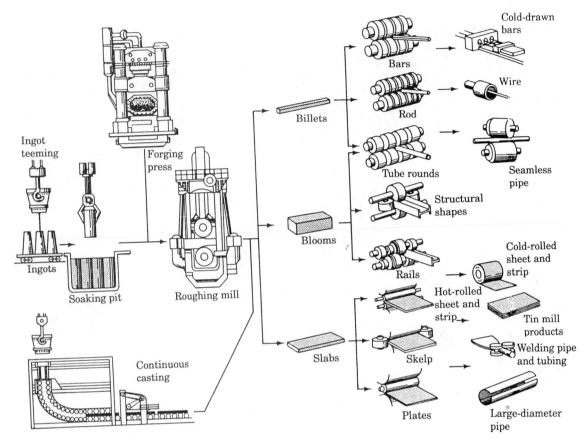

Figure 3-7   Blooms, billets, and slabs *(Courtesy of American Iron and Steel Institute)*

NONFERROUS METALS

The most commonly used nonferrous materials are aluminum, copper, lead, tin, zinc, magnesium, gold, silver, and platinum. Many of the nonferrous metallic materials are used as alloys, but few are used in their pure form. Since this book is primarily interested in materials used in manufacturing and construction industries, this section will concentrate on a discussion of the more commonly used industrial nonferrous metal materials.

Aluminum

Approximately 8% of the earth's crust is made up of aluminum. Although almost all rocks and clays contain some aluminum, it is not practical to mine such material unless the ore has an aluminum oxide content of at least 45%. These ores are known as *bauxite,* most of which are located outside the United States. Some deposits of bauxite are being mined in Arkansas, Georgia, and Alabama.

Bauxite is generally mined using the strip or open-pit method, then conveyed to a processing plant where it is ground, washed, screened, and dried to form a powder. It is then transported to a refining plant where the aluminum oxide is separated from the other impurities and reduced to form pure aluminum, which is alloyed and processed into ingots.

The commercial refining of bauxite is known as the **Bayer process** and is illustrated in Figure 3–8. The steps in this process include the following:

1. The bauxite powder is charged into large tanks (digesters) of caustic soda, which convert it into sodium aluminate.
2. The sodium aluminate is filtered and mixed with a crystalline substance known as alumina hydrate, which creates large amounts of aluminum.
3. The alumina hydrate is then heated in kilns to about 1,800 degrees Fahrenheit. This removes the water, leaving a powder alumina.
4. The alumina is then separated into aluminum and oxygen by melting the powdered alumina in large electrolyte cells. Powdered alumina is charged at the top of the cell; and an electric current is passed between the anode and cathode, causing the aluminum to separate from the oxygen. The melted aluminum falls to the bottom of the cell where it is further treated and poured into ingots.

The process of fabricating aluminum mill products from ingots is similar to that used in fabricating steel products. Standard plate, strip, tube, structural shapes, and others are produced and used extensively in the manufacturing and construction industries.

Copper

Copper is a reddish metallic element. It is malleable, ductile, and has high electrical and thermal conductivity. It is corrosion resistant and has a low melting point as well.

Copper and its alloys constitute one of the major groups of commercial nonferrous metals. The copper industry is divided into two areas: The producing companies are responsible for the mining, smelting, and refining of copper; and the manufacturing companies manufacture copper-based products from the refined copper. The end product of their effort is refined cathode copper.

Copper ore is mined by the open-pit or underground method then crushed, ground, and concentrated to produce an ore containing 20–25% copper. Most copper ores are sulfide ores which are a composition of copper sulfate and significant amounts of iron sulphide plus recoverable traces of silver and gold.

The concentrated ore is smelted in a reverberatory furnace to produce a copper sulfide–iron sulphide matte. The matte is oxidized in a converter to change the iron sulphides to iron oxides, which are separated from the copper sulfate in the form of slag. This process also burns off the sulfur from the copper sulfate, producing *blister copper*—a name derived from the appearance of the copper after the liberation of the sulfur and other gases.

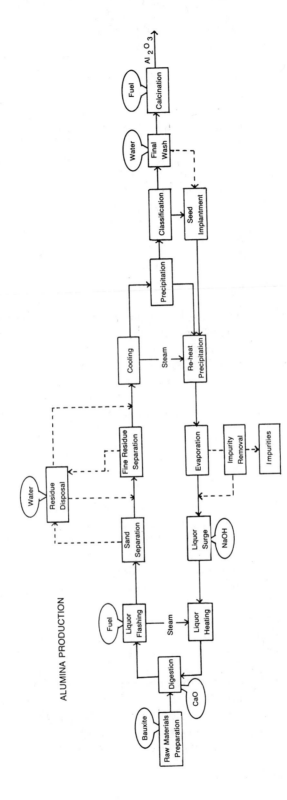

Figure 3–8   The Bayer process is used for refining aluminum

Blister copper contains at least 98.5% copper. Further refining of blister copper removes most of the oxygen and other impurities, leaving 99.5% pure copper which is cast into anodes (electropositive poles). Anode copper is electrolytically refined using a cathode (negative cell) of pure copper and an electrolyte of copper sulfate. Direct current of high amperage and low voltage dissolves the anode and deposits the pure copper on the cathode.

This copper is 99.5% pure and is further refined to 99.95% using an electrorefining process which leaves gold and silver as a by-product. The refined copper is then shipped to the fabricators and transformed into mill products.

Fabricators

There are four major classes of copper fabricators:

1. Wire mills   44% of copper fabrication
2. Brass mills   44% of copper fabrication
3. Foundries   9% of copper fabrication
4. Powder plants   1% of copper fabrication

In addition to the copper, the copper alloys are extremely important in the manufacture of usable goods. Brass and bronze are alloys whose basic constituent is copper. Brasses consist of zinc and copper and have improved strength and machinability yet maintain ductility. Bronzes consist primarily of tin and copper and have increased strength, workability, hardness, and ductility.

Magnesium

There are two main production methods for metallic magnesium. One is electrolytic, and the other is thermal.

The electrolytic process uses seawater in the electrolysis process. The first step in the process is the production of magnesium chloride, which is refined from seawater and other brines.

The electrolytic cell is an iron or steel box with carbon electrodes. The charge consists of the anhydrous magnesium chloride together with alkaline earth chloride. In the reduction process, chloride is given off at the anodes and collected for reuse in the production of more magnesium chloride. The liberated metal floats on the surface of the cell and is removed for refining in a melting furnace and then cast into ingots.

The thermal reduction process begins with pulverizing the ore and mixing it with ferrosilicon. The mixture is then charged into a steel furnace and heated to approximately 1,200 degrees Celsius. The magnesium vapors that form are condensed at the cool end of the processing furnace. The condensed magnesium is then further refined and cast into ingots.

Magnesium is a light metal with a specific gravity lower than aluminum. It has excellent casting characteristics and is highly machinable. Magnesium alloys have a high strength/weight ratio and good anticorrosion qualities.

WOOD

Harvesting

Wood is used to manufacture the lightest paper and tissue products as well as huge laminated beams capable of supporting massive structural shapes. Wood not only provides us with the raw materials for furniture and buildings; but, through continuous improvements in science and technology, organic chemicals and other manufactured wood-based products are being developed to improve the quality of life.

The early history of the forest industry was not as promising. The natural forests which were subject to insect infestation and disease were exploited by the early lumberjacks, leaving the land barren. During the 1800s, the industry was an extractive industry only. Little or no attempt was made at reforestation, and only about 20% of the volume of the tree was used. Attempts to stop the practice of "cut and run" lumbering were not initiated until the 1930s.

During the 1930s and early 1940s, the concept changed to the idea of timber as a crop; and the industry focused its efforts on reforestation and more extensive utilization of the raw material. New pulping processes were developed which permitted a greater volume of the tree to be converted into useful materials such as pressboard, flakeboard, and particleboard.

In 1941, Weyerhauser initiated the tree farm certification system, which supported reforestation through genetic engineering studies and efficient forest management. Today, the forest industry can truly be called a harvesting industry; and we can rest assured that the generations to come will enjoy the warm aesthetic characteristics of wood products just as past generations have.

Processing the Raw Material

The boards, beams, plywood, and other standard shapes and sizes of wood products that are used in the construction of a house or a fine piece of furniture begin with the timber. Timber can be classified as softwood timber or hardwood timber. Softwoods have been identified as *conifers* or *coniferous*. With few exceptions, the softwood trees are cone-bearing trees which have needlelike leaves that are retained throughout the year. Approximately 75% of the total lumber production is from softwood trees. The lumber obtained from softwood trees is primarily used in the construction industry for a variety of purposes. However, some softwoods are extremely popular for the construction of certain furniture styles (e.g., colonial, early American).

There are three major producing regions for softwood in the United States. The *Western Woods Region* is the largest, comprising thirteen states located in the western third of the country. Douglas fir, white pine, hemlock, and red cedar are among the species of softwood taken from this region. The *Southern Pine Region,* which covers the entire South, produces various species of pine which are sold commercially as short-leaf and long-leaf pine. The smallest of the three regions is the *Redwood Region,* encompassing a handful of counties in northern California. Half of the lumber production in this region is redwood production. The rest is various softwood species which grow in the region. The hardwood timbers come from trees known as *deciduous*. They are broad-leaf trees such as oak and maple; and unlike the conifers, they generally lose their leaves at the end of each growing season. Approximately 25% of the total lumber production is a result of the manufacture of hardwood products. Hardwoods are sometimes used in the construction industry but are primarily processed for use in the manufacture of furniture and cabinets.

Most of the hardwood species grow east of the Great Plains region of the United States. The *Northern Forest Region,* which extends from Maine to Wisconsin along the Canadian border, is primarily responsible for providing maple, birch, ash, cherry, and beech. The *Central* and *Southern* regions, which stretch from the East Coast to beyond the Mississippi and in the West from the Great Lakes in the North to the Gulf of Mexico in the South, are responsible for production of oak, yellow poplar, basswood, walnut, gum, and hickory.

From Timber to Commercial Product

The timber, once selected for harvesting and felled, is removed from the growing site primarily with the use of large tractors and specially designed equipment which grip the logs and drag them to a point from which they can be transported to the processing facility. The transportation to the sawmill is generally in the form of logging roads, railway, or river.

The purpose of the sawmill is to convert the log into a form which can be commercially marketed and used or into a form which will require further processing. The operations performed at the sawmill are debarking, edging, trimming, sorting, and grading.

The processing of the log at the sawmill begins with the debarking (Figure 3–9). Debarkers and debarking equipment remove the outer bark through the use of high pressure water jets, thus revealing the shape and irregularities of the log to the sawyer. The log is then transported by conveyor to the headrig (either a large circular saw or a band saw) and is sawed into timber known as rough cuts, planks, and board. This conversion, however, leaves the lumber with rounded, often rough edges.

The second step in the process is edging. This process utilizes a series of small saws to rip the lumber into desired widths and remove the rounded edges. From the edger, the lumber is conveyed to the trimmer, which removes rough ends and defects and produces desirable lengths. The final step is to sort the lumber according to size, species, and grade.

Grading and Standard Classification—Softwoods

Standards of quality have been established in both the hardwood and softwood industries. These standards are the industries' method of providing uniform quality in their product, whether purchased on the East or West coast.

The commercial grading rules for the various species of softwoods have been established in the *American Softwood Lumber Standard PS 20-70.* The grading rules for each region of the country conform to Standard PS 20-70. The basic purpose of these rules is to separate the wide range of qualities into usable classifications.

Softwood lumber is classified according to two basic groups: yard, and factory and shop. *Yard lumber* is generally intended for construction and general building purposes and is

Figure 3–9A   The bark is first removed from the log *(Photo courtesy of USDA, Forest Products Laboratory)*

subdivided into three major categories: boards, dimension lumber, and timbers. Boards are 1 to 1½ inches thick and 2 inches or more wide. They are typically used for sheathing, flooring, siding, trim, and paneling.

Dimension lumber refers to material 2 to 5 inches thick and 2 inches or more wide. Dimension lumber is primarily used as structural members of residences. They consist of rafters, planks, joists, and studs. *Timbers* are 5 inches or more in the smallest dimension. They are primarily used as heavy structural members.

Factory and shop is primarily for remanufacturing. The factory and shop lumber will be used to manufacture commercial wood products used in residential and commercial construction. Examples include windows, doors, trim, and moldings.

Figure 3–9B   The slab from the circular portion of the log is removed *(Photo courtesy of USDA, Forest Products Laboratory)*

Grading of softwoods has been divided into two major categories: select and common. Select boards are classified by letters A to D. However, Grade A (practically clear-face grain) and Grade B (some imperfections) have been combined to form the top grade of B and better which is used for high-quality interior-finish carpentry. Grades C select and D select are lower grades which have defects but are basically sound and are more economical.

Common grades are much more economical than select grades. They also have more defects and are primarily used for construction purposes. There are four grades of common lumber: 1, 2, 3, 4. Numbers 1 and 2 should be free of defects which would affect the strength of the board. This lumber may have knots, but it should be able to be used without waste. Number 3 utility and Number 4 economy may have loose knots, checks, and may be warped. Numbers 3 and 4 are normally used for sheathing, subflooring, and other rough construction purposes.

Figure 3–9C   The log is sawed into boards *(Photo courtesy of USDA, Forest Products Laboratory)*

Grading and Standard Classification—Hardwoods

The standards of quality used for grading *hardwood* lumber have been established by the *National Hardwood Lumber Association* (NHLA). The first two grades of hardwood lumber have been combined to form the grading classification known as First and Seconds (FAS). Lumber in this grade classification must be 6 inches wide and at least 8 feet long. It also must be 85–90% face clear on the poorest side.

Select grades are similar to FAS except that the best side must be 85–90% face clear.

Common grades are approximately 66²/₃% face clear on the better side. They are generally 2 to 3 feet long and 3 to 4 inches wide.

Because of its lack of availability, hardwood is generally cut in lengths and widths that are most economical.

Seasoning

When timber is felled in the forest, it is extremely high in moisture content. This moisture must be reduced considerably in order to make the wood a more stable product. The process of reducing moisture content is known as *seasoning*. Moisture content must be reduced from its green state of 200–300% to between 4 and 19% depending on its intended use and area of the country. One of two methods is used for seasoning lumber: kiln or air-drying (see Figure 3–10). *Kiln* drying is the more popular of the two and produces seasoned lumber in a few days. With the kiln method of seasoning lumber, the atmosphere inside the kiln is controlled for humidity, heat, and air circulation. The kiln method allows the manufacturer to reduce the moisture content in just a few days to desired levels.

Lumber that is being air-dried is stacked in such a manner as to allow air to flow through the stacks, thus reducing the percentage of moisture by evaporation. This process normally takes three to four months and usually will not reduce moisture content below 12–15%. Most cabinet-grade lumber has a moisture content, when seasoned, of between 4 and 12%, while construction-grade lumber is used at between 12 and 19%.

The reduction in the size of the lumber as a result of seasoning is as much as 10%. Figure 3–11 illustrates the percentage of loss of dimension from drying in the cross-sectional, tangential, and longitudinal directions of a board.

Plywood, Hardboard, and Flakeboard

Plywood, hardboard, flakeboard, and other manufactured panels represent the best use of natural wood and wood fibers, combined with adhesives to make an extremely strong, stiff panel with excellent qualities for finish carpentry, cabinetmaking, and construction applications.

Plywood has the advantage of being equally strong in both directions. This is a result of the individual veneers being placed so the grain and layer is perpendicular to the one before it. Plywood also is available in a wide range of thicknesses and quality. It represents the maximum use of raw materials and reduces the burden of construction and manufacturing of wood products.

The transformation of lumber to plywood begins with the selection of peeler logs for processing. The peeler logs are cut into blocks about 8½ feet long and 12 to 20 inches in diameter. The large blocks are then placed on a giant lathe and rotated against a long knife which peels the log into long continuous veneer sheets. The sheets are then transported to clippers and cut to desired widths. Then, the sheets are transported by conveyors to ovens to be dried. This process reduces moisture content. Next, the graded veneer goes through the glue spreaders, and both sides are coated evenly. The coated panels are then laid up perpendicular to one another, and pressure (150 psi) is exerted to form the veneer and adhesive into plywood panels. The final operations involve trimming the panels to exact size, sanding, and grading.

Figure 3–10A Air-drying of lumber *(Photo courtesy of USDA, Forest Products Laboratory)*

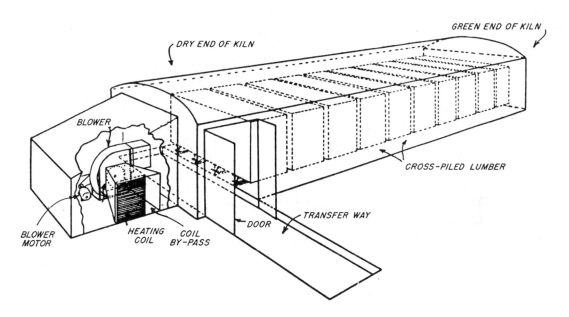

Figure 3–10B   Kiln drying of lumber *(Courtesy of USDA, Forest Products Laboratory)*

Standards of Quality

All plywood products are divided into one of two major categories: construction or decorative. The standards established for governing the manufacture of plywood products are PS 1-66 for softwood plywood—construction and industrial, and PS 51-71 hardwood and decorative plywood.

Hardwood plywoods are inspected and placed into one of six grade classifications:

1. Premium—face of specified hardwood, such as walnut or mahogany, which have been carefully matched as to color and grain
2. Good—for natural finish; similar to premium, but veneers are not matched as accurately
3. Sound—provides a base for smooth painted surfaces; face free from open defects
4. Utility—discoloration and knotholes up to ¾ inch in diameter
5. Backing grade—unselected for grain or color; defects
6. Specialty—used for architectural plywood; matched grain panels for special purposes

Softwood Plywood

Softwood plywood grades are identified by capital letters A, B, C, D, and N (Figure 3–12). Capital A indicates that the face is free of knots and it is sanded. B means that knots

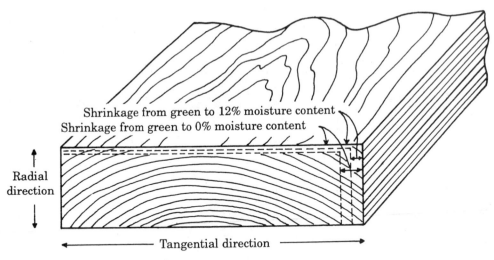

Shrinkage from green to 12% moisture content
Shrinkage from green to 0% moisture content

Radial
direction

←——————— Tangential direction ———————→

Figure 3–11   Dimensional loss during drying process

are permitted, but they must be tight. Capital C indicates that knotholes less than 1 inch in diameter are permitted. D represents interior plywood, and N means special order free from all defects.

Hardboard

Hardboard is an all-wood panel made from wood fibers. Logs are cut into small wood chips and reduced to fibers by steam or mechanical processes. The fibers are then compressed and refined under heat and pressure to produce a sturdy, strong material. There are five basic grade classifications of hardboard (Table 3–2). All classes of hardboard are manufactured with two types of surfaces, either S1S (sanded one side) or S2S (sanded two sides).

Particleboard

Particleboard is made by combining wood chips, scraps, flakes, and other wood fragments with an adhesive to form a large panel. It was first manufactured in 1948 and represents excellent utilization of wood and adhesives. Particleboard is formed by heat and pressure; and its properties can be changed by varying the kinds and amounts of adhesive, pressure, size and shape of chips, and methods of forming. It is used widely in both the manufacture of furniture and cabinets, and the construction of homes and commercial buildings. Figure 3–13 illustrates and describes the transformation of particleboard from raw material to finished product.

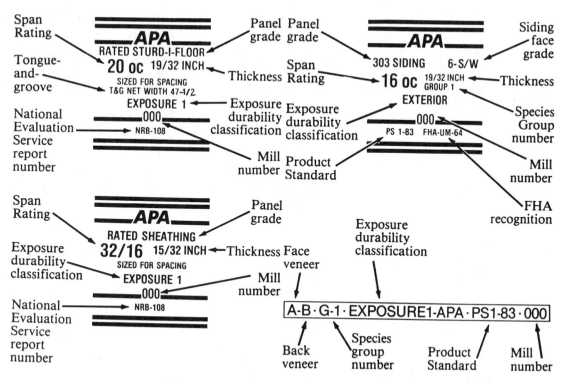

Figure 3–12   Grading marks on plywood

PLASTICS

Compared to wood and metals, plastics are relatively new materials used in manufacturing and construction. Plastics are considered synthetic materials and are based on the carbon atom in which the covalent bond predominates. Although the first synthetic plastics were discovered more than 100 years ago, they did not emerge as a significant material for manufacture until after World War II. Shortages of metal and natural rubber during the war caused greater emphasis in the development of synthetic materials that could be used as a substitute for dwindling supplies of natural materials. In 1995, millions of tons of plastics were being used and transformed into an almost infinite number of manufactured products. Today, engineers and technologists have approximately 5,000 chemically different types of plastics from which to select for manufacturing purposes.

Plastics are classified into one of two large groups whose structures are very similar until heated. **Thermoplastics** are a group of plastics which can be heated and shaped or formed repeatedly. **Thermosets** are cured by heat and, once set, cannot be melted and reshaped. The molecular structure of the thermosetting plastics, after curing, is different from that of thermoplastics, thus preventing remelting. Cross-links between molecules prevent the thermosets from slipping and flowing when heat is reapplied.

Class	Surface	Nominal thickness (inches)	Water resistance, max av per panel (percentage)				Modulus of rupture (min av per panel) (p.s.i.)	Tensile strength (min av per panel)	
			Water absorbtion based on weight		Thickness swelling			Parallel to surface	Perpendicular to surface
			S1S	S2S	S1S	S2S			
	S1S	1/12	30	—	25	—			
1 Tempered	S1S & S2S	1/10	20	25	16	20	7,000	3,500	150
		1/8	15	20	11	16			
		3/16	12	18	10	15			
		1/4	10	12	8	11			
		5/16	8	11	8	10			
		3/8	8	10	8	9			
2 Standard	S1S & S2S	1/12	40	40	30	30	5,000	2,500	100
		1/10	25	30	22	25			
		1/8	20	25	16	18			
		3/16	18	25	14	18			
		1/4	16	20	12	14			
		5/16	14	15	10	12			
		3/8	12	12	10	10			
3 Service-tempered	S1S & S2S	1/8	20	25	15	22	4,500	2,000	100
		3/16	18	20	13	18			
		1/4	15	20	13	14			
		3/8	14	18	11	14			
4 Service	S1S & S2S	1/8	30	30	25	25	3,000	1,500	75
		3/16	25	27	15	22			
		1/4	25	27	15	22			
		3/8	25	27	15	22			
		7/16	25	27	15	22			
		1/2	25	18	15	14			
	S2S	5/8	—	15	—	12			
		11/16	—	15	—	12			
		3/4	—	12	—	9			
		13/16	—	12	—	9			
		7/8	—	12	—	9			
		1	—	12	—	9			
		1-1/8	—	12	—	9			
5 Industrialite	S1S & S2S	3/8	25	25	20	20	2,000	1,000	35
		7/16	25	25	20	20			
		1/2	25	25	20	20			
	S2S	5/8	—	22	—	18			
		11/16	—	22	—	18			
		3/4	—	20	—	16			
		13/16	—	20	—	16			
		7/8	—	20	—	16			
		1	—	20	—	16			
		1-1/8	—	20	—	16			

Table 3–2 Kinds of hardboard

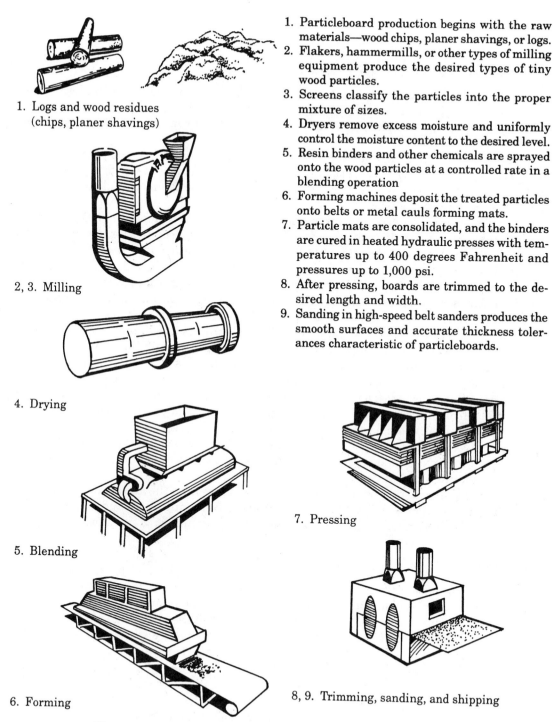

1. Logs and wood residues
(chips, planer shavings)

2, 3. Milling

4. Drying

5. Blending

6. Forming

7. Pressing

8, 9. Trimming, sanding, and shipping

1. Particleboard production begins with the raw materials—wood chips, planer shavings, or logs.
2. Flakers, hammermills, or other types of milling equipment produce the desired types of tiny wood particles.
3. Screens classify the particles into the proper mixture of sizes.
4. Dryers remove excess moisture and uniformly control the moisture content to the desired level.
5. Resin binders and other chemicals are sprayed onto the wood particles at a controlled rate in a blending operation
6. Forming machines deposit the treated particles onto belts or metal cauls forming mats.
7. Particle mats are consolidated, and the binders are cured in heated hydraulic presses with temperatures up to 400 degrees Fahrenheit and pressures up to 1,000 psi.
8. After pressing, boards are trimmed to the desired length and width.
9. Sanding in high-speed belt sanders produces the smooth surfaces and accurate thickness tolerances characteristic of particleboards.

Figure 3–13 Particleboard—from raw material to finished product

From Raw Materials to Primary Plastics

Most plastics are formed from natural materials, which are carbon based. They are developed through chemical as well as mechanical processes. Heat, pressure, and chemical action are used to separate the atoms and molecules of the raw materials, which are recombined to form new materials.

The most common thermoplastics include cellulosics, acrylics, styrene plastics, nitrile barrier copolymers, vinyl plastics, polyolefins, polymides, acetal resins, ionomer plastics, and thermoplastic rubbers. Many plastics fall within one of these families, and each is physically different from the others.

The principal raw material sources for the thermosets and thermoplastics are petroleum, natural gas, coal, limestone, sulfur, salt, water, and air. The products formed from these materials are known as intermediates and are transformed into plastic resins.

For example, the result of linking benzene, which is extracted from coal (as a by-product of coke ovens), and ethylene, which is extracted from coal or natural gas, is ethylbenzene. Ethylbenzene is formed when the ethylene is forced through benzene in the presence of a suitable catalyst such as aluminum chloride. Ethylbenzene is converted to styrene by splitting hydrogen away from the rest of the molecule. This process is accomplished by a decomposition of the material under heat. The styrene produced is a liquid which will boil at 292 degrees Fahrenheit. This liquid is then converted to a solid plastic through polymerization.

The widely used families of thermosets include the phenolics, the aminos, polyester plastics, silicone plastics, epoxies, epoxy-graphite compounds, and casein plastics. When phenol and formaldehyde are combined in the presence of an alkaline catalyst, they form resins which are permanently soluble and fusible. Heat is used to start the reaction in a resin kettle, and an agitator stirs the mixture. Once the reaction begins, the heat must be removed. Water forms during the process and must be removed through condensation reaction. This process is known as **condensation polymerization.**

An amber material, the consistency of molasses, comes from the kettle and forms into a solid. This material is then ground into powder and mixed with fillers and pigments. The material is then formed into balls, kneaded, cooked, and ground into granules.

Additives

Plastic resins have little use without additives to color them or change their properties. The list of additives includes fillers, extenders, antioxidants, colorants, lubricants, plasticizers, and stabilizers. Other special-purpose materials such as biodegradable resins, antistatic agents, and flame retardants are also added. Fillers reduce the cost of plastics and add body to the material. Aluminum powder, kaolin, and graphite are among the many materials used as fillers. Antioxidants are used to inhibit the degradation of plastic materials within the specification given for the material. Plasticizers are additives which cause the material to improve in flexibility and softness. These additives and others play an important part in the manufacture of plastics and reduce the cost of manufacturing them.

CERAMICS

The purpose of this chapter is to describe the transformation of raw materials to primary materials, that is, the conversion of materials found in their natural state to materials which have been refined and processed for use in the manufacture of usable goods. In some cases, that transformation process is quite extensive and involved (e.g., metals and plastics). The preprocessing or transformation of ceramic materials is perhaps not as involved as other materials.

Excluding oxygen, ceramic materials are the most abundant chemical elements on earth. They are among the hardest materials known and exhibit excellent strength and thermal properties. Ceramic materials and products are classified as

1. Crystalline ceramics
2. Clay products
3. Glasses

Crystalline Ceramics

Crystalline ceramics include those materials commonly known as **refractories,** a group of high-melting-point oxides or a compound of oxides and aluminum (Figure 3–14). The most common refractory materials used today are the aluminum oxides and silicon carbides. These refractories are formed by carefully mixing the constituent mineral powders into clay bodies which are then pressed into desired shapes and fired. The temperature at which they are fired is generally over 1,000 degrees Celsius, but it varies according to desired strengths and properties.

Clay Products

Clay products have been used since the dawn of civilization. After the discovery of fire, the desire to cook food required the development of cooking vessels and utensils. Whether consciously or subconsciously, humans discovered that the heating of the clay caused it to harden; and so, the first vessels for cooking and storing food were shaped of ceramic materials.

The principal elements in natural clay are oxygen, silicon, and alumina. As a result of the reaction of water and carbon dioxide on feldspar, a mineral called kaolin is formed. *Kaolin* is a binder which cements the particles of mineral elements together. These clays, when wet, exhibit plasticity, which enables the material to be molded easily. Most clay products can be grouped according to their use. They are primarily used for *manufacturing* of dinnerware and other commercial and industrial products or as *structural* components in residential and commercial construction projects.

Figure 3–14   Refractory clay products *(Photo courtesy of Batelle Columbus Labs)*

Glass

Glass is a material formed as a result of the fusion of silica, basic oxides, and other compounds that react with silica or basic oxides. Its structure is amorphous, which means that it solidifies from a liquid state without crystallizing. It is actually a supercooled liquid.

Raw Materials

Clay, flint, and feldspar have been the more widely used natural raw materials in the production of ceramic bodies. Alumina is probably the most widely used nonsilicate material used in the production of clay bodies. The vast majority of ceramic products contain more than one constituent. For example, porcelain is a blend of feldspar and flint, ball, and china clays. Following are some of the typical raw materials used in the production of clay bodies:

Fire clay—plastic clays of high refractoriness

Flint clays—hard clays almost devoid of plasticity

Ball clays—fine, dark sedimentary clays of excellent plasticity

Kaolin—a fine, white clay which is used in making china and porcelain

Feldspar—alkali-alumina-silicate which is formed from igneous rock and used as a fluxing agent

Processing

Clays as found in the earth are generally unsuited for immediate fabrication into products. The mined materials must be screened, crushed, and cleaned before further processing can take place. A *ball mill* is used to grind the raw materials into very fine particles. The ball mill is lined with a wear-resistant material, and wear-resistant balls provide the grinding surface. Although the materials are wear resistant, some abrading generally takes place; thus, the lining and balls must be compatible with the raw material. Normal practice is to grind the materials wet to avoid dust in the mill.

A number of these finely ground powders are then mixed with water to develop a desired level of plasticity. The mix is conveyed to a *pug mill* where the material is kneaded and vacuum pumps are used to remove air bubbles. The pug mill extrudes the mix in clay to preform clay bodies which are ready to be shaped into one of many ceramic products.

Ceramics have a number of uses that correspond to their excellent properties and characteristics. In addition to traditional ceramic products such as dinnerware, they are used as coatings over metals and as insulators in the electronics industry; they are also among the hardest abrasive materials known. In the construction industry, they are used in the form of concrete, concrete blocks, brick, tile, and glass panels. Their value is becoming more and more evident since being used on the space shuttle Columbia to protect the vehicle from burning up during reentry. Although traditionally thought of as a material primarily used in the making of pots and vases, it has recently been identified as a modern space-age material which may transform the next generation of manufactured products just as plastic did the last. Chapter 4 discusses the new generation of space-age ceramics and ceramic processes in greater detail.

SUMMARY

From about 90 chemical elements, some of which cannot be efficiently recovered, scientists and technologists have been able to refine, combine, and take apart and recombine elements to produce thousands of primary metals, wood products, plastics, and ceramics that are chemically different and have distinctively different characteristics. The many shapes and forms of these primary materials have provided the basis from which an almost infinite number of manufactured goods have been developed. These essential goods and nonessential goods have increased our standard of living and helped us to adapt to a changing world.

REVIEW QUESTIONS _____

1. How many chemical elements have been identified?
2. The chemical elements and the primary materials produced from them can be divided into two broad classifications. List the two classifications and describe the difference between them.
3. Give examples of the two material classifications listed above.
4. Metals are classified as *ferrous* or *nonferrous*. Describe the difference between the two and list examples of each.
5. What are the primary raw materials used in the refining of iron? Check library references for major sources of these raw materials.
6. What is the difference between iron and steel?
7. What gas is produced when the oxidation of carbon occurs during the steel-making process?
8. What is the principal alloy in *specialty* steels?
9. What forming process is used to produce standard shapes of steel products?
10. What are the most commonly used nonferrous metals?
11. Describe the process used in refining bauxite to form aluminum.
12. List the two important copper alloys and the constituents of those alloys.
13. Describe the movement from lumbering as an extractive industry to one of harvesting and reforestation.
14. What are the major softwood and hardwood regions in the United States?
15. What historic event brought about great advancements in the plastics industry?
16. List the two broad categories of plastics and provide examples.
17. What are the two most commonly used refractory ceramics today?

SUGGESTED ACTIVITIES _____

1. Search for articles related to our rapidly diminishing natural resources.
2. Plan a project using totally recycled materials. If you have a foundry, collect aluminum cans for a cast project.
3. Several good films and filmstrips are available which describe the mining and refining of iron. View one of these and discuss the importance of iron and steel to the human race.
4. Locate examples of the various standard mill shapes formed in steel mills. List products that use the shapes.
5. Promote reforestation in your neighborhood by planting trees in empty lots.

6. Cut a small block of wood and weigh it. Place it in an oven on low heat overnight and reweigh it in the morning. Determine its percentage of moisture with the following formula:

% M.C. = $(W_w - W_d)/W_d$ x 100
   M.C. = Moisture Content
      W_w = Wet weight (weight before oven-drying)
      W_d = Dry weight (weight after oven-drying)

CHAPTER 4

Synthetics and Space-Age Materials

The ability of the scientist and technologist to develop new materials, each with its own characteristics, from the limited natural elements available appears to be unlimited. The parade of new materials developed and described in the journals related to plastics, metals, ceramics, and composites is a continuous occurrence.

It is estimated that there are as many as 2,000 types of steel, 5,000 types of plastics, and 10,000 kinds of glass. Add to this the phenomenal growth in synthetic ceramic materials and composites and the choices seem almost infinite. The development in outer space and the advantages it offers in the area of research and development of new materials and processes will certainly further extend the phenomenal growth in the area of synthetics.

This chapter describes some of the developments that have taken place with respect to space-age materials technology.

COMPOSITES

Composites consist of two or more materials or material phases combined to produce a material that has superior properties to those of its individual constituents. These materials, when combined, do not merge into one another but are linked together to combine the properties of the separate constituents. Composites are included in a class of materials generally referred to as engineered. The term **engineered** is used to distinguish materials designed for specific applications from conventional materials such as steel, iron, and wood. All composites consist of a **reinforcement,** a **matrix,** and a boundary between the two known as an **interface.**

Composite materials have existed in nature for a long time. For example, wood is a natural composite material whose constituents consist of cellulosic fibers in a matrix of lignin, and leather is a composite of a matte of fibers in a matrix of protein. While both of these examples can theoretically be classified as composite, most developments in this area are related to the ceramic, plastic, and metal matrix composite materials. The result of these developments is materials known as advanced composites.

Composites are often classified as **polymer matrix composites** (PMCs), **metal matrix composites** (MMCs), and **ceramic matrix composites** (CMCs). Although the PMCs dominate the advanced-composites applications at the current time, significant advancements and applications are occurring in the areas of advanced CMCs and MMCs.

The form and size of the reinforcement used in the composite is often used as a method of classifying or categorizing composites as well. **Particle-strengthened composites** include those reinforced by spheres, flakes, and other reinforcements of approximately equal dimension on all axes. **Fiber-strengthened composites** have reinforcers whose length is greater than their width. **Laminar composites** are made of layers (two or more) of reinforcement. The reinforcing material used in laminar composites has two dimensions significantly greater than the third. Figure 4–1 illustrates the types of reinforcement used in each of these composites.

Composites are being used in a wide variety of defense, space, and consumer products, ranging from space shuttles and jet aircraft to tennis racquets and golf clubs. Aerospace applications of advanced composites account for approximately 50% of current sales. Golf clubs, tennis racquets, skis, and other sporting goods account for another 25%.

The use of composites is growing rapidly. Products made of composite materials in 1988 valued approximately $2 billion. This figure is projected to increase to $20 billion by the year 2000.

Carbon-Fiber Composites

Carbon-fiber composites are being used today in the production of airplanes, boats, golf clubs, tennis racquets, and fishing rods. In addition, they have recently been used in the robotics industry because of their excellent strength/weight ratio.

Carbon-fiber composites are formed by heating carbon-based fibers like rayon and orlon to temperatures of about 2,000 degrees Fahrenheit. This process burns everything but the pure carbon threads which are soaked in plastic resins such as epoxy or polyester. The soaked fibers are then woven into a cloth for further processing or wound around giant molds and allowed to cure and form rigid structural shapes.

Carbon-fiber composites have several advantages over metals:

1. They do not rust.
2. Huge assemblies can be made one piece at a time.
3. They are wear resistant.
4. Structural strength can be built into each part in the desired direction.

Synthetic graphite is a mixture of crystalline graphite cross-linked with intercrystalline carbon, thus producing increased strength at high temperatures. Synthetic graphite can be

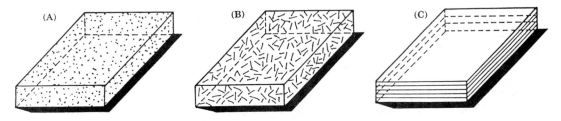

Figure 4–1 **(A)** particle, **(B)** fiber, and **(C)** laminar composites

made from almost any organic (carbon-based) material that leaves a high carbon residue upon heating to temperatures of 2,500–3,200 degrees Celsius.

The carbon fibers are filamentary forms of carbon offered in the form of yarn containing from 1,000 to 500,000 filaments per strand. They are flexible, lightweight, thermally and chemically inert, and good thermal and electrical conductors.

High-performance graphite cloth has been used recently in the development of a new jet aircraft. The lightweight cloth has made these aircraft extremely fuel efficient.

Union Carbide now has a 4,000-filament continuous-length graphite cloth called Tharnel P-55 that is aimed at the aerospace industry and robotics control arms. This high-strength yet ultralight material is the first to sell at under $20.00 per pound. Tharnel T-300 PAN-based fiber cloth, which also is produced by Union Carbide, has a tensile strength that exceeds 600,000 psi.

Cermets

Cermets are a group of composite materials consisting of ceramic and metallic constituents. They are fabricated by mixing finely divided constituents of powders and fibers, compacting them under pressure, and sintering the compact to produce combined properties. Sintering causes a glass bond to occur, thus combining the constituent materials.

Ceramic constituents include metallic oxides, carbides, borides, silicones, nitrides, or a mixture of these compounds. These composite materials have increased strength and hardness, wear resistance, and resistance to corrosion. Cermets are commonly used as cutting material for lathe, mill, and drill bits.

Powder Metallurgy

Powder metallurgy, commonly referred to as P/M, is a process used to form metal parts from powder exactly or close to final dimension and finish, with little or no machining required. Modern P/M technology began during World War II and has been growing ever since. It is an economical, high-production method of producing parts of various shapes and sizes.

There are four basic steps used in the production of powdered-metal products:

1. Production of metal powders and additives
2. Blending and mixing of powders
3. Shaping blended powders
4. Sintering the shaped product

One of three basic methods is employed to produce the powdered metals. These methods are either physical, chemical, or mechanical. Following are some examples of these methods:

Atomization (physical): dispersion of metal into particles by impact force of high-velocity air or inert gas

Reduction (chemical): chemical decomposition of a compound of a metal into a powder (e.g., the contact of tungsten oxides with hydrogen at temperatures below their melting points causes reduction to powders)

Electrolytic: special kind of reduction by chemical decomposition of a metal by electric current from an electrolyte containing a metal salt

Mechanical: pulverization by crushing, grinding, and milling of metals

Blending or mixing is done to distribute the powdered constituents and additives uniformly in the mix. Powdered graphite is often added to the mix as a lubricating agent.

Shaping of powdered-metal products is accomplished by one of several methods. These methods include compaction, extrusion, and rolling. Compaction is the most commonly used technique. It involves forming the metal powders in a closed-die arrangement using pressures between 10 and 60 tons (Figure 4–2). Extrusion, roll compacting, and isostatic pressing are other fabrication techniques used to shape powdered-metal parts.

Sintering is the heating of the shaped part in a controlled-atmosphere furnace to about 70% of the melt temperature of the constituents. Sintering is often done in a hydrogen atmosphere at a temperature of 3,000 degrees Fahrenheit (1,800 degrees Celsius). This process bonds the constituents together into a strong cohesive mass and eliminates oxidation as oxygen is replaced by hydrogen.

Powdered-metal fabrication can be used to manufacture a wide range of products for manufacturing as well as construction industries. Its applications extend from agriculture to aerospace as the industry continues to develop and grow (Figure 4–3). Among the advantages of powdered-metal fabrication are these:

1. Machining is eliminated.
2. Material losses are eliminated.
3. Close dimensional tolerances are maintained.
4. Unique properties such as controlled porosity or variations in composition can be achieved.

Ceramics

Ceramics have been used for over 10,000 years by humans to make usable products. Perhaps the first came after the discovery and use of fire for cooking. Cooking required vessels for holding food, and the first such containers were made from the bountiful supplies of clay.

Today, ceramics are being used experimentally in the production of engines (Ford, GM, Cummins), gene-splicing equipment, office machinery, body replacement parts, and others (Figure 4–4). Ceramics are easy to shape, durable, and able to withstand corrosion and high temperatures. They are also ultralight and have amazing insulating properties. It has been predicted that by the year 2000, approximately 0.75 ton of ceramic parts per person will be produced.

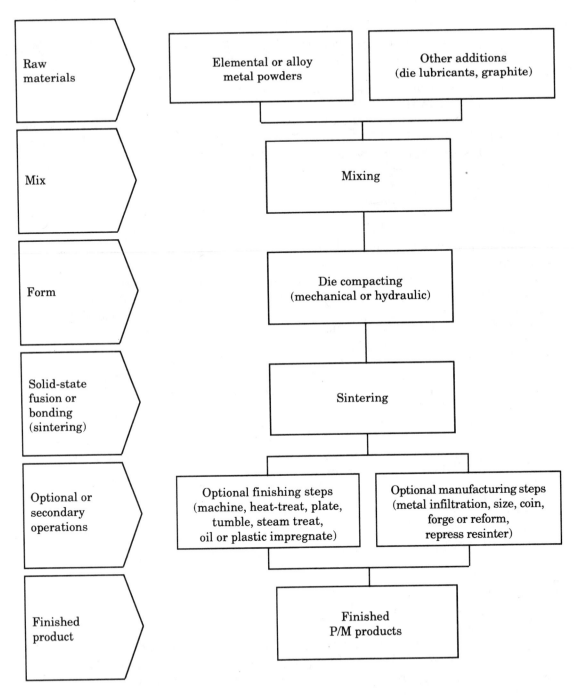

Figure 4–2 Processing powdered metals

Figure 4–3   Powdered metal products *(Courtesy of Metal Powders Industries Federation, Princeton, NJ)*

Ceramics are cheap, compared to competing materials. Al_2O_3 alumina ceramic powder sells for as little as $0.40 per pound, compared to polyethylene plastic at $0.43 per pound and steel at $0.80 per pound. Silicon nitride engine-component parts are one-half to one-third less in cost than metal alloys.

Ceramics are made from raw materials, primarily in the form of silicon and aluminum; after oxygen, they are the most abundant elements in nature.

Fabrication

In addition to the gains made in the development of new forms of ceramic materials, and perhaps more important, are the new techniques developed to fabricate the ceramic parts. **Injection molding,** a process first developed for ceramics and then abandoned only to be picked up by the plastics manufacturers and successfully applied to plastics fabrica-

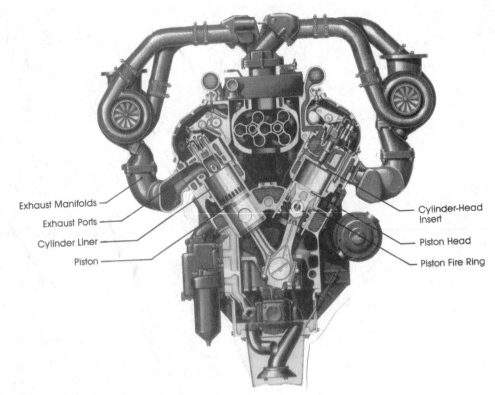

Exhaust Manifolds

Exhaust Ports

Cylinder Liner

Piston

Cylinder-Head Insert

Piston Head

Piston Fire Ring

Figure 4–4   Ceramic engine *(Courtesy of Cummins Engine Company)*

tion, is now being used in the ceramics industry. The process, developed by Ray Weich, uses equipment designed for injecting plastics, with little modification, to produce ceramic parts that need little or no machining. This process reduces brittleness and allows the fabrication of complicated shapes.

Weich's process immobilizes ceramic powders to 100 microns in size in a thermoplastic binder and injects them into a die cavity (Figure 4–5). The die is then heated and pressure applied so that the powders flow into desired shapes. These shapes are then fired twice— once to evaporate the binders, and once to fuse the powders.

Perhaps the most publicized use of ceramic materials in recent years has been the tiles attached to the Columbia space shuttle (Figure 4–6). The 34,000 tiles used to shield the Columbia from extreme reentry temperatures are 90% air by volume. The rest is ultrapure silica fibers capable of withstanding temperatures of 1,540 degrees Celsius (2,800 degrees Fahrenheit) without expanding. These tiles are able to withstand thermal shock. They can be heated with a blow torch and dipped in an ice-water bath without damage.

The tiles were developed by Johns Manville Corporation and required 34,000 different computer-controlled programs to machine each one to specifications. The programs were

Figure 4–5   Injection molding of ceramic parts *(Courtesy of Batelle Columbus Labs)*

designed to control a high-speed diamond-tipped cutter, and each tile cost $1,000. Sixty percent of this price was for the machining and quality control. The rest was for the silica fibers.

Closer to earth are the developments taking place in the truck and automobile industries. Ford and General Motors, with the help of other related industries, are working toward the development of a ceramic gas turbine powerful enough to propel 3,000-pound vehicles at 40 miles per gallon.

Cummins Engine Company is primarily responsible for the development of a diesel that uses ceramics for the pistons, piston caps, cylinder liners, exhaust gas manifold, and valve train components (Figure 4–7). Cummins has already developed a prototype of an uncooled turbo compound diesel designed to power a five-ton army truck. Cummins is hoping to begin commercial production of an adiabatic engine (occurring without loss or gain of heat) which will operate on synthetic lubricants and at temperatures up to 1,200 degrees Celsius (2,210 degrees Fahrenheit).

If this type of engine is to be a reality, then a bearing capable of the same extreme heats will be needed. S.K.F. industries is under way in developing a bearing that will meet these requirements. The cost will be close to $100, compared to $0.50 for a steel bearing.

Figure 4-6 Ceramic tiles on the nose of the space shuttle Columbia *(Photo courtesy of NASA)*

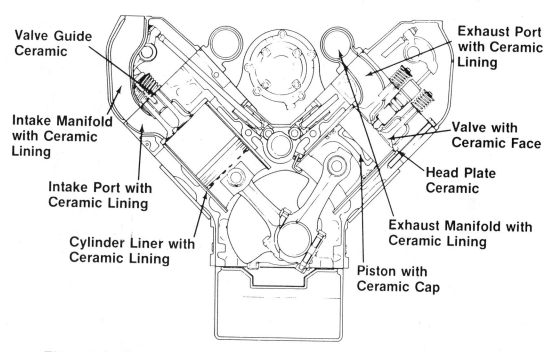

Valve Guide
Ceramic

Exhaust Port
with Ceramic
Lining

Intake Manifold
with Ceramic
Lining

Valve with
Ceramic Face

Head Plate
Ceramic

Intake Port with
Ceramic Lining

Cylinder Liner with
Ceramic Lining

Exhaust Manifold with
Ceramic Lining

Piston with
Ceramic Cap

Figure 4–7   Experimental ceramic engine *(Courtesy of Cummins Engine Company)*

Ceramics are ideal for numerous applications because of their excellent thermal properties; however, they are often too brittle to be used where any vibration or shock is applied. The ideal space-age material is one that has the thermal properties of ceramics and the toughness and ductility of metal. Although this appears to be unrealistic, United Technologies Research Center has been developing silicon carbide fibers embedded within a ceramic or glass matrix. The fiber diverts the path of any fracture toughness five to six times greater than conventional ceramics. However, temperature capabilities are limited to 1,000 degrees Celsius (1,823 degrees Fahrenheit).

Semiconductors

Semiconductors are a family of materials which structurally lie between ceramics and polymers and have some of the characteristics of each. Their use in electronics has enabled scientists and technologists to miniaturize almost anything that operates electronically.

Semiconductors are located between conductors and resistors. They are materials with filled valence bonds but with a small energy gap between the upper filled bond and the overlaying vacant energy bond. Electrons can jump across these energy gaps if stimulated by photons or thermal energy.

Group IIIA	Group IVA	Group VA
Non-metals		
5 B 10.811	6 C 12.011	7 N 14.007
13 Al 26.98	14 Si 28.086	15 P 30.97
31 Ga 69.72	32 Ge 72.59	33 As 74.92
49 In 114.82	50 Sn 118.69	51 Sb 121.75
81 Tl 204.37	82 Pb 207.19	83 Bi 208.98

Table 4–1   Group III, IV, and V elements from the periodic table

Semiconducting materials are compounds made of elements in Groups III, IV, and V of the periodic table (Table 4–1). Group III elements include such nonmetals as boron and indium. These elements have three electrons in their outer valence shell. Group IV elements include silicon and germanium, which are the primary materials used in the production of semiconductors. These elements have four electrons in their outer valence shell. Group V elements include such materials as arsenic and phosphorous, and they have five electrons in their valence shell.

Group III and Group V elements, when compounded with Group IV elements, alter the conductance of Group IV elements. Group III elements, when added to silicon, create an imbalance of electrons, which causes the formation of electron holes. These holes remain in the valence bond as positive carriers for P-type semiconductors.

Group V elements, when added to silicon, create additional electrons which have a negative charge, thus forming n-type semiconductors.

At low temperatures, there is little activity between the energy bonds of a semiconductor material. However, heat—whether radiant or photonic—causes the semiconductor material to become active and the electrons to begin moving across energy bonds. The n and p junctions are formed to P-N-P or N-P-N transistors. Current can be made to flow in one direction or the other.

The fact that both the n-type and p-type semiconductors can be made is the basis for present-day solid-state electronics. Semiconductors are the basis for solar cells and light emitting diodes, used in electronic displays on everything from home computers and calculators to video arcade games and wristwatches.

SUMMARY

New and continuing developments in materials science and technology have provided the engineer and technician with an ever-widening choice of materials for application in manufacturing. The developments of composites and new forms of synthetic ceramics and plastics have provided better materials for our space-age demands.

With the already-developing technology in space manufacturing and experimentation, the list of new and better synthetic materials is certain to continue at a runaway pace. It will then be up to the engineer and technologist to apply new materials in a way that benefits humanity.

 REVIEW QUESTIONS _____

1. Define *composite materials*.
2. List two natural composite materials. Describe the constituents of those listed.
3. How are cermets commonly used?
4. List the four basic steps used in the production of powdered-metal products.
5. What is the most commonly used shaping technique in forming powdered-metal products?
6. List two or three advantages of powdered-metal fabrication.
7. List examples of products made from carbon fiber composites.
8. What advantages does injection molding of ceramics have over other forming processes?
9. Describe n and p junctions and the formation of P-N-P and N-P-N transistors.

SUGGESTED ACTIVITIES _____

1. Write to the Metal Powder Producers Association for a copy of *The Many Uses of Metal Powders*. Address: Metal Powder Producers Association, 105 College Road East, Princeton, NJ 08540
2. Obtain materials for reinforced plastics (fiberglass) and make a small reinforced plastic mat. Test the strength of this mat and compare it to the strength of its constituent materials (fiberglass cloth and plastic resin).
3. Discuss the present uses of ceramics and potential future uses in replacing traditional materials in the manufacture of products.

PRE PROCESSING

PROCESSING

GENESIS

SECTION
2

RECOVERY

POST PROCESSING

INDUSTRIAL MATERIALS

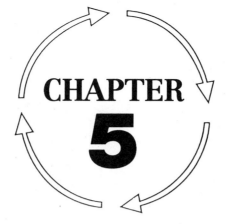

CHAPTER 5

Metals: Structure, Properties, and Characteristics

Most authoritative sources define *metal* by describing its properties or atomic structure. The combination of properties that distinguish metals from other materials are hardness, ductility, tensile strength, and opacity. Metals are also good conductors of heat and electricity. This is a result of their atomic structure.

The conductivity of metallic materials is the result of an atomic structure that has free electrons in the valence (outer) shell of each atom. Atoms are neutral, which means the number of positive protons and negative electrons are equal. However, elements with few electrons in the outermost shell tend to give up those electrons to atoms that need them to fill their valence shell (Figure 5–1). This process is referred to as **ionization.** When the atom gives up electrons, it becomes a positively charged ion; and when it accepts electrons, it becomes a negatively charged ion. Most metals have few electrons in the outer shell and tend to give them up, thus becoming positive ions.

When large quantities of atoms from an element of metal come together, the electrons from the valence shell form what is referred to as an **"electron cloud."** This electron cloud moves freely among the atoms. The mobility of the electrons is the characteristic that causes metals to be such excellent conductors of electrical and thermal energy.

ATOMIC STRUCTURE OF METALS

The smallest particle of an element is known as an **atom.** Atoms are often referred to as the building blocks of all materials. An understanding of the properties and characteristics of metal must begin with a general understanding of atomic structure.

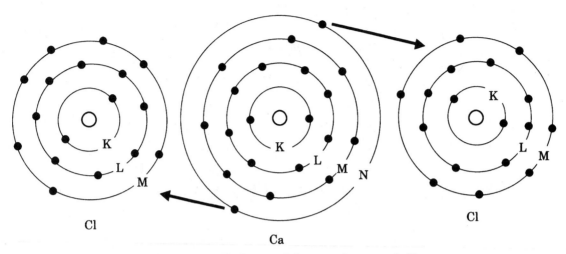

Figure 5-1   Exchange of electrons in outer shell

The atom consists of neutrons, protons, and electrons (Figure 5–2). The neutrons and protons form what is referred to as the nucleus. Protons have positive electric charges, and neutrons have no charge.

The neutrons and protons are much heavier than the electrons; therefore, the **atomic weight** of the atom is determined by the sum of the weights of the protons and neutrons. The sum of the protons and neutrons in the nucleus of metal atoms is generally very high compared to lighter material and gases.

Iron (Fe) has an atomic weight of 56, zinc (Zn) 65, and lead (Pb) 207. In comparison, helium (He) has an atomic weight of 4. The dense nuclei of the metal atoms is a contributing factor to the physical properties of metal.

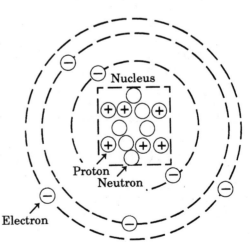

Figure 5-2   The atom is the basic building block of metals

GRAIN AND GRAIN BOUNDARIES

Metals, with few exceptions, are polycrystalline aggregates of many individual crystals or grains. The size of the individual grains or crystals varies with the type of metal or alloy. Some metals have individual grains that can be seen with the unassisted human eye. Others require the use of a microscope or other magnifying device.

Grain size and the area between adjoining grains, known as the grain boundary, affect the properties and chacteristics of metals. For example, as grain size becomes smaller, yield strength increases. Surface finish also improves as grain size decreases. Metals with large-grain structure are generally easier to machine and harden than those with small- or fine-grain structure. However, the machined surface of coarse-grained metals is not as smooth as fine grained. Coarse-grained metals also fracture more easily than fine grained.

Grain size is determined by several factors. The type of metal and the alloy are among the factors determining grain size. Grain size established during melting and solidification processes is a product of the cooling rates. Metal cooled quickly will form finer grains. Slow cooling rates produce larger more stable grains.

Grain boundaries also influence the properties of metals. For example, during the refining process, impurities sometimes collect in the grain boundaries and may be a source of weakness to the material.

PROPERTIES OF METAL

The properties of metal are generally classified as mechanical, thermal, electrical, and chemical. Each of these properties are considered carefully by the engineer or technologist when searching for a material that meets the specifications for a new part or product.

MECHANICAL PROPERTIES

Mechanical properties describe behavior of a material when static and dynamic loads (forces) are applied to that material. Static loads are loads at rest. The weight of a hammer resting on a board or the weight of a house on the foundation are examples of static loads. The force exerted when a nail is driven into a board is dynamic. Dynamic loads refer to forces in motion. The forces exerted when an automobile engine is running also represent dynamic forces (e.g., the movement of pistons, crankshaft, valves, etc.). The following represents some of the important mechanical properties of metals.

Hardness

The **hardness** of metal refers to the ability of the material to resist penetration. Hardness is related to mechanical strength of metal and therefore is an important property to consider when selecting the best material for the part or product.

The most commonly used indexes for hardness are the *Rockwell hardness number* (R) and the *Brinell hardness number* (BHN). Although the scales for each system are different, each is based on the indentation made in the metal by an indenter (usually a steel ball).

The Rockwell hardness test is accomplished by forcing a $1/16$-inch steel ball into the sample under a load of 60 to 100 kg. For harder materials, a diamond indenter and an increased load (150 kg) are used. Two scales, R_c and R_b, are commonly used in determining hardness using the Rockwell system.

The Brinell hardness test is accomplished using a larger-diameter indenter (0.39 inch) and a load of 3,000 kg. A load of 500 kg is used for nonferrous metals. The diameter of indentation is the reference for the Brinell scale. The softer the material, the deeper the indenter is forced into the surface and the larger the diameter of the indentation.

The Brinell hardness number is calculated using the following formula:

$$BHN = \frac{F}{(\pi D/2)[D - \sqrt{(D^2 - d^2)}]}$$

where

BHN = Brinell Number = Force/Unit Area
F = load
d = indentation diameter (mm)
D = indentor ball diameter (mm)

Table 5–1 compares Rockwell and Brinell hardness for some common metals.

	BHN	Rockwell (B)
Cold-rolled steel	150	80
Stainless steel	150	80
Cast iron	200	95
Wrought iron	100	55
Aluminum	100	55
Brass	120	55
Magnesium	60	14

Table 5–1   Brinell and Rockwell hardness scales

Tensile Strength

The **tensile strength** of metal is defined by the resistance to a force or load that pulls the material apart. The response of the metal to the applied load is normally described as the elastic and plastic behavior. The initial response to an applied load is an elastic response. The load, when first applied, will cause the metal to stretch or elongate. If the response of the metal to the removal of the load is to return to its original shape, the strain on the metal is referred to as **elastic behavior.** This elastic behavior is equivalent to stretching and then releasing a rubber band.

If the load is increased and the strain is great enough to stretch the metal beyond its elastic limits, thus causing permanent deformation, it is referred to as **plasticity** or **plastic behavior.** This behavior is often desirable, especially when primary shapes such as ingots are stretched to form larger structural shapes.

Tensile strength is determined by conducting a *stress–strain tensile* test using standards outlined by the American Society for Testing and Materials (ASTM). A variety of mechanical properties can be determined as a result of the test and data received from this test.

In preparing for the actual test, a standard specimen is machined to size and shape. The standards for the specimen are defined by ASTM and vary by the type of material being tested. A commonly used metal specimen is referred to as a ".505" specimen. The ".505" refers to the diameter of reduced section of the specimen (Figure 5–3).

The convenience of the .505 diameter is that it produces a 0.20 cross-sectional area, which makes the calculations for ultimate strength and breaking strength much simpler. The following terms and formulas are commonly used in determining tensile strength:

Proportional Limit is the point beyond which stress is no longer proportional to strain. As the diagram in Figure 5–4 illustrates, stress and strain are linearly proportional to the point known as proportional limit.

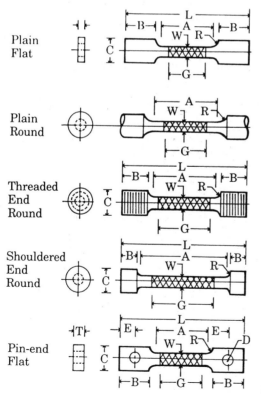

Figure 5–3   ASTM .505 standard tensile specimen

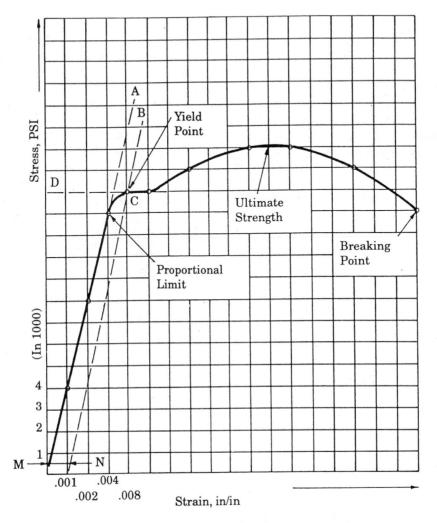

Figure 5–4   Typical stress–strain curve of mild steel. Brittle materials may have steeper curves. To determine the yield strength by the "offset" method, on the stress–strain curve lay off MN equal to the specified value of the set (0.2%). Draw NB parallel to MA and locate C. The intersection of NB with the curve gives the strength value as read at point D.

Yield Point is defined as the point at which strain occurs without a concurrent increase in stress. Plastic deformation is observed at this point.

Breaking Strength (BS) is a point at which the specimen fractures. It is equal to the breaking load divided by the cross-sectional area (CSA).

BS = Breaking Load (psi)/CSA

Ultimate Tensile Strength (US) is the maximum load force per unit of original cross-sectional area.

US = Ultimate Load/CSA

Elongation is total plastic strain before fracture measured as a percentage of strain along the axis of the specimen. *Note:* Gauge length must be specified in calculating percentage of elongation. A 2-inch gauge length is commonly used.

% Elongation = [(g.l. after – g.l. before)/g.l. before] × 100

Reduction of Area (AR) is total plastic strain before fracture measured as the percentage decrease in the cross-sectional area. The difference between the original cross-sectional area and the cross-sectional area at the point of rupture is generally expressed as a percentage of the original area.

% AR = [(CSA before – CSA after)/CSA before] × 100

The following example and calculations should provide further clarification of tensile strength concepts.

Specimen: 0.505
Material: SAE-AISI No. 1040 Hot-Rolled Steel
Breaking load: 12,500 psi
Ultimate load: 15,200 psi
Original gauge length: 2 inches
Area of reduction after fracture: 0.12
Gauge length after fracture: 2.4 inches

Calculate the following:

BS = 12,500/0.2 = 62,500
US = 15,200/0.2 = 76,000
% Elongation = [(2.40 – 2.0)/2.0] × 100 = (0.40/2.0) × 100 = 0.20 × 100 = 20%
% AR = [(0.20 – 0.12)/0.2] × 100 = (0.08/0.2) × 100 = 40%

The percentage of elongation and the area of reduction are factors used to determine the ductility of the material. Ductility is the characteristic which enables the metal to be bent, twisted, drawn out, or otherwise shaped without fracturing.

Compressive Strength

Compressive strength is the ability of the material to withstand a squeezing or pressing force or load. Compression tests are generally done with concrete or brittle metals such as cast iron. The values for stress and strain in compression testing are very similar to those found in tensile testing. This is due to the fact that forces required to cause stress required to pull material apart are usually the same as the stress required to push them together.

The proportional limit, the elastic range, and the yield strength are generally the results sought in compression testing. Tensile strength and compression strength for the 1040 hot-rolled steel used in the previous example would be the same, approximately 50,000 psi. However, in the compression test, the cross-sectional area of the specimen increases rather than decreases. Compressive strength is calculated using the following formula:

$$\text{Compressive Strength} = \text{Breaking Load/CSA}$$

Stiffness

Stiffness is a term used to describe the resistance of a material in the elastic range of the stress–strain diagram (see Figure 5–4). It is the ratio of stress to elastic strain. The elastic range is the straight-line region from the origin to the elastic limit.

The measure of stiffness for ductile materials as metal is **modulus of elasticity** or **Young's modulus.** The letter E is used to designate modulus of elasticity or Young's modulus. E is found by dividing stress in pounds per square inch by strain. The result is pounds per square inch. E is calculated by dividing any stress value along the elastic range of the stress–strain diagram by the corresponding strain. The following formula is used to calculate modulus of elasticity or stiffness:

$$E = \frac{(S_1 - S_2)}{(e_1 - e_2)}$$

where

E = modulus of elasticity (stiffness), psi
S = stress, psi
e = strain, in./in.

For Figure 5–3 the following values are used to compute E:

$$S_1 = 36,000$$
$$S_2 = 25,000$$
$$e_1 = 0.0012$$
$$e_2 = 0.0008$$

$$E = \frac{36,000 - 25,000}{0.0012 - 0.0008} = \frac{11,000}{0.0004} = 27.5 \times 10^6 \text{ psi}$$

Shear Strength

Compression and tensile strength relate to the resistance of a load or force that is axial. The areas or planes resisting the load are perpendicular to the force. Shearing stresses occur when the area or plane resisting the load is parallel to the force or load (Figure 5–5). Shearing stress is determined by dividing the shear force by the cross-sectional area.

$$S = F/CSA$$

Shear strength is often considered when two pieces are fastened together in either an overlapping or perpendicular configuration. For example, when a vertical and horizontal member of a large steel-framed building are riveted together, shear strength of the fasteners must exceed the shear load or stress.

Fatigue Strength

Fatigue strength is the ability of a material or part to resist failure during repeated loads or stresses. For example, the ability of a shock absorber in an automobile to withstand the continual vibration from the road without failing is a product of its fatigue strength.

Generally, a material that is subjected to conditions whereby loads are continually applied and withdrawn will fail at stresses lower than the material's ultimate strength.

Impact Strength

Impact strength is the ability of a material or part to withstand the shock of a sudden applied force. For example, a bowling ball striking a bowling pin illustrates an impact force.

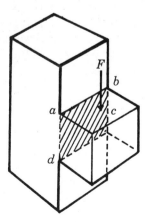

Figure 5–5   Shear stress

Other examples include a hammer striking a nail, a jackhammer breaking up concrete, or a wrecking ball knocking down a masonry building.

Several materials have high tensile and compression strengths but low impact strength. Ceramic and glass products are good examples of this. Glass has a theoretical tensile strength that is well above that of the strongest steels but has extremely low impact strength. Wood, many metals, and plastics have high impact strength. However, brittle metals like cast iron and some plastics will not withstand even moderate impact forces.

THERMAL PROPERTIES OF METAL

Metal has been a very popular manufacturing material for the centuries for many reasons, some of which relate to the mechanical properties described earlier. Other reasons relate to the wide variety of both ferrous and nonferrous metals and alloys available to the design engineer as a result of metallurgical science. The variety of steel available to industry today has been estimated to exceed 2,000 different types.

However, the various types of metals and their different mechanical properties are only part of the reason they have been so popular. Another part of its formula for success relates to thermal properties of metals, that is, the behavior of metal to conditions of heat and cold. The ability of metals to readily conduct heat has, for the most part, been an advantage to product designers and engineers.

The ability of metal to conduct heat energy can be easily demonstrated. The tip of a steel rod placed in a hot flame will quickly conduct the heat energy from the heated end to the unheated end. Within minutes, the end not in the flame will be too hot to handle.

The high thermal conductivity of metals is a property that can readily be used in products that are commonly found in homes and industry. For example, certain types of heating units, thermostats, radiators, and cooking surfaces rely on the high thermal conductivity of metals.

Heat-Treating

The heating and cooling of metals at different rates has been used for centuries to alter the properties of metal. Before metallurgy was a science, the principles of heat-treating metals to make them harder and stronger was used to condition swords, agricultural tools, and industrial machine parts. Within the past 100 years, science has been able to explain why the heating and cooling of metal changes its properties.

Crystallization

In order to understand basic heat-treating methods, it is important to have a basic understanding of metal structures. As mentioned earlier, cooling rate has an effect on the structure of metals. Heating of metals to liquid or near liquid form is commonly used as a primary or secondary manufacturing process. For example, steel and cast iron ingots are heated to melt or near-melt temperatures to further process into standard structural shapes (e.g., bar and sheet or cast into a mold).

As metal cools from a liquid state, crystals slowly begin to form. The term *crystal* is often used interchangeably with grain. As solidification continues, groups of crystals form in geometric patterns. As crystal growth continues, the front of the formation begins to branch out in a tree-like fashion (Figure 5–6). These tree-like crystalline structures are called **dendrites.**

Each crystal is composed of individual cells repeated in a regular pattern to form three-dimensional lattice structures. Metal is an aggregate of thousands of these interlocking three-dimensional crystals or grains. Three types of three-dimensional lattice structures are commonly associated with metals (Figure 5–7):

1. Body-centered cubic
2. Face-centered cubic
3. Close-packed hexagonal

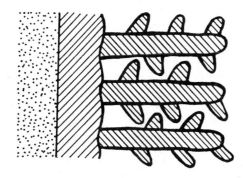

Figure 5–6 Dendrite crystal formation

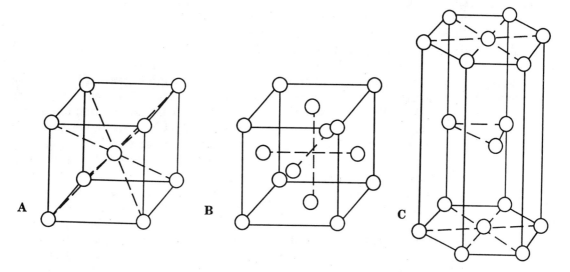

Figure 5–7 (**A**) Body-centered cubic, (**B**) face-centered cubic, and (**C**) close-packed hexagonal

Iron, in its normal state, has a body-centered cubic crystalline structure. There is a relationship between the form of the structure and the property of the metal. Metals with face-centered and body-centered cubic lattice structures are relatively ductile. In contrast, metals with close-packed hexagonal structures are brittle.

It is possible to change the structure of metal crystals, thereby changing the properties. This is accomplished by heating the metal to transformation points and cooling it at different rates. Transformation temperatures are the points at which the basic crystal structure changes. For example, as stated previously, iron has a body-centered unit-cell structure under normal conditions. When heated to 1,670 degrees Fahrenheit, its unit-cell structure changes to face-centered cubic. Therefore, 1,670 degrees Fahrenheit is a transformation temperature for pure iron.

Physical Metallurgy

An understanding of the principles of how the property of metal changes during heating and cooling cycles should begin with an introduction to the equilibrium or phase diagram of carbon steels. Figure 5–8 illustrates the various phases and transformations that carbon steel goes through during a heating and cooling cycle.

The percentage of carbon and the temperature range are the critical elements in the iron–carbon equilibrium diagram. Certain relationships between carbon content and temperature are readily observable by examining this diagram. For example, the melting point of metal generally decreases as the carbon content increases until it reaches a carbon content of 4.3%. This point is known as the **eutectic point,** which means *melts easily.* As the carbon increases beyond 4.3%, the melting temperature increases.

It is also readily observable that certain transformations take place as the temperature and carbon content change. As the temperature of metal increases, it goes through transformations of its basic grain structure. Carbon steel or iron at ambient temperature has a crystalline grain structure described as body-centered cubic lattice structure and is known as alpha iron or ferrite (see Figure 5–7).

As the temperature increases to higher levels, the atoms begin to realign themselves; and the crystalline structure changes from a body-centered to a face-centered cubic lattice structure (see Figure 5–7). During this process, ferrite makes a transformation to austenite. Austenite is nonmagnetic and is often referred to as gamma iron. At higher temperatures, 2,552 degrees Fahrenheit for pure iron, the atoms again realign themselves to form a body-centered cubic lattice structure. Pure iron at temperatures between 2,552 degrees Fahrenheit and its melting point (2,800 degrees Fahrenheit) is known as delta iron.

These transformations are described as **allotropic** changes. *Allotropic* refers to the ability of a material to exist in two or more crystalline forms. The temperatures at which these changes take place are referred to as transformation temperatures. The lower transformation temperature defines the point at which ferrite starts to change to austenite or the structure realigns itself to form a new cubic lattice structure. The upper transformation temperature is the point at which the transformation from ferrite to austenite is completed. These lines are clearly marked on Figure 5–8. Notice that as the percentage of carbon changes, the upper and lower transformation temperatures also change.

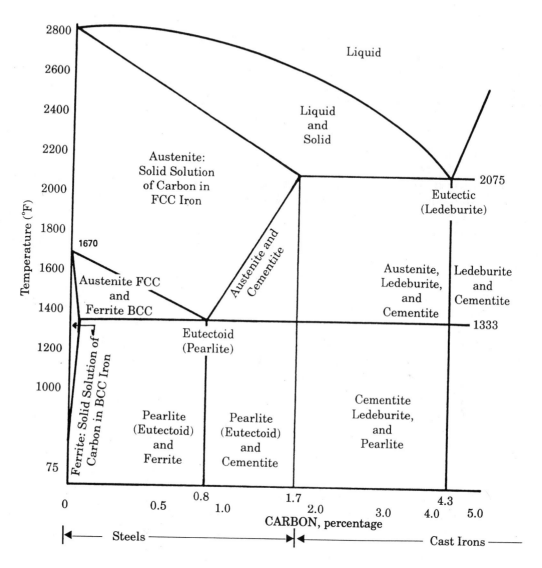

Figure 5–8   Iron–iron carbide diagram

HEAT-TREATING

Hardening

The hardness of metal was defined previously as the ability to resist penetration. The hardness of most carbon steels can be increased through controlled heating and cooling cycles. However, the hardenability (ability to be hardened) is related to the amount of carbon in the iron. The higher the carbon content, the greater is the hardenability.

Hardening process begins by raising the temperature of the metal to approximately 50–100 degrees Fahrenheit above the upper transformation temperature. This temperature will vary with the percentage of carbon. You may recall that this temperature represents the transformation phase from ferrite to austenite. Hardening takes place when the metal becomes austenitic (nonmagnetic) and is then quenched (dipped in water or other cooling solution). Water, salt water (brine), and oil are the most commonly used quenching media. Water cools more rapidly than oil and is less expensive; however, it can cause internal stresses. Therefore, oil, which cools more slowly, is used for critical parts having thin sections.

When the temperature is brought above the upper transformation temperature, the carbon in the iron begins to dissolve into the molten iron much like sugar dissolves in water. This action causes the iron and carbon to become a solid solution referred to as austenite. If the temperature is reduced quickly, the metal will become martensite. Martensite is the hardest and most brittle form of steel and has little practical use in this state.

Its structure is a body-centered tetragonal lattice formation, and it has higher internal stress than most steels. The geometric pattern of the lattice structure of martensite is illustrated in Figure 5–9. As one can see, the structure is similar to that of the body-centered cubic lattice structure of room temperature iron or austenite.

Because martensite has little practical use, additional heat-treating must take place to reduce hardness and brittleness and increase toughness. The following heat-treating processes are used to reduce hardness and brittleness and produce desirable properties in steel and iron.

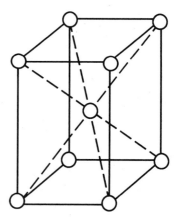

Figure 5–9   Body-centered hexagonal lattice formation

Annealing

Annealing is a softening process that relieves the internal stresses created during the hardening process. The hardening process is often a result of a manufacturing process that was used to transform the metal into a part or product. For example, metal that is welded, cold worked, or hot worked often becomes hardened and develops internal stresses. Even the larger primary metal shapes (slabs, sheets, bars) that come through the rolling mills are hardened from the rolling process and must be annealed.

The following steps are required for annealing to occur:

1. Heat metal to a temperature of 10–100 degrees Fahrenheit above upper critical or transformation point. This temperature will vary with percentage of carbon in the steel or iron.
2. Soak the metal at this temperature for an extended period of time. A good rule of thumb is to allow the metal to soak one hour for every inch of thickness. The purpose of soaking the metal for an extended period of time is to ensure that the carbon completely dissolves in the iron to form homogeneous austenite.
3. The metal is then slowly cooled in the annealing oven. A cooling rate of about 100 degrees Fahrenheit per hour is common. The slower rate of cooling reduces the internal stress, hardness, and brittleness.

Spheroidizing

Spheroidizing is a form of annealing that reduces hardness and increases machinability of the metal. The term *spheroidizing* refers to the appearance of the grain structure of the metal after this heat-treatment process has been completed. The process causes the carbon to form as small spheres that are dispersed throughout the microscopic grain structure.

Spheroidizing is accomplished by raising the temperature to slightly below or at the lower critical temperature and then slowly cooling it.

Normalizing

Normalizing, like annealing and spheroidizing, is done to relieve internal stresses often caused by work hardening. Normalizing is accomplished by raising the temperature of the metal to approximately 50–100 degrees Fahrenheit above the upper critical temperature, soaking it at that temperature to form austenite, and then cooling it in still air.

Cooling the metal in still air rather than furnace cooling makes this process faster than annealing. However, the end results are slightly different. Normalizing will not soften the hardened metal as much as annealing does. Normalizing is often used to refine the grain, improve machinability, and increase ductility in low- and medium-carbon steels.

Tempering

The purpose of **tempering** is very similar to the other heat-treating processes described earlier. Tempering, sometimes referred to as *drawing,* is done to reduce hardness and increase toughness. The following procedure is used in the tempering process:

1. Tempering begins with hardening. Using temperatures and procedures described earlier, harden the steel by raising the temperature above the upper critical temperature and quenching it in a solution of water, brine (water and salt), or oil.
2. Reheat the steel to a temperature slightly lower than the lower critical temperature. Table 5–2 is useful in identifying appropriate tempering temperatures for carbon and tool steel.
3. Quench or allow the steel to air-cool to room temperature.

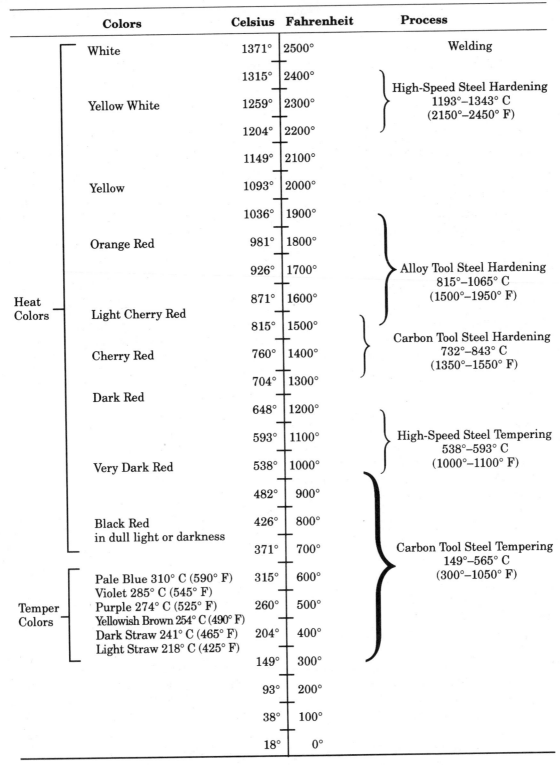

Colors	Celsius	Fahrenheit	Process
White	1371°	2500°	Welding
	1315°	2400°	
Yellow White	1259°	2300°	High-Speed Steel Hardening 1193°–1343° C (2150°–2450° F)
	1204°	2200°	
	1149°	2100°	
Yellow	1093°	2000°	
	1036°	1900°	
Orange Red	981°	1800°	Alloy Tool Steel Hardening 815°–1065° C (1500°–1950° F)
	926°	1700°	
	871°	1600°	
Light Cherry Red	815°	1500°	Carbon Tool Steel Hardening 732°–843° C (1350°–1550° F)
Cherry Red	760°	1400°	
	704°	1300°	
Dark Red	648°	1200°	
	593°	1100°	High-Speed Steel Tempering 538°–593° C (1000°–1100° F)
Very Dark Red	538°	1000°	
	482°	900°	Carbon Tool Steel Tempering 149°–565° C (300°–1050° F)
Black Red in dull light or darkness	426°	800°	
	371°	700°	
Pale Blue 310° C (590° F) Violet 285° C (545° F)	315°	600°	
Purple 274° C (525° F) Yellowish Brown 254° C (490° F)	260°	500°	
Dark Straw 241° C (465° F) Light Straw 218° C (425° F)	204°	400°	
	149°	300°	
	93°	200°	
	38°	100°	
	18°	0°	

Heat Colors (White through Black Red)

Temper Colors (Pale Blue through Light Straw)

Table 5–2 Temperatures, steel colors, and related processes

SUMMARY

Approximately 75 percent of the 90 naturally occurring elements have been identified as metallic. Metals have been the materials of choice for the production of a wide variety of product categories, from aerospace to household. There are more than 10,000 varieties of pure metals and alloys available for commercial use, each with individual properties and characteristics. The sheer number of metals available and the fact that properties of metals can be altered by heat treatment make them the ideal material choice for the design engineer.

Metals have been the material of choice for thousands of years. Their importance can easily be affirmed by the names given to various historical periods (e.g., Iron Age, Bronze Age).

 REVIEW QUESTIONS _____

1. List physical properties of metals that distinguish them from other materials.
2. Explain why the atomic structure of metals increases their ability to conduct electricity.
3. What is the difference between an ion and an atom?
4. How is the weight of an atom determined?
5. Describe the difference between static loads and dynamic loads. Give examples other than those identified in the textbook.
6. Define the following physical properties:

   Hardness
   Tensile Strength
   Plasticity
   Elasticity
   Compressive Strength
   Shear
   Fatigue
   Impact Strength

7. Given the following information, determine the Ultimate Strength and % Elongation.

   Diameter of original reduced area = 0.505
   Ultimate Load = 18,500 psi
   Area of Reduction after fracture = 0.15
   Original gauge length = 2.00
   Gauge length after fracture = 2.35 inches

8. Describe the transformation that takes place as carbon steel changes from alpha to delta iron.

9. Define *allatropic changes*.
10. Describe the procedure used in annealing steel or iron.

SUGGESTED ACTIVITIES

1. Using $1/8$-inch dowel rods and 2½-inch diameter styrofoam balls, build models of body-centered and face-centered cubic lattice structures.
2. Acquire several different samples of steel and iron. Test for hardness using a Rockwell or Brinell hardness testing machine. Record the type of metal and corresponding hardness number.
3. Select a medium-carbon steel rod ¾ inch in diameter and approximately 8 inches in length. Using Figure 5–3, machine a .505 specimen from the steel rod.
4. Using the .505 specimen from the previous activity, complete the following procedures and record data.
   a. Test hardness of specimen and record.
   b. Harden the specimen using the procedure described in this chapter. *Note:* Use water to quench. Test hardness again and record.
   c. Anneal the hardened .505 specimen and test for hardness. Record.
   d. If a tensile testing machine is available, pull a tensile of the specimen and record the breaking load and ultimate load.
   e. Using the breaking load and ultimate load, calculate breaking strength and ultimate strength.

CHAPTER 6

Polymers: Structure, Properties, and Characteristics

Classifying the millions of different types of materials available for use today is no easy task. There are a number of different systems that have been used and accepted by scientists, engineers, and technologists. One system—organic and inorganic—was used earlier in this text to distinguish between materials such as wood and plastics that are derived from organic (derived from living organisms) substances and those inorganic materials such a metals and ceramics.

Such a classification system makes it easy to simply separate plastics, woods, metals, and so on into easily defined categories. However, when discussing structures, properties, and characteristics, more narrowly defined categories are needed. In Chapter 5, the structure, properties, and characteristics of materials commonly referred to as metals were discussed. The term *metal* is easily recognizable because of its common everyday use. **Polymer**, however, is not as easily recognizable. This term refers to the basic structure of materials generally referred to as plastics and woods.

Metals, as discussed in Chapter 5, are made up of atoms, which were described as the basic building blocks of any substance. However, the building blocks of polymer materials are **molecules**. *Molecules* are groups or units of different atoms that are joined together either naturally or artificially by humans. For example, water is a molecular material. We know that atoms of hydrogen and oxygen are joined together to form H_2O, or water.

Wood and plastic are also examples of molecular materials because their basic structure is formed from units of atoms. Most of the materials discussed in this chapter are classified as **organic polymers**. Polymer is a Greek derivation and means many *(poly-)* parts *(mer)*.

The building blocks of organic polymers are referred to as **mers**. *Mers* are molecules formed from atoms of different elements. Polymers are larger building blocks made from the bonding of many mers. These molecular structures can contain millions of atoms.

VISCOELASTICITY

The majority of the mechanical properties of plastics and woods are a result of a characteristic in these materials known as viscoelasticity. The term *viscoelasticity* refers to the behavior of polymers when a load is applied. *Visco-* refers to the viscous flow of polymeric materials. Under a constant load, over time, this characteristic will cause a permanent deformation. *Elasticity* refers to a reversible response plastics have to a load. Therefore, when stress (load) is applied to a polymer, the initial strain is recoverable; and no permanent deformation of the material will result when the load is removed. However, a more permanent load will result in a flow of the material, which will cause permanent deformation. The degree of deformation is time and temperature related.

This knowledge can be used by engineers and technologists who are responsible for identifying appropriate materials for specific uses. Because of the viscoelastic nature of plastics, they would not be appropriate to use as a foundation for a heavy piece of equipment. Why? The heavy load and heat generated by such a machine would most likely cause considerable deformation of the foundation.

There are many common examples that can be used to illustrate viscoelasticity in molecular materials. For example, a straight pine board that is placed between two vertical members and used for a bookshelf will exhibit viscous flow and elastic deformation. If several heavy books are placed in the center of the shelf, it will deform. If the books are removed within a short period of time, the shelf will return to its original shape. However, if the books are left on the shelf for an extended period of time, the shelf will become permanently deformed.

This same behavior is common in rubber and plastic rollers for spreading ink over plates that is ultimately transferred to paper. When the presses are operating, a certain amount of pressure is applied to the rollers. If this pressure is not released at the end of the day, the rollers will deform and flat spots will develop. This may cause the spread of ink to be uneven. This same behavior will occur in children's wheeled toys. A wagon or tricycle that sets for an extended period of time will develop a flat spot on the surface of the tire that is resting on the ground or pavement.

Figure 6–1 illustrates the viscoelasticity of polymers. When the stress is applied, the elastic strain is immediate. However, the viscous flow occurs over time. When the load is removed, only the elastic component is recovered.

PROPERTIES OF PLASTICS

Plastic has become an extremely important material during the last half of the twentieth century. The raw materials (carbon based) are readily available, they are easily formed, and they exhibit excellent mechanical and thermal properties for many applications. Although a relatively weak material, they are easily formed and can be strengthened by adding such

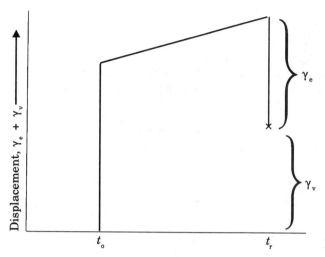

Figure 6–1   Viscoelasticity of polymers

material as glass fibers. The mechanical properties of plastics can vary greatly. This is one of the advantages of plastics over other materials.

Weight

Plastics are generally considered to be low-density materials. **Density** is the mass per unit volume of a substance under specified conditions of pressure and temperature. Low-density materials weigh less per volume than high-density materials. Another indicator of weight is **specific gravity.** *Specific gravity* is the ratio of the weight of a given volume of a material to an equal volume of water. Material with a specific gravity less than water will float.

Several types of plastics will float. Polyethylene and polystyrene are among the lightest plastics. The heaviest, fluorocarbons, have a specific gravity of 2.3. These plastics are still lighter than the lightest metals.

Tensile Strength

Tensile strength is the ability of a material to resist a pulling force. Generally, plastics are considered to have low tensile strength. However, reinforced plastic resins can develop strengths of 100,000 psi or more. Most thermoplastics do not exceed tensile strength of 10,000 psi. Thermosets will achieve tensile strengths that exceed 20,000 psi. Many plastics will elongate 500% or more without fracturing. Low-density polyethylene will elongate as much as 700% and polypropylene 500%.

Compressive Strength

Compressive strengths of plastic materials can vary greatly. Typically, plastics are not known for their compressive strength. However, plastics reinforced with glass fibers can resist compressive or crushing forces in excess of 70,000 psi. This compares favorably with steel, which has a compressive strength of 50,000 psi.

The range of compressive strengths for thermoplastic and thermoset plastics is typically less than 25,000 psi. Thermosets have greater resistance to compressive forces than thermoplastics. Table 6–1 shows typical compressive and tensile strengths for some of the more commonly used plastics.

Plastic	Tensile psi (thousands)	Compressive psi (thousands)
ABS	4–8	9–11
Acetals	8.5–10	4.5–5.0
Acrylics	6–10	12–18
Fluorocarbons	2.5–8.5	1.6–14.0
Nylons	7.5–12.5	9–20
Phenolics	5–10	15–36
Polyester	1–10	12–37
Polyethylene (high-density)	3.2–5.0	3.0
Polyethylene (low-density)	1.5–2.5	3.0
Olefin	2.0–4.0	2.5–10.0
Polypropylene	2.8–5.5	4.4–6.5
Polystyrenes	2.8–10	4–16
PVC	1–8	10.0

Table 6–1 Tensile and compressive strengths

Hardness

Deformation hardness is the resistance to a penetration force. Even the hardest plastics are softer than the softest metals. However, hardness can be increased by adding other nonplastic materials. For example, glass fibers added to plastic resins will increase hardness.

Hardness is tested by forcing a round rod into the plastic surface. This test is generally conducted on a Rockwell hardness tester or a Shore durometer tester. The test results in each case are a measure of how far the indenter penetrates into the sample.

The Shore durometer test is conducted using standard test specimens that are flat and ¼-inch thick. These test specimens can be cut from standard sheet stock, or they can be molded. The specimen is placed on a hard flat surface; and the spring-loaded indenter is released, forcing the rod into the specimen. A gauge on the front of the tester indicating the depth of penetration is read and each measurement recorded. At least five measurements should be made on each sample at locations at least ¼-inch apart. The readings should then be summed and averaged and the arithmetic mean recorded. The step-by-step procedure is described in ASTM standard 2240.

Two types of indenters are used on the Shore testing instrument: Type A and Type D. Type A is used for softer plastics and Type D with harder materials. Remember, even the hardest plastics are softer than the softer metals.

Table 6–2 lists Rockwell and Shore durometer hardness readings for the more commonly used plastics. Rockwell is typically used for the harder materials.

Plastic	Durometer		Rockwell	
	Type A	Type D	M	R
ABS				75–115
Acetals			80–90	
Acrylics			80–100	
Fluorocarbons			110–115	
Nylons				108–118
Polyester				90–120
Polyethylene (high-density)		65		
Polyethylene (low-density)		45		
Olefin		35		
Polyporopylene				80–100
Polystyrenes			45–75	
PVC (nonrigid)	50–100			
PVC (rigid)		70–85		110–120

Table 6–2   Rockwell and durometer hardness of plastics

Impact Strength

Impact strength is the ability of a material to resist sharp blows or shocks. It is often a measure of the toughness of a material. Toughness is the ability of a material to absorb applied energy. The impact strength of some reinforced plastics is greater than that of steel.

The ability of plastics to resist an impact force is referred to as **impact resistance**. Impact resistance varies widely among the different types of plastics. Rigid or stiff polymers like polystyrenes and acrylics are brittle and fracture easily. Flexible plastics such as polyvinyls and polyethylenes have high impact strength.

Impact strength can be altered through the addition of certain types of rubber and other plastic compounds. However, these additives usually result in a loss of rigidity. Environment can also affect impact strength. Increased temperatures will significantly increase impact strength. Conversely, lower temperatures will reduce impact strength.

Impact strength is measured in foot-pounds per inch (ft.-lbs./in.) or inch-pounds per inch (in.-lbs./in.). Several standard types of impact tests have been developed for use by the plastics industry. Two different tests have generally provided the standards for impact strength in polymers. The procedures for the IZOD and the Charpy impact tests are described in detail in ASTM standard D 256.

Creep

The phenomenon of deformation under load, over time, is referred to as **creep**. When a load is applied to a plastic material, it will deform quickly according to its stress–strain modulus. Then it continues to deform slowly under a constant load which can eventually lead to excessive distortion or failure.

Creep, under normal conditions, is slow (10^{-3}%/hr.); however, this rate can be affected by environmental conditions. Increased temperature can cause the rate of creep to increase. Creep is not restricted to plastics. Wood will creep also. An example of creep in wood is easily illustrated through the distortion that occurs in bookshelves over time and under load. The permanent bow in the shelf after an extended period of time is a result of creep.

Plastics are generally not considered as appropriate foundation materials for heavy equipment because of their creep properties, although some reinforced plastic resins are now being used for light-machine bases. The glass fibers in the reinforced plastics prevent slippage of the molecules and, therefore, limit creep.

Thermal Properties

Thermal properties refer to the behavior of a material to temperature increases and decreases. This behavior is generally a major concern to designers and engineers who must select materials that will perform according to the customer's specifications. Typically, plastics have low melting temperatures and high coefficients of expansion. However, because of their molecular bond, they do not conduct heat readily and, therefore, are excellent insulators against heat and cold.

Thermal Conductivity

Thermal conductivity is a measure of a material's ability to transmit heat. The letter K is used for thermal conductivity; and the lower the K value, the better insulating quality the material has.

Thermal conductivity is measured in British thermal units (BTUs) per hour conducted per square foot of material 1-inch thick per 1 degree Fahrenheit temperature differential. The foamed plastics such as polystyrene and urethane are the best insulators, with K values between 0.01 and 0.03. The covalent bonding of the molecules which resist transmission of heat and the voids in the foamed plastics are key factors in their excellent insulating qualities.

Thermal Expansion

Temperature changes cause materials to expand and contract to different degrees. Ceramics typically expand the least, followed by metallic materials. Polymer materials have the highest coefficients of thermal expansion.

This property is especially important when designing parts that must fit together tightly. Sufficient clearance must be designed into the product to allow for expansion and contraction of the material.

PROPERTIES OF WOOD

The forests have provided wood for shelter, fuel, and tools for thousands of years. Its abundance and the ease with which it can be shaped made it an ideal material for our early ancestors.

Wood has many advantages over other materials, including the fact that it is a renewable source. Scientists have provided society with methods for growing better trees at faster rates than nature can, that is, unassisted. Their efforts, in combination with better forest-management techniques, should assure an endless supply of lumber to meet present and future needs.

The progress that has been made in the area of wood science during the twentieth century has enabled the manufacturers to build better and more durable wood and wood composite products. Much of the research related to mechanical and other properties of wood has come from the Forest Products Laboratory located in Madison, Wisconsin.

The Forest Products Laboratory is an agency of the Forest Service of the U.S. Department of Agriculture (USDA). Each year, the Forest Products Laboratory conducts or sponsors research related to a wide range of topics. Sample areas include mechanical properties, thermal properties, adhesives, composites, finishes, processes, and economics.

This research has helped guide the designer and manufacturers in selecting the best materials at the greatest economy to the customer and producer. Through the ongoing research sponsored by the Forest Products Laboratory, new and better uses of an age-old resource have been discovered. The warmth, strength, and aesthetic appeal of wood will continue to make it an extremely desirable material for manufacturing and construction.

Structure

The principal molecular species in wood is cellulose. Hemicellulose and lignin are the other constituents in wood. *Hemicellulose* is a low-molecular-weight polymer formed from glucose and other sugars. *Lignin* is a hard amorphous polymer. The amount of lignin is complementary to the amount of hemicellulose. Lignin accounts for 12–28% of wood and hemicellulose from 25–33%. When the hemicellulose is high, the lignin is lower and vice versa. In general, hardwoods contain less lignin and more hemicellulose than softwoods do.

Cellulose is manufactured from units of glucose that are formed through photosynthesis. Glucose anhydride units are linked end to end to form long-chain polymers that form cellulose. Most hemicellulose molecules are branch-chained rather than straight-line (or linear) polymers. Figure 6–2 illustrates the difference between branching and straight-line or linear polymer molecules. The branch-chained molecules do not crystallize as readily; however, when forced to bond, they are more stable than other molecules.

Mechanical Properties

Many of the strength properties described in Chapter 5 and the section on plastics are equally important for wood. However, the structure of wood and its chemistry are considerably different than other materials; and a number of factors can substantially influence the strength qualities.

Moisture content and specific gravity are two of the more important factors to consider when describing strength qualities of wood. The cellular structure of wood is directly related to its moisture content as well as its specific gravity. You may recall from Chapter 3 that the

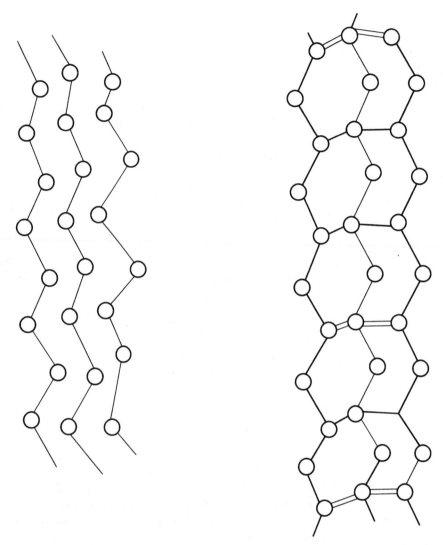

Figure 6–2   Branch-chained and straight-line or linear polymer molecules

moisture content of wood is very high when it is in the green state. The water in the cell and that absorbed by the cell walls or fibers can be as high as 200%.

As the wood dries, it becomes stronger. This is due to the stiffening of the individual fibers in the cell walls and to the increased compactness of the wood; that is, the wood becomes more dense as it dries. Much of the increase in strength takes place after the wood reaches its fiber saturation point. This is the point at which the wood cells have given up all free water and the fibers are fully saturated; this occurs at about 30% moisture content.

Specific gravity and density are often used interchangeably when describing the mass of a material per unit volume. However, although they are related, they are different. Density is the mass or weight per unit volume expressed as pounds per cubic foot or kilograms per cubic meter. Specific gravity is the ratio of density of a material to the density of water at a standard temperature.

The higher the specific gravity of wood, the more dense and, therefore, the stronger. Through research conducted by the U.S. Forest Products Laboratory, fairly accurate predictions of strength can be made using specific gravities of certain wood species. Table 6–3 shows the relationship between specific gravity and strength for some popular species of U.S. woods.

Species	Specific Gravity	MOR* (psi)	Compression (psi)
Hardwoods			
White oak	0.68	15,200	7,440
Yellow birch	0.62	16,600	8,170
Maple	0.63	15,800	7,830
Walnut	0.55	14,600	7,580
Ash	0.60	15,400	7,410
Basswood	0.37	8,700	4,730
Cherry	0.50	12,300	7,110
Locust	0.69	19,400	10,180
Yellow poplar	0.42	10,100	5,540
Softwoods			
Red cedar	0.47	8,800	6,020
Douglas fir	0.48	12,400	7,240
Hemlock	0.45	11,300	7,110
White pine	0.35	8,600	4,800
Ponderosa pine	0.40	9,400	5,320
Spruce	0.44	10,400	5,650
Redwood	0.40	10,000	6,150
Larch	0.52	13,100	7,640

*MOR, Modulus of Rupture

Table 6–3   Relationship of specific gravity strength for common woods

Bending Strength

Bending strength is a property that was not discussed in the other sections of the book. It is a strength property that is especially important for structural wooden members that have loads placed on them while in a horizontal position and supported at each end. Examples of applications where bending strength is important include roof and floor joints, beams used to support walkways and bridges, a balance beam used for gymnastics, and a shelf that is supported at both ends and used for books.

Stress–strain in a uniformly loaded beam
Stress distribution

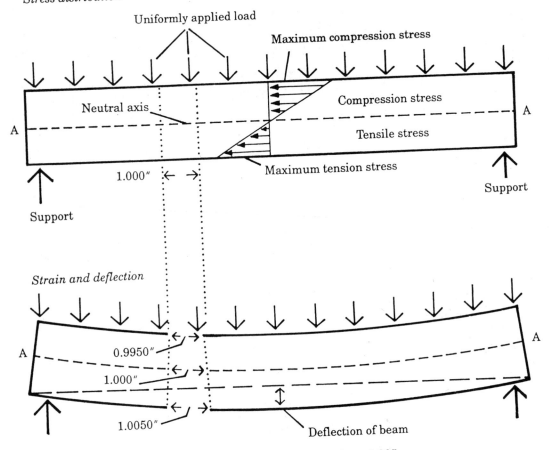

Strain at the surface = 0.005 in. ÷ 1 in. = 0.005.

Figure 6–3 Standard test for bending strength of wood

Bending strength is a resistance to a bending stress, which is a combination of compression and tensile stresses. Figure 6–3 illustrates the combination of stresses that are exerted on a horizontal structural member when a load is applied at midpoints. As one can see, the stress applied to the top surface is a compressive stress; and that at the bottom edge is in tension.

Modulus of elasticity (MOE) and **modulus of rupture (MOR)** are important principles in discussions related to strengths of wood. Although the bending strength is generally expressed in terms of MOR, the relationship between MOE and MOR is important to define.

The MOE is a ratio of stress to elastic strain. As stated previously in the discussion of tensile strength in metals, stress is proportional to strain until it reaches its proportional limit. Until this point is reached, the material will have elastic qualities. When a load is removed, it will return to its original shape.

Modulus of elasticity is tested in bending by measuring the amount of deflection and load. The load is applied at the midpoint of the beam, and the deflection of the beam is measured and recorded. Modulus of elasticity is calculated using the following formula:

$$MOE = PL^3/48\ ID\ (psi)$$

where

P = Load in pounds

D = Deflection at midspan with load applied (inches)

L = length of span (inches)

I = Moment of inertia (a function of beam size)

I = Width × Depth^3/12 (inches)

Example: An ASTM Standard Specimen 2 × 2 × 30 inches of Ponderosa pine is to be tested for bending strength. The test piece is placed between support, so there is a 28-inch span. A load of 1,500 pounds is applied and a deflection of 0.3125 is recorded. Calculate the modulus of elasticity.

$$MOE = PL^3/48ID = (1,500)(28)^3/(48\{[2(2)^3]/12\}0.3125) = 1,646,400\ psi$$

The MOE for woods ranges from approximately 500,000 to 2,800,000. The higher the MOE, the higher is the resistance to bending stresses.

The MOR is calculated using the maximum load at the time of rupture. The same testing procedures are used except the proportional limit is exceeded and the load is increased until failure occurs. The MOR is calculated using the following formula:

$$MOR = 1.5PL/bd^2$$

where

P = Breaking load

L = Span (distance between supports)

b = width (inches)

d = depth (inches)

Example: The specimen used in the previous example is loaded to failure. The recorded breaking load is 1,800 pounds. Calculate the MOR.

$$MOR = 1.5\ PL/bd^2 = [(1.5)(1,800)(28)]/[2(2)^2] = 9,450\ psi$$

Tensile Strength

The method of testing the tensile strength of wood is similar to that used in testing metals. The specimen has a reduced center section, and the load is applied parallel to the direction of the grain (Figure 6–4). It is important that the specimen is not crushed during the testing procedures. Any deformations of the ends of the test specimen could affect its ultimate strength.

In most applications where wood is used, tensile strength is not a factor; bending strength, compression strength, and shear strength are. Compressive strength in wood ranges from approximately 2,000 to 10,000 psi.

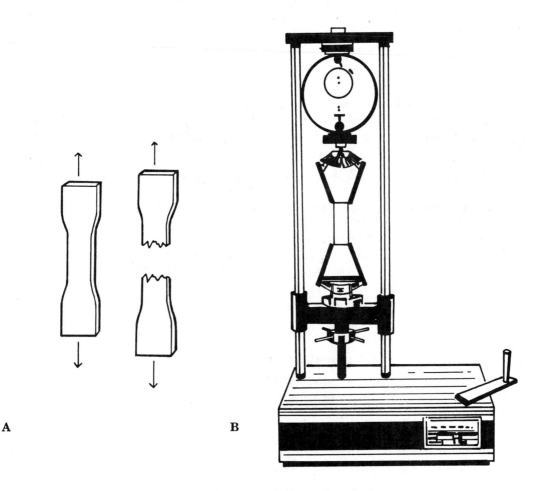

A   B

Figure 6–4   Tensile testing of wood

Shear Strength

Shear strength refers to the ability of the wood to resist forces which cause the fibers to tear or slip past one another. Figure 6–5 illustrates the standard method of determining the shear strength of wood. Horizontal and vertical shears are of importance to designers who plan to use wood in any application. Shear strength is tested parallel to the grain of the specimen. Wood is generally low in shear strength parallel to the grain but higher perpendicular or across the grain.

Horizontal shear is illustrated in the section on bending strength (see Figure 6–3). The load on the specimen causes the top fibers to move horizontally with respect to bottom fibers. Tensile strength of wood is generally in the range of 100 to 1,000 psi.

Compression

Compression strength in wood is usually tested in two directions: parallel and perpendicular to the grain. In the majority of applications, wood-compression strength parallel to the grain is a factor in applications such as vertical supports for decks, piers, roofs, and the like.

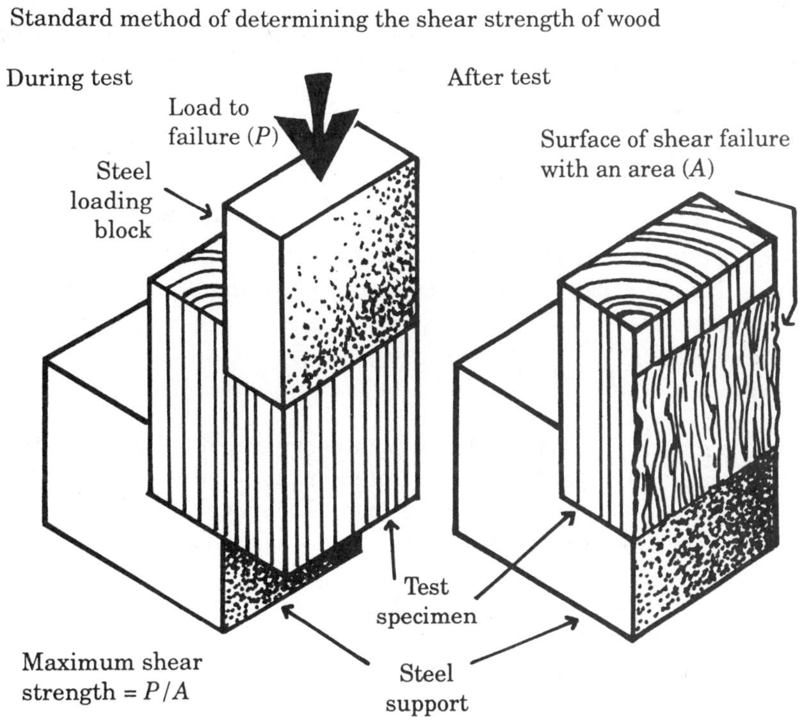

Standard method of determining the shear strength of wood

Figure 6–5   Testing shear strength of wood

Compressive strength is determined by dividing the maximum load (breaking load) by the cross-sectional area of the specimen.

$$\text{Compressive Strength} = \text{Breaking Load/Cross-Sectional Area}$$

The standard test specimen is a $2 \times 2 \times 8$ inches.

Hardness

Hardness as a mechanical property of wood is defined as resistance to wear, marring, or indentation. It is tested using a 0.444-inch steel ball that is forced into the wood. Hardness ranges from 250 to 1,700 psi. Sugar pine is one of the softest woods, and locust is one of the hardest.

Thermal Properties

Wood is considered an insulator of heat rather than a conductor. Thermal conductivity in wood is influenced by a number of factors, including moisture content and density. The lower the moisture content and the lower the density, the better is the thermal conductivity. These qualities make it a good insulator against the cold and heat.

Thermal conductivity is a measure of the amount of heat that travels through a material. It is expressed in BTUs and is measured using a standard specimen that is 1-inch thick and 1 square foot in area.

Thermal Expansion

Thermal expansion refers to the property that causes a material to expand and contract with temperature differences. Wood that is dry (free of high moisture content) is a relatively stable material. It does not expand and contract to any significant degree when temperature changes occur.

SUMMARY

Wood and plastic materials are included in a category called polymers. The basic building blocks of polymer materials are molecules—groups or units of different atoms joined together by nature or artificially by humans. The majority of the mechanical properties of polymers is a result of a characteristic known as *viscoelasticity*. *Visco-* refers to viscous flow, and *elasticity* refers to recoverable deformation.

Polymer materials typically have low tensile strength when compared to metals. They can be reinforced with materials such as glass fibers to improve their mechanical properties. Polymers are insulators of heat and electrical current and have low melt or combustion temperatures.

Polymers are generally easy to form and inexpensive in comparison to metals. The exception to this is some hardwoods and exotic species of wood.

REVIEW QUESTIONS _____

1. What is the difference between organic and inorganic materials?
2. What is the difference between the structure of molecular materials and metals?
3. The majority of the mechanical properties of plastics are a result of a condition known as viscoelasticity. Define *viscoelasticity*. Give an example of how viscoelasticity affects a product around the home or in industry.
4. Define *density*. What is the difference between density and specific gravity?
5. How does the compressive strength of plastics compare to that of metals?
6. How does the tensile strength of plastics compare to that of metals?
7. Describe the testing procedures for determining the hardness of plastics.
8. What is the relationship between temperature and impact strength of plastics?
9. What is meant by a property known as *creep*?
10. What agency has been most responsible for research in wood science and technology?
11. What is the chemical composition of wood?
12. What two important factors affect the mechanical and thermal properties of wood?
13. Explain why wood becomes stronger as it dries.
14. Define *bending strength*. Give examples of where bending strength is important.
15. An ASTM standard specimen of 2 × 2 × 30 inches is tested for bending strength, and a deflection of 0.375 is recorded. A load of 1,350 pounds is applied to a span of 28 inches. What is the modulus of elasticity (MOE)?

SUGGESTED ACTIVITIES _____

1. Select several specimens of different types of plastics and woods. All specimens must have the same volume. Determine the weight of an equal volume of water and use the following formula to determine the specific gravity for each specimen:

   Specific Gravity = Dry weight of specimen/weight of water (equal volume)

2. Select four species of hardwood and four species of softwood. Prepare ASTM standard samples for compression and bending strength testing. Test samples using a universal testing machine and prepare a chart comparing the strengths.

CHAPTER 7

Ceramics: Structure, Properties, and Characteristics

Ceramic products and materials have served the needs and desires of society for thousands of years. The development and use of brick by the Sumerians in Mesopotamia around 3000 B.C. allowed them to build temples and fortifications that gave them supremacy over their neighbors for centuries. Their strength, heat, and electrical properties make them excellent materials for a wide range of product applications. Their compressive strength, wear resistance, and high degree of chemical stability are attributes that engineers and technologists look for when specifying materials for manufacturing and construction projects.

The structural clay products (brick, clay pipe, etc.) used every day in construction and the ceramic products around the home are generally very familiar items to most people. However, the advanced ceramics and nuclear ceramics used as high-speed cutting tools, optical fibers, semiconductors, and fuel for nuclear power generation are not as readily identifiable to the average individual. Just as the clay products have helped advance civilization through the present, the advanced ceramics will do so into the future.

Ceramics have become a multi-billion-dollar industry, and the growth of this industry should continue well into the future. New ceramic materials are being introduced to the market in significant numbers, and new applications are being developed just as quickly. This chapter explores how the thousands of ceramic materials are classified, describes their structure, and discusses the properties that make them so popular.

CLASSIFICATIONS

Ceramics are hard, brittle compounds of metallic and nonmetallic elements that have high melting temperatures and are chemically inert. As a class of materials, ceramics are much more stable in chemical and thermal environments than metals and molecular materials.

Ceramics are solid materials that have crystalline and noncrystalline or amorphous atomic structures. The properties, composition, and characteristics of ceramics are widely diverse. However, the thousands of materials can generally be classified under the following structure:

Structural clay products	Refactories	Abrasives
Whiteware	Glass	Cement and concrete

Structural Clay Products

Products classified under this heading include such widely used items as brick, drain tile, flues, and floor tile. Structural clay products represent a large industry today, with millions of dollars in sales. Products made from clay were among the first used by humans.

Clay is a term applied to a natural mixture of substances found in the earth. The principal elements of clay are oxygen, silicon, and aluminum, which form compounds referred to as *alumina-silicate*. Brick clay is a composition of 10% magnesia, 15% lime, 8% iron oxide, 10–25% alumina, and 35–65% silica. These materials are among the most abundant on earth. Oxygen and silicon, which are commonly found in clays and other ceramics, make up nearly 75% of earth's mass.

The plasticity of clay is the quality that makes it such an excellent material to form into a variety of shapes for structural and artistic purposes. The plasticity in clay is a result of a mineral known as kaolinite. Kaolinite is composed of small thin platelets that adhere to one another when wet and at the same time slide over one another when minimal pressure is applied.

The plasticity of clay will increase as water is added to a point. Excessive water will cause the clay material to adhere more to other materials than to each other. The correct proportions of water will prevent the clay from being too brittle or crumbly at the lower end and sticky and weak at the upper end.

Once the clay substance is molded into a desired shape, it will hold that form upon drying. Controlled uniform drying will prevent cracking and warping. If the clay is subjected to higher temperatures, the clay will lose its ability to be remolded.

Ceramic Whiteware

Whiteware is a category of ceramics that includes products such as ceramic bathroom tile, china, and earthenware. These products have essentially the same substance as the structural clays, but the ratios are different. In most cases, the ratio of feldspar to clay is higher in whiteware than in the structural clay products discussed earlier.

Refractories

Refractory ceramics are high-melting-point oxides sometimes referred to as crystalline ceramics. Most refractory ceramics are mixtures of alumina and silicon that are mixed in various proportions, dry pressed to shape, and fired to develop strength. Refractory brick used in kilns of various shapes and sizes is perhaps the most popular use for this type of ceramic material. Crucibles for melting metals are also made of refractory ceramics (Figure 7–1).

Figure 7–1 Refractory crucible used for melting metals

Glass

Glass is noncrystalline ceramic that is made from mixtures of silica and basic oxides such as boron. There are more than 10,000 kinds of glass in use today. The four basic types of commercial glass are lead, soda-lime, borosilicate, and high-silicate.

Soda-lime glass is the type most individuals are familiar with. It is used for such commonly used products as window glass, bottles, and light bulbs (Figure 7–2). Theoretically, glass is a super-cooled liquid since it has solidified from a liquid state without crystallizing.

Figure 7–2   These products are made of soda-lime glass *(Courtesy of GE Lighting)*

This characteristic gives glass its optical quality of transparency. Other mechanical properties are listed in Table 7–1.

Abrasives

Abrasives are generally classified as natural and synthetic. The *natural abrasives* are made of materials that are found in nature: flint, crocus, diamond, garnet, and emery. These materials have generally been replaced by the synthetic or manufactured abrasives.

The most popular *synthetic ceramics* currently being used as abrasives in manufacturing are aluminum oxide, silicon carbide, boron carbide, and artificial diamond. Many of these synthetic abrasive materials were discovered before the turn of the twentieth century. Today, these ceramics are used in the production of abrasive paper and wheels as well as cutting bits for lathes and machining centers (Figure 7–3).

Cement and Concrete

Cement and concrete have been among the most important construction materials for a long time. Their use in providing the foundations, floors, walls, stairs, streets, and other

Property	Soda-lime	Soda-lead	Borosilicate	Alumina-silicate	96% silica	Fused silica
Young's modulus, psi	10×10^6	9×10^6	9.5×10^6	12.8×10^6	9.6×10^6	10.5×10^6
Poisson's ratio	0.24	—	0.20	0.26	0.18	0.17
Specific gravity	2.47	2.85	2.23	2.53	2.18	2.20
Linear coefficient of thermal expansion, cm²/ C	92×10^{-7}	91×10^{-7}	32.5×10^{-7}	46×10^{-7}	8×10^{-7}	5.6×10^{-7}
Service temperature, C						
Annealed						
Normal	110	110	230	200	800	900
Maximum	460	380	490	650	1,100	1,200
Tempered						
Normal	220	—	260	400	—	—
Maximum	250	—	290	450	—	—
Volume resistivity, ohm-cm						
At 250°C	2.5×10^6	8×10^8	1.3×10^8	3.2×10^{13}	5×10^9	1.6×10^{12}
At 350°C	1.3×10^5	1×10^7	4×10^6	2×10^{11}	1.3×10^8	2.5×10^{10}

Source: Corning Glass Works

Table 7–1　Properties of glass

Figure 7–3　Abrasive products of synthetic abrasive materials *(Courtesy of Norton Company)*

components of residential and commercial structures has been invaluable (Figure 7–4). As with other ceramic materials, cement and concrete are essentially hard, brittle materials that are chemically inert.

Concrete is a mixture of cement, water, fine aggregate (sand), and course aggregate (rock). While wet, concrete can be poured or formed into any basic shape. However, once the moisture is removed, the concrete forms a hard, brittle substance that cannot return to its original form.

Portland cement is the binder in the concrete mixture and is a finely pulverized mixture of lime, silica, alumina, and iron. There are several types of portland cement available for different specialized needs. For example, high-early-strength portland will provide the user with high strength in a shorter time than normal portland.

STRUCTURE OF CERAMICS

The mechanical, thermal, optical, and electrical properties of ceramics are a result of the microstructure of ceramics. Ceramics are compounds of metals and nonmetals, and the bond between the metal and nonmetal elements is primarily ionic.

Figure 7–4   Concrete is used for footers, floors, highways, and other construction products

In ceramic compounds, nonmetallic and metallic elements are bonded by either acceptor or donator electrons from their outer or valence shell in order to fill their shells and become stable. The metal atoms generally have few electrons in their outer shell and donate them to a ceramic atom that accepts these electrons in order to fill its valence or outer shell. By giving up electrons, metal atoms are no longer electrically neutral but become positively charged ions.

The ceramic atoms accepting the free electrons from the metal atoms fill their outer shells and become negatively charged ions. The attraction between the positively charged metal atom and negatively charged ceramic atom forms a tight ionic bond that makes the ceramic compound strong. This bond also is the reason ceramics are excellent thermal and electrical insulators.

The basic structural unit in ceramics is the silicate SiO_4 tetrahedron (Figure 7–5). The SiO_4 tetrahedral structure is found in many ceramic materials including brick, cement, glass, whiteware, and electrical insulators. SiO_4 is a silicate ion that has four atoms of oxygen and one silicon atom. The silicon atom gives up its four valence-shell electrons to each of the four oxygen atoms.

In this structure, the silicon atom is satisfied; but the oxygen atoms are still one electron short in their valence shell. In order to reach a stable condition, the oxygen atom will link with other silicon atoms through covalent bonding (sharing of electrons). The covalent bonding of the structural units of the ceramic materials form chain-like units called silicate sheets. These silicate sheets are the basis for clay ceramics.

PROPERTIES OF CERAMICS

The mechanical, thermal, optical, and electrical properties of ceramics are a product of their structure, processes employed to manufacture them, and their chemical composition. In general, ceramics are hard, brittle, strong materials that are poor conductors of heat and electricity and are chemically inert.

Many of these properties are developed as the ceramics give up moisture through regulated drying and firing (heating to high temperature) processes. The rate and temperature are important to the development of strength properties.

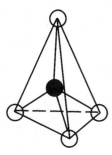

Figure 7–5 SiO_4 tetrahedral structure of ceramics   Silicon   Oxygen

During the firing process phase, changes occur much like those described in Chapter 5 for metal materials. The phase diagram in Figure 7–6 is the alumina-silicon phase diagram. This diagram illustrates the extremely high temperatures required to reach a liquid state in ceramics.

As solidification occurs, ceramic bonding takes place, causing a smooth glassy surface to form on the ceramic material.

Mechanical Properties

As a class, ceramics are the most rigid of all materials. Ceramics are stiffer than metals; their MOE runs as high as 50–65 million psi, compared to 29 million psi for steel. Table 7–2 shows mechanical properties for several commonly used industrial ceramics.

Hardness

Ceramic materials are the hardest materials on earth, diamond being the hardest. The Knoop and Mohs hardness scales are generally used for ceramic materials. Table 7–3 is a listing of some of the more commonly used ceramics and a comparison of hardness on the Knoop and Mohs scales.

Hardness makes ceramics especially useful for wear-resistant parts, abrasives, and cutting tools. Silicon carbides, nitrides, zirconia, and SIALON are becoming popular materials for cutting tools for superalloys and other hard metals. Silicon nitride–based cutting tools have captured applications in cast iron machining.

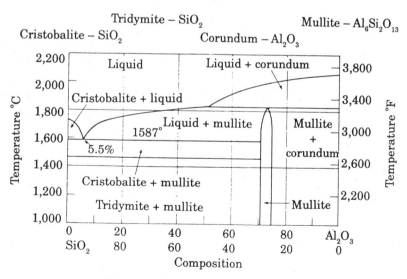

Figure 7–6 Al_2O_3-SiO_2 phase diagram

Property	Oxide Ceramics			Refractory Ceramics		
	Alumina	Magnesia	Zirconia	Carbides	Nitrides	Borides
Melting point, °F	3,700	5,070	4,710	2,800–6,300	1,200–4,900	2,000–5,800
Modulus of elasticity in tension, 10^6 psi	65	40	30	26–65	8–50	32–72
Tensile strength, 1,000 psi	38	20	21	—	—	—
Hardness, Mohs scale	9	6	9	9	7–9	9
Compression strength, 1,000 psi	320	120	300	20–420	40–300	—

Table 7-2 Properties of ceramics

Ceramic	Knoop scale	Mohs scale
Talc	20	1
Gypsum	50	2
Quartz	1,000	7
Aluminum oxide	2,000	9
Tungsten carbide	2,100	—
Titanium carbide	2,800	—
Silicon carbide	3,000	—
Boron carbide	3,500	—
Diamond	8,000	10

Table 7-3 Relative hardness values for ceramics

Tensile Strength

In theory, ceramics have extremely high tensile strengths. It is possible to get 700,000 to 1,400,000 psi in tension from newly drawn glass fibers. However, the smallest flaw in the fiber filament will cause the tensile strength to decrease significantly.

The manufacturing process used to make glass and other ceramic products usually can not prevent small flaws from occurring. Conventional window glass typically has so many small flaws that its tensile strength is reduced to 30,000 psi or less.

Specialty glass such as borosilicate reinforced with glass fibers can achieve much higher tensile strengths than conventional window glass. Tensile strengths of borosilicate glass have been recorded at 120,000 psi and higher. Fracture toughness—the ability to resist sudden impacts—is also increased in this type of reinforced glass.

Compression Strength

Ceramics are considerably stronger in compression than they are in tension. This strength property is important to many applications of structural ceramics. The concrete foundations and clay drainage pipes must withstand the weight of compressive loads in the form of earth and rock or heavy building materials like brick, steel, and wood (see Figure 7–4).

The compressive strength of concrete is a product of the ratio of water to portland cement. It is generally desirable to have a water/cement ratio of less than 0.55 by weight for maximum strength.

The maximum strength of concrete occurs over a period of days. Testing for compression strength of various mixtures is done after a period of twenty-eight days. Compression strengths will range from approximately 3,500 psi to 8,000 psi.

Thermal Properties

The ionic bond, which is the principal bond in ceramics, causes the electrons to be in a stable condition. This means that the electrons are immobilized and the atoms are at a low energy level. This condition is responsible for the thermal properties exhibited by ceramics.

If you recall from the unit on metals, both heat and electrical current are readily conducted in metals because of the electrons that freely move from atom to atom. These moving electrons freely conduct both heat and electric current readily. The electrons in ceramics are stable or at a low energy level and try to remain in that configuration. Therefore, ceramics do not conduct heat or electricity like metals do.

Ceramics have the highest-known melting temperatures of all known materials. Specialty ceramics have melting temperatures that exceed 7,000 degrees Fahrenheit; metals melt well below 3,000 degrees Fahrenheit. Compared to metals and plastics, thermal expansion of ceramics is very low. Ceramics are much more stable when exposed to a wide variation in temperatures than either plastics or metals.

Space shuttle tiles of ultrapure silica fibers transfer heat so slowly that they can be held by the human hand seconds after being removed from a hot oven. Shuttle tile can experience temperatures of 2,300 degrees Fahrenheit (1,260 degrees Celsius), while the shuttle skin beneath never exceeds 350 degrees Fahrenheit (180 degrees Celsius).

Chemical Properties

Ceramics are highly resistant to chemical attack. This characteristic is due, in part, to the high surface hardness that resists abrasion, thereby retarding chemical corrosion.

SUMMARY

Ceramics have been used for thousands of years to make vessels to carry liquids; to cook with; and to construct buildings, walls, and roads. Within the last twenty years, new applications and processes have been developed that have experts referring to ceramics as the material of the future.

New applications and processes related to ceramics have increased cutting efficiencies of cutting tools for machining and provided the medical community with improved prosthetic devices and implants. Ceramics have been used extensively in the electronics industry and show great promise for increased use in the automotive and aerospace industries.

Ceramics are already being used in the production of some engine components, and experimental engines with major ceramic components are being tested. Ceramics are chemically inert, among the hardest materials on earth, and have thermal properties ideal for many applications involving temperature extremes.

REVIEW QUESTIONS

1. List three applications of advanced or nuclear ceramics.
2. What characteristics distinguish ceramics from polymers and metals?
3. What classifications are used to identify the variety of ceramic products?
4. Describe the chemical composition of clay.
5. Why is it important to have the correct proportion of water in a clay mixture?
6. List the four basic types of commercial glass. Which of the four is commonly used in products such as windows, bottles, and light bulbs?
7. Give three examples of *natural* abrasives and three examples of *synthetic* abrasives.
8. What is the composition of portland cement?
9. Describe the molecular bonding between atoms of ceramic mixtures.
10. What term is used to describe the basic structural unit found in most ceramics?
11. What hardness scales are used in testing ceramics?
12. What causes the difference between the theoretical tensile strength and practical tensile strength to be so great?

SUGGESTED ACTIVITIES

1. Locate ASTM Standard C 31 in the ASTM Standards in your library. Mix enough concrete (one part portland, two parts sand, three parts shale, and water) to fill three cylinders as described in ASTM C 31. Perform a compression test as described in C 31 on the samples after 3, 7, and 28 days. Develop a chart to show your findings.
2. Prepare a seven- to ten-page paper on the subject of "Advanced Ceramics in Aerospace and Automotives."

CHAPTER
8

Composites: Structure, Properties, and Characteristics

New requirements for high-performance defense and space systems have led to the development of advanced engineered materials. Service requirements that could not be met with conventional materials are now being met through the recent development of advanced composites and ceramics. Recently developed and tested composites are able to retain strength at high temperatures and are often lighter in weight than the conventional materials they replace. More than 10,000 pounds of advanced composites are used in the space shuttles, and they will be used extensively in the construction of the space station.

Although advanced ceramics and composites have been used primarily in the aerospace industry, experts predict that they will be a primary manufacturing material in the automotive, machinery, sporting goods, construction, and aircraft industries in the 1990s and on into the next century. The sporting goods industry has taken the lead in the commercial market. Advanced composites have been used to manufacture a variety of sporting goods, including golf clubs, fishing rods, tennis racquets, and track and field equipment.

Research is being conducted around the world to develop composites that operate at higher temperatures and lower costs. The U.S. government currently spends about $170 million a year (Table 8–1) for research and development related to advanced ceramics and composites. This is more than any other nation spends in this area. Japan, however, has been very aggressive in developing and using advanced ceramics and composites in commercial products.

The value of components produced from advanced materials in 1988 was approximately $2 billion. This figure is projected to increase to $20 billion by the year 2000. Most of the advanced ceramics and composites are currently being used by the government in projects related to space or advanced weapon systems. Aircraft and sporting goods manufacturers have accounted for the bulk of the commercial use of these materials. This should change as

Agency	Ceramics and ceramic matrix composites	Polymer matrix composites	Metal matrix composites	Carbon-carbon composites	Total
Department of Defense	21.5	33.8	29.7	13.2	98.2
Department of Energy	36.0	—	—	—	36.0
NASA	7.0	5.0	5.6	2.1	19.7
National Science Foundation	3.7	3.0	—	—	6.7
National Bureau of Standards	3.0	0.5	1.0	—	4.5
Bureau of Mines	2.0	—	—	—	2.0
Department of Transportation	—	0.2	—	—	0.2
Total	73.2	42.5	36.3	15.3	167.3

Source: U.S. Congress, Office of Technology Assessment

Table 8–1   U.S. government agency funding for advanced structural materials, fiscal year 1987 (millions of dollars)

the materials become less expensive and manufacturing processes for high-volume production improve.

Currently, materials can account for as much as 30–50% of the production cost of manufactured goods. Advanced materials are significantly more expensive than conventional materials and usually require a longer time to process. Therefore, many analysts believe that profitable commercial use of advanced ceramics and composites is still 10 to 15 years away.

The development of composites is also changing the design process. Traditionally, engineers would select materials that most closely met specific service requirements. In many cases, a single monolithic material could not meet the requirements. Now, properties can be designed into the materials to meet special service requirements of high-performance products. As a result, composites and other advanced engineered materials are often referred to as "tailored" or "tailorable materials."

DEFINITION AND STRUCTURE

Composites consist of two or more materials or material phases that are combined to produce a material that has superior properties to those of its individual constituents. Although the focus of this chapter is on composites of a more advanced nature, it is important to understand that they have been used for centuries and that they do exist in nature.

Wood is perhaps the best example of a natural composite. Wood consists of fibers made of cellulose in a matrix of lignin (Figure 8–1). Some composites, manufactured by humans, have been used for thousands of years. Plaster, concrete, plywood, and flakeboard are a few examples examples of conventional materials that may be classified as composites.

Composites are based on the controlled distribution of one or more materials in a continuous phase of another. All composites consist of a *reinforcement, matrix,* and a boundary

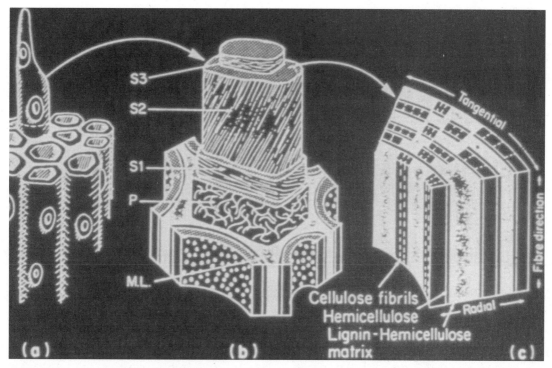

Figure 8–1   Wood, a natural composite, consists of cellulose fibers in a matrix of lignin *(Courtesy of USDA, Forest Products Laboratory)*

between them known as the *interface* (Figure 8–2). The reinforcement, matrix, and interface contribute unique properties that give the composite improved characteristics over conventional materials.

Reinforcement

The *reinforcement* is primarily responsible for the structural properties of a composite, that is, strength and stiffness. Materials used as reinforcements in composites are placed in one of three categories:

Fibers
Whiskers
Particulates (flakes, spheres)

Fiber reinforcement has dominated the composite industry since the 1950s. The largest volume of structural composites are reinforced with glass fibers, which were first commer-

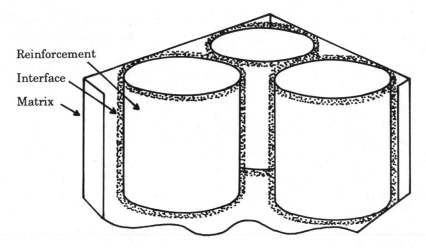

Reinforcement

Interface

Matrix

Figure 8–2 Composites consist of a reinforcement, a matrix, and a boundary between called the interface

cialized in 1939. Currently, fiberglass accounts for approximately 90% of the reinforced plastics market. Their light weight, high strength, and nonmetallic characteristics helped them gain acceptance during World War II. Boat hulls, corrugated roofing, fishing rods, and a wide range of other products are manufactured from fiberglass.

The two types of glass fibers commonly used as reinforcements are E glass (calcium aluminoborosilicate) and S glass (soda-lime-silica). E (electrical) glass provides excellent dielectric properties; resistance to chemicals, corrosion, and fatigue; and a high strength/weight ratio. It retains 50% of its tensile strength to 650 degrees Fahrenheit. E glass is available as continuous filament, chopped fiber, and fiber cloth.

S (high-strength) glass is used for components that require higher strengths. It has higher compressive and tensile strength, a higher MOE, and a slightly lower density. It is also significantly more expensive than E glass.

Other fiber-reinforcement materials include carbon-graphite (CG), boron, and aramid. CG and aramid fibers have received considerable attention in the production of advanced composites used in the aerospace industry.

Aramid organic fibers were introduced to commercial markets in the early 1970s by Du Pont. Introduced as Kevlar, this aramid fiber has been used in a wide variety of products, including helmets, jet fighters, and commercial aircraft.

Approximately 2,500 pounds of Kevlar was used in the fabrication of the Lockheed L1011-500 aircraft. Fairings, panels, and control surface components are fabricated from Kevlar 49 epoxy laminate. Kevlar 49 has excellent electrical properties, high tensile strength and toughness, but inferior compressive strength compared to CG. Du Pont's Nomex, Teijin's HM50, and Sumitomo Chemical's Ekonol are examples of other commercial aramid fibers on the market.

Boron fibers have been used extensively in the aerospace industry. First introduced in 1959, their relatively high cost has kept them from being used in high-volume applications. Boron fibers have excellent compressive strength and stiffness, and they are extremely hard. Boron fibers are manufactured by vapor deposition of boron onto a thin filament of tungsten.

CG fibers have become the most prevalent reinforcement material for composites used in advanced systems. CG fibers are produced from organic precursor fibers such as polyacrylonitrile (PAN) or by means of pyrolysis (burning of rayon or other polymers in a high-temperature furnace). The product of this process is a strong, stiff fiber that is a few microns in diameter.

By controlling the process, the tensile strength and stiffness can be raised or lowered for different applications. CG fibers with intermediate strength and stiffness are generally produced for aircraft applications. Spacecraft applications generally require higher-stiffness fibers, while higher-strength fibers are used in missile applications. Graphite epoxy is currently used or proposed for a wide range of aerospace structures, including the Boeing 767, F-15 and F-18, and the Advanced Tactical Fighter.

Table 8–2 compares various properties of selected fiber-reinforcement materials used in advanced composites.

Ceramics dominate the whisker reinforcements used in advanced composites. Although brittle by nature, ceramic whiskers retain their strength at high temperatures. Ceramic whiskers are generally single crystals with an **aspect ratio** (length/diameter) of 50 or more. Silicon carbide (SiC), silicon nitride (Si_3N_4), and high-purity alumina are common types of whisker-reinforcement materials used in ceramic matrix composites (CMCs). SiC is the dominant material used as a whisker reinforcement.

Whisker-reinforced ceramic composites have excellent insulating properties, high strength/weight ratio, and a low coefficient of thermal expansion. Their thermal stability has made them the material of choice in electronic, military, and aerospace applications where high temperatures are a problem.

Glass, ceramics, and metal are typically used to form particulates for reinforcing metal, plastic, and ceramic matrix composites. Flake reinforcement is primarily used when uniform properties are required. Particulate-reinforced composites are **anistropic** (different properties, depending on the direction of measurement). Particulates, depending on their

| Property | Fiber | | | | | |
	Boron	Aramid	E Glass	S Glass	Carbon-Graphite	PEEK
Tensile strength (10^3 psi)	510	400	500	660	300	75
Young's modulus (10^6 psi)	58.0	18	10.5	12.5	32	—
Elongation (percentage)	19	2.4	4.8	5.4	1.2	20
Coefficient of thermal expansion (10^{-6} °F in./in.)	2.5	−1.1	1.6	2.8	−.55	—
Specific gravity	2.58	1.44	2.54	2.49	1.8	1.3

Table 8–2 Properties of selected fiber reinforcements for advanced composites

composition, can provide excellent electrical resistance or conductivity. Ceramic flake–reinforced composites make excellent insulators, and metal flake–reinforced composites make good conductors.

Spherically shaped reinforcements aid in resisting shock and have excellent compressive strength. They are dimensionally stable and exhibit "low coefficient of thermal expansion." These properties make them extremely useful for a wide range of products. Cutting tools, seals, bearings, valves, heat engines, and medical implants are only a few of the current and proposed uses for particulate-reinforced composites.

Matrix

Bonding fibers, whiskers, or particulates together in one component is the primary function of the matrix. However, the function of the matrix extends beyond that of providing the "glue" that holds the reinforcement together. Among other tasks, the matrix also transfers the loads between reinforcing fibers, whiskers, or particulates. Protecting the reinforcement is yet another function of the matrix. The matrix protects the reinforcement against abrasion and corrosion caused by the environment (e.g., moisture, oxidation, chemical corrosion).

The importance of the matrix is easily suggested by the fact that composites are generally classified by the type of matrix used in their production. Composites are classified as polymer matrix composite (PMC), metal matrix composite (MMC), and ceramic matrix composite (CMC). Although the polymer matrices dominate current composite applications, both metal and ceramics have what appears to be a bright future.

Polymer Matrices

Organic polymer matrices have dominated the composite industry since World War II. Polymer resins used as matrices in the production of composites are classified as thermoset and thermoplastic. Until recently, the use of polymers was restricted to thermosets. The thermoplastics were considered to have inferior properties when compared to thermosets. Recent advances in thermoplastic technology, however, have led to the development of several new resins that compare favorably to the thermosets.

Epoxies are the dominant thermosetting resin in current use. Epoxies are high-strength adhesives that are combined with a variety of reinforcements to form high-strength materials for the aerospace and other industries. When combined with graphite, they produce a strong, tough, dimensionally stable material.

The service temperature limit for epoxies generally does not exceed 350 degrees Fahrenheit. However, specialty epoxies are commercially available with temperature limits to 475 degrees Fahrenheit (246 degrees Celsius). Epoxies are relatively expensive when compared to other resin systems.

Other commonly used thermoset resins include polyesters, polyimides, vinylesters, and bismaleimides. These thermosetting polymers are used primarily in the production of PMCs. Polyesters are used for high-volume production of auto parts and for large marine structures.

Thermosetting resins tend to have higher temperature and chemical resistance and dimensional stability than thermoplastics. This is a result of the structural difference between thermosetting and thermoplastic resins. During the curing process, the polymer chains in thermosetting plastics cross-link to form a three-dimensional network that resists heat and chemicals (Figure 8–3).

Thermoplastics do not cross-link and, therefore, are not as resistant to heat and chemicals. This difference can be an advantage or a disadvantage. Thermoplastics can be processed

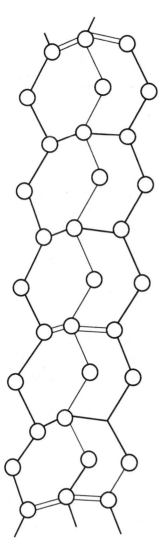

Figure 8–3   Polymer chains in thermosetting plastics cross-link, forming three-dimensional networks

faster than thermosets and tend to be tougher, that is, more resistant to cracking and impact damage. Also, processing is reversible for thermoplastics; that is, they can be formed a number of times if the initial forming process fails. Thermosets cannot be recycled or formed more than one time.

Once considered to be greatly inferior to thermosets for the production of composites, thermoplastics are receiving considerable attention in the composite industry. Thermoplastics are especially attractive to high-volume industries such as the automotive industry. The fact that thermoplastics can be formed more quickly than thermosets is a great advantage to high-volume industries.

Thermoplastic resins include polyesters, polyetherimide (PEI), polyimide, polyphenylene sulfide (PPS), thermoplastic polyimide (TPI), and polyetheretherketone (PEEK). PEEK and some of the other newly developed thermoplastics compare favorably with thermosets in high-temperature strength and chemical resistance. These resins are generally used with glass, graphite, boron, and aramid fibers. Table 8–3 compares general characteristics of thermoset and thermoplastics matrices.

Resin Type	Process temperature	Process time	Use temperature	Solvent resistance	Toughness
Thermoset	Low	High	High	High	Low
Toughened thermoset					
Lightly cross-linked thermoplastic					
Thermoplastic	High	Low	Low	Low	High

Source: U.S. Congress, Office of Technology Assessment

Table 8–3   Comparison of general characteristics of thermoset and thermoplastic matrices

Metal Matrices

Metals used as a matrix in composites generally include low-density metals such as aluminum or magnesium. These metals are typically reinforced with fibers or particulates of a ceramic material such as silicon carbide or graphite. These composites are high in cost and are used primarily for military and space applications.

Compared to conventional, unreinforced metals, MMCs have significantly greater strength (Table 8–4) but at the expense of lower ductility and toughness. MMCs have been developed for use in aircraft, missiles, and the space shuttle.

Although commercial use of MMCs is limited, an aluminum piston selectively reinforced with a ceramic fiber is being produced for diesel engines. MMCs will most likely be developed for a number of commercial, military, and space applications. Spacecraft structures, fighter aircraft engines, high-speed mechanical systems, and electronic packaging are suggested applications for MMCs. Commercial applications include diesel engine pistons, golf clubs, tennis rackets, skis, and other sporting goods and recreational products.

Property	Unreinforced		Reinforced				
	Aluminum 6061	Titanium Ti-6A1-4v	Graphite-Aluminum	Boron-Aluminum	Silicon Carbide Aluminum	Alumina Aluminum	Silicon Carbide Titanium
Tensile strength (axial) MPa	290	1,170	690	1,240	1,040	620	1,720
Tensile strength (transverse) MPa	290	1,170	30	140	70	170	340
Stiffness (axial) GPa	70	114	450	205	130	205	260
Stiffness (transverse) GPa	114	114	34	140	99	140	173

Table 8–4   Comparison of conventional metals to metal-matrix composites (MMCs)

MMCs exhibit high strength and stiffness (see Figure 8–4), excellent dimensional and thermal stability, and resistance to environmental factors when compared to conventional metals and plastic matrix composites. Their fracture toughness is generally lower than that of conventional unreinforced metals—a tradeoff for higher strength and stiffness.

Fiber-reinforced MMCs are excellent conductors of heat and electricity. The heat or electrical current is transmitted from fiber to fiber much more efficiently because of the higher aspect ratios of fiber-reinforced MMCs. High-thermal graphite fibers in a matrix of aluminum or copper are especially good conductors.

The major disadvantages of MMCs are costs associated with the constituent materials and the processing used to produce the composite structure. In general, these costs restrict their use to high-performance military and space applications.

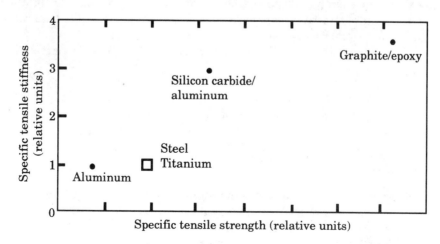

Figure 8–4   Strength of stiffness of selected advanced materials

Ceramic Matrices

The most commonly used ceramic matrix is cement. Portland cement, used as a binder in concrete, is a ceramic matrix that was first used long before "advanced composites" were even theorized. Advanced ceramic composites consist of glass or ceramic matrices reinforced with fibers, whiskers, or particulates.

Glass and ceramic matrix composites have improved toughness, compared to unreinforced ceramics. The inherent brittleness of ceramics is the primary obstacle that must be addressed before advanced ceramics and ceramic composites enjoy the same popularity and range of application as polymer composites.

Perhaps the most popular glass-ceramic matrices are lithium-alumina-silicate (Li_2O-Al_2O_3-SiO), commonly referred to as LAS; BMAS (BaO-MgO-Al_2O_3-SiO); and magnesia-alumina-silicate (MgO-Al_2O_3-SiO). These matrices and others are typically reinforced with fibers, whiskers, or particulates of silicon carbide (SiC), silicon nitride (Si_3N_4), zirconium oxide (ZrO_2), aluminum oxide (Al_2O_3), or other engineered ceramics.

Research in CMCs focuses on methods of increasing their toughness. Ceramics are, by nature, a very brittle material. They do not withstand impact forces or shock very well. Researchers, however, have been able to increase fracture toughness in ceramic composites over monolithic ceramic structures.

In general, ceramics have superior high-temperature strength and low thermal conductivity. They also are extremely hard. By reducing the strength of the bond between the matrix and the reinforcement, researchers have been able to increase fracture toughness of ceramic composites. The presence of long fibers in the ceramic or glass matrix has increased toughness by a factor of two over monolithic ceramics. This is a result of the apparent ability of the fiber to stop the propagation of cracks or flaws caused by shock or impact. Table 8–5 compares physical and mechanical properties of various ceramics and CMCs.

Compatibility between the matrix and the reinforcement is especially important in ceramic composites. For example, if the coefficient of thermal expansion is significantly different, the component could be seriously flawed. Ideal compatibility is generally assured when the matrix and reinforcement are from the same material.

A wide range of applications for CMCs exist or are being proposed. A partial list includes the following:

Cutting tools	Bearings	Seals
Valves	Orthopedic implant	Heat exchangers
Heat engines	Diesel engines	Gun-barrel liners

Interphase–Interface

The boundary between the matrix and the reinforcement in a composite is known as the *interphase* or *interface*. The interface generally does not receive the attention that the matrix and reinforcement do; however, it is an important element in engineering composite structures. The bond between matrix and reinforcement affects the overall performance of

Property	Alumina 99.5	SiC Reaction Bonded	Si_3N_4 Reaction Bonded	Al_2O_3	ZrO_2
Compressive strength, psi $\times 10^3$	380	100	112	350	300
Tensile strength, psi $\times 10^3$	30	20	—	—	21
Flexural strength, psi $\times 10^3$	50	37	30	—	100
Modulus of elasticity, psi $\times 10^6$	55	56	24	38	30
Impact resistance, in./lbs. Charpy	6	< 0.8	< 0.8	—	—
Hardness Mohs	9	9	9		
Knoop				1,470	1,100
Density, kg/m^3 $\times 10^{-3}$	3.5	3.21	3.44	4.64	5.78

Table 8–5   Properties of selected ceramics and ceramic-matrix composites

the composite. The interface is especially important in controlling the mechanical behavior of composites.

Although it is generally more common to increase the strength of the bond between the matrix and reinforcement, there are composites that perform better with a weaker bond. As discussed earlier, the toughness of ceramics is increased if the bond between the matrix and reinforcement is weaker. Polymer reinforcements, on the other hand, are often treated with a coupling agent to increase the bond between the constituent of the composite component.

Coupling agents, used to improve the interface between the matrix and the reinforcement, can perform one of several functions in fabricating composite materials. Among others, coupling agents improve wetability (ability of any solid to be wetted when in contact with a liquid). This increases the bond strength between the reinforcement and the matrix.

PROPERTIES OF COMPOSITES

Any attempt to describe the various properties of PMCs, CMCs, or MMCs would require volumes rather than pages of information. The wide range of properties made possible by almost-infinite combinations of matrices and reinforcements make a detailed description of properties a nearly impossible task.

A discussion of the elements of composites that affect their properties is not quite as difficult, however. As mentioned previously, each constituent of the composite (matrix and reinforcement) and the interface contribute to the thermal, mechanical, and electrical properties of the composite. For example, an aluminum or copper matrix with long carbon fibers conducts electricity and heat much more efficiently than an aluminum matrix with carbon flakes. The length of the fibers, in this case, affects the property of the composite.

The aspect ratio is also important when high strength is required. Rather than focus on properties of individual composites, this section will deal with a more general description of how the size, shape, composition, and configuration of the constituents affect various mechanical, thermal, and electrical properties.

The following is a partial list of the factors that affect the properties of composites:

Composition of components
Size of components
Shape of components
Proportion of reinforcement and matrix
Orientation of reinforcement
Interphase adhesion

Composition of Components

Several generalizations can be made about the properties of PMCs, MMCs, and CMCs. PMCs are lighter in weight and tougher than either metal or ceramic composites, but their service temperature is generally much lower. MMCs have higher service temperatures than polymers, and they conduct heat and electricity better than either polymers or CMCs. Ceramic composites are brittle but high in compressive strength, and they have a low coefficient of thermal expansion.

Polymers melt or char at temperatures above 600 degrees Fahrenheit (316 degrees Celsius), while metals generally retain their strength to about 1,900 degrees Fahrenheit (1,038 degrees Celsius) and ceramics can retain their strength to well above 3,000 degrees Fahrenheit (1,649 degrees Celsius). Figure 8–4 compares specific strength and stiffness of some advanced materials to conventional metals.

Wear resistance is much higher in MMCs and CMCs than it is in PMCs. The alumina-silica–reinforced aluminum piston used in the diesel engine demonstrated an 85% improvement over the conventional-type piston.

Thermal expansion can be reduced to near zero by choosing the proper matrix and reinforcement. Silicon carbide particulates in aluminum result in a much lower coefficient of thermal expansion when compared to the conventional metals.

Size and Shape of Reinforcing Components

The concept most commonly used when referring to the physical dimensions of the reinforcements used in composites is aspect ratio. *Aspect ratio* is the relationship of the length to the diameter of the reinforcing material. Aspect ratio is calculated by dividing the length of the individual reinforcement by its diameter. Aspect ratios can easily exceed 500 in fiber reinforcement or as little as 10 in particulate. The ratio of length to width is important in determining the properties of the composite.

Longer, continuous fibers in PMCs increase their strength. Longer fibers (higher aspect ratios) in MMCs will improve their electrical and thermal conductivity. An aspect ratio of 300 in E-glass fibers will increase its tensile modulus by 65% over shorter fiber reinforcements (aspect ratio of 50).

As explained earlier, longer fibers in CMCs help increase the toughness of ceramics. The longer fibers tend to stop the propagation of cracks in CMCs caused by shock or impact.

Reinforcements with lower aspect ratios (particulate) are also used to design specific properties into composites. Spheres and flakes generally improve the compressive strength, shock resistance, and dimensional stability of composites. They also provide uniform properties in all directions; unlike the fiber reinforcements. Flakes and spheres also provide high electrical and thermal resistance.

Proportion of Constituents

The proportion of the matrix and reinforcement is critical to the ultimate strength of the composite. The relationship is curvilinear. The strength and stiffness of the composite will increase as the reinforcement increases to a point.

In fiber-reinforced PMCs, the strength will fall drastically if the volume reinforcement exceeds about 60%. A volume of 40% silicon carbide particulate in aluminum increases its strength by about 65% over that of 6061 aluminum. The composite loses strength when the volume of reinforcement is too high because the matrix cannot thoroughly wet all the fibers causing the bond to be weak.

Orientation of the Reinforcement

Composites can be designed to have different properties in different directions. The direction of the strength of a composite is related to the orientation of the reinforcement. Aligning all fibers parallel to one another will provide the highest strength when stressed parallel to the direction of the fibers. This directional character is referred to as *anisotropy*.

To obtain a degree of **isotropy** (strength in all directions), fibers would have to be arranged in multiple directions. However, this reduces the unidirectional strength of a composite. For example, a unidirectional-fiber composite mat may have a tensile strength of 300,000 psi. If the fibers are oriented in two directions, 0 degrees and 90 degrees, the tensile strength in either direction is about one-third, or 100,000 psi. If three plies are laid at an orientation of 0, 45, and 90 degrees, then the tensile strength will be approximately one-sixth of a single-dimensional mat.

Interphase Adhesion

The interphase was discussed in detail earlier. Its relationship to the properties of the composite component is a very important one, as explained earlier. The wetability of the reinforcement is critical in determining the mechanical properties of the composite. The more readily the matrix is able to wet the reinforcement, the better the bonding between reinforcement and matrix. This is especially important for PMC fiber composites.

However, a weaker bond is more desirable in CMCs. The weaker bond increases the toughness of the ceramic by causing the reinforcement to pull loose from the matrix when impact or shock causes the matrix to crack. This tends to stop the crack from continuing.

STRUCTURAL SHAPES

Composites or "advanced materials" take many forms, depending on specific applications. This is one of the major advantages of composites over other conventional materials. In the past, the discrete properties of the material would greatly influence the design of the final product. Composites and other advanced materials allow the designer and engineer to tailor the material to meet the desired final performance requirements. The materials and the structure are created in an integrated manufacturing process.

Metals, woods, and plastics—as discussed in other chapters—are available in shapes and sizes that have been standardized by professional organizations. Composites and other advanced materials, however, are available in a wide variety of forms depending on the specific application.

Preformed composites are often classified by the shape of the reinforcement: fiber, whisker, or particulate (Figure 8–5). Fiber reinforcements have dominated the composite industry to date. Each type of reinforcement offers its own special property enhancements to the composite.

Fabrication

A wide variety of fabrication processes are used to transform primary composite materials into finished products. Many of the processes are adapted or the same as those used to process conventional materials (e.g., metal and plastic). One of the earliest processes used and the least complicated is hand lay-up, commonly used for fiber-reinforced fabric (Figure 8–6). During this process, fiber-reinforced fabric is placed in a mold; and resin is poured, sprayed, or brushed into the fibers to provide stiffness.

Other processes include compression molding, injection molding, and cold stamping (Figure 8–7). Figure 8–8 shows a sheet-molding process used by Ford Motor Company to fabricate PMC automotive body components.

Standards

The field of composites is relatively new, and much of it has been experimental in nature. Therefore, standards for this field are not yet well established. Several groups, however, are beginning to collect historical data on performance levels of well-established composites. Committees are being formed, and the groups responsible for establishing standards are being identified.

Several organizations and associations (private and public) have begun establishing standards for composites. The American Society for Testing and Materials (ASTM) has estab-

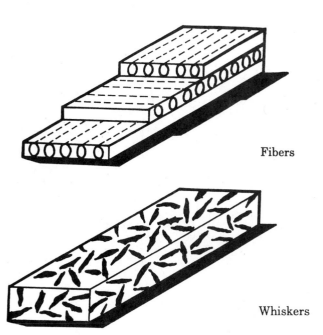

Fibers

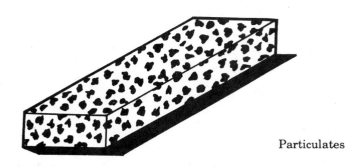

Whiskers

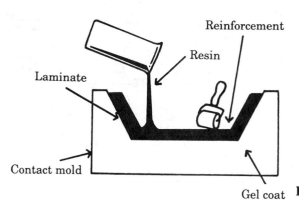

Particulates

Figure 8–5 Structural composites are classified as fiber, whisker, or particulate

Reinforcement

Resin

Laminate

Contact mold

Gel coat

Hand lay-up

Figure 8–6 Hand lay-up method used to manufacture reinforced fabric composite

145

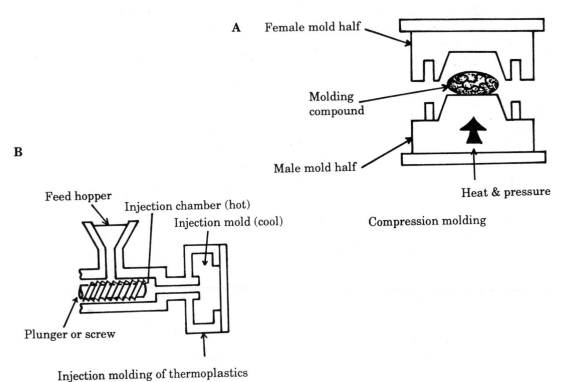

A — Female mold half

Molding compound

Male mold half

Heat & pressure

Compression molding

B

Feed hopper

Injection chamber (hot)

Injection mold (cool)

Plunger or screw

Injection molding of thermoplastics

Feed hopper

Injection chamber (cool)

Part

Plunger or screw

Injection mold (hot)

Injection molding of thermosets

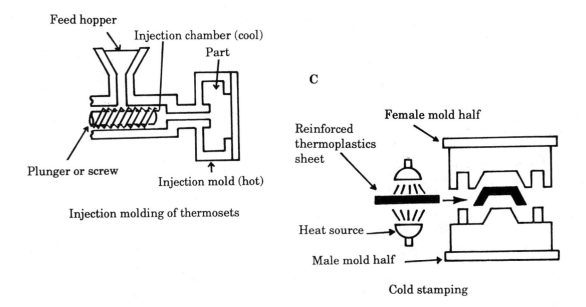

C

Reinforced thermoplastics sheet

Female mold half

Heat source

Male mold half

Cold stamping

Figure 8–7 Typical process used to manufacture composite parts and products: **(A)** compression molding, **(B)** injection molding, **(C)** cold stamping

146

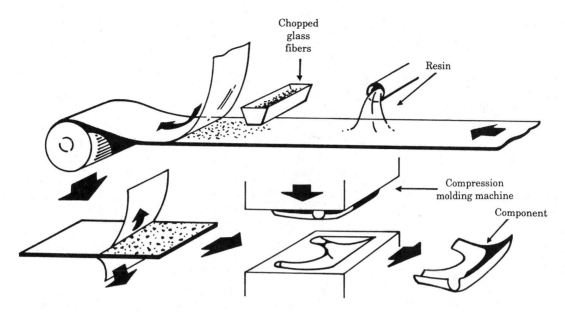

Figure 8–8   Sheet-molding process used by Ford to fabricate PMC automotive body parts

lished an advanced ceramics committee (C28) which is reviewing properties, performance, characterization, processing, terminology, design, and evaluation. ASTM's Committee on High-Modulus Fibers and their Composites (D30) and Committee on Plastics (D20) are the primary sources for standardized test methods for PMCs.

Trade associations, such as the U.S. Advanced Ceramics Association (USACA) and the Suppliers of Advanced Composite Materials Association (SACMA) are also working with ASTM to establish standards. The Department of Defense (DOD) has initiated a new program for standardization of composite technology. The DOD, through the Army, has established the Composite Materials Program Standards (CMPS). It is responsible for integrating diverse standards for composites by gathering standardized test methods into Military Handbook 17 (Mil 17) and developing separate test methods when necessary.

Other groups involved with establishing standards for composites include the Composites Group of the Society of Manufacturing Engineers (CoGSME), American Society for Metals (ASM), Society for the Advancement of Material and Process Engineering (SAMPE), and Society of Plastics Industry (SPI). Several international organizations are also pursuing advanced materials standards.

SUMMARY

Technologies related to composites and other advanced materials are in their infancy. The research in this area has been funded primarily by the federal government; however, many

many private companies recognize the advantages of composites and are supporting research and development as well. The majority of advanced materials and composites have been used in high performance military and aerospace systems such as missiles, fighter aircraft, and space shuttles.

Commercial manufacturers of a wide variety of products have recognized the potential of these materials for existing and proposed products. Efforts to reduce the costs of the primary materials and to improve the manufacturing processes for fabricating composite components are being made by many public and private organizations.

REVIEW QUESTIONS

1. List factors that created a need for composites.
2. How have composites changed the design process?
3. What department of government is spending the most money on research related to advanced ceramics and composites?
4. Define *composite*.
5. List the three phases of a composite material and describe the unique contributions of each to the composite.
6. What year were glass fibers introduced to commercial markets?
7. Describe the physical property differences between E glass and S glass.
8. What is the commercial name of the first aramid fiber material used in composites? What company developed it?
9. How are composites generally classified?
10. Describe the basic physical differences between thermosetting and thermoplastic resins?
11. What potential advantages do thermoplastics have over thermosetting resins?
12. How have researchers increased the toughness of ceramic matrix composites over conventional ceramics?
13. Define *wetability*.
14. Define *aspect ratio* and describe its effect on properties of composites.
15. How does the proportion of the constituents of a composite affect its strength?

SUGGESTED ACTIVITIES

1. Research four composite materials not listed in Table 8–5 to determine their mechanical properties. Prepare a chart comparing the properties of the four materials selected. Use the same format as Table 8–5.

2. Select a product made of wood, plastic, or metal. Determine an appropriate composite that could be substituted for the wood, plastic, or metal and prepare a list of advantages and disadvantages of replacing the original material with the composite.
3. Prepare a composite of your own, using available resins and fibers or laminate plastic to wood or metal. Compare the mechanical properties of the composite to its constituent materials.

SECTION 3

PROCESSING · POST PROCESSING · RECOVERY · GENESIS · PRE PROCESSING

DESIGNING AND DEVELOPING MANUFACTURED GOODS

CHAPTER 9

Product Design

Product design is the process of creating goods that have a pleasant appearance and function effectively. When we design a product, it is possible to break down the process into elements and principles. *Elements* are what we work with, and *principles* are what we do with them.

Before the design process can begin, we need to know something about the product and what it is expected to do. To accomplish this task we need to conduct a product research study. This study will identify basic information about the variables (changeable data) and the constraints (limitations) concerning the suggested product.

PRODUCT RESEARCH

Most product designers agree that form should follow function. Therefore, a good place to start a product research study would be to determine exactly what the product is expected to do. This procedure can be set up as a design problem and written as a problem statement. For example, a desk organizer is needed that will hold four pencils, a pad of notepaper, and fifty paper clips and provide a place to attach important memos.

This is an example of a problem statement written in terms of performance and quantity. It lacks specific information such as size, type of material, costs, and design features. The designer must be able to break the problem down and create several design solutions. In doing so, the designer must have more information and may require the answers to several questions. For example:

1. What is the desired finished size of the product?
2. What type of materials should be used?
3. Is there a limit on the cost of the product?

4. Will the product be used in a home or office environment?
5. Is there a preferred style for the design such as modern, colonial, or Danish?
6. What are the manufacturing requirements and limitations?
7. What type of finish will be needed to make it durable yet pleasing to the eye?

Product Profile

Once the designer has obtained the answers to these and perhaps other questions, the **product profile** will begin to develop. The product profile is a collection of known facts about the requirements of the product based on the problem statement and research information which is used to clarify any unknowns.

Not all product designs begin with a problem statement. Sometimes products are *redesigned* or represent an adaptation or imitation of an existing successful product. This concept is discussed in further detail in Chapter 11. Because the product already exists, the only questions which need to be answered are concerned with the desired changes or modifications. The basic product remains essentially the same, but design changes enhance its looks or may even help the product function more efficiently.

Design Variables

In the process of developing the product profile, the designer must be aware of the various variables that are possible when developing the product. Indeed, the questions which provide the basic information for product research are developed with the design variables in mind. Regardless of the type of product to be processed, the designer and design engineer must deal directly with eight basic design criteria:

1. Function
2. Form
3. Ergonomics
4. Aesthetics
5. Ease of manufacture
6. Standardization
7. Durability
8. Cost

The process of selecting the best combinations of design criteria is an important activity for developing a good product profile. As you begin to design your product in class, you will also need to define your problem and develop a product profile. Let us first review some basics in the design process.

Basic Design

In addition to gathering data about the product for a good profile, you will also need to understand some basic rules concerning good design. As mentioned earlier in this chapter, design can be divided into two basic areas: the elements and the principles. The elements of design are form, space, line, texture, and color. The principles of design are balance, continuity, and emphasis. These basic areas should be considered when developing the design solution. The creative designer is one who is able to combine and function effectively and gracefully without compromise.

ELEMENTS OF DESIGN

Every product exhibits form in either a single-, double-, or three-dimensional plane. Form includes both the inside and outside shape of the object. Some forms are simple single-plane ideas, while others are complex and confusing to the beginning designer. For example, a house is merely a combination of different geometric shapes which together create a single form (Figure 9–1).

Figure 9–1   The use of geometric shapes can render a beautifully designed home *(Courtesy of Photri, Inc.)*

By combining straight, angled, and curved forms, designs can exhibit different feelings and values. For example, if we wanted to create the feeling of formality we would use straight, 90-degree sharp-cornered shapes. Rectangles, squares, and triangles suggest rigidness and, therefore, a more formal feeling. If overdone, a feeling of monotony will be created, particularly if these formal shapes are repeated.

If you add angles to the straight formal forms (particularly a tiled square or rectangle), a sense of motion will be felt. Using the formal straight forms with angular forms can provide exciting and interesting designs. When the angles are changed into triangles and pyramids, an additional feeling of stability will begin to occur. Multiple angles and sides (more than four) have a softening effect, and this feeling continues to grow as the shape begins to take on the circle form.

Curves and circles are considered to be more natural forms which can provide pleasant feelings. When curves and circles like those in Figure 9–2 are placed among the other geometric shapes, they tend to display the feeling of dominance. In addition, the three-dimensional circle called a sphere also conveys the feeling of motion.

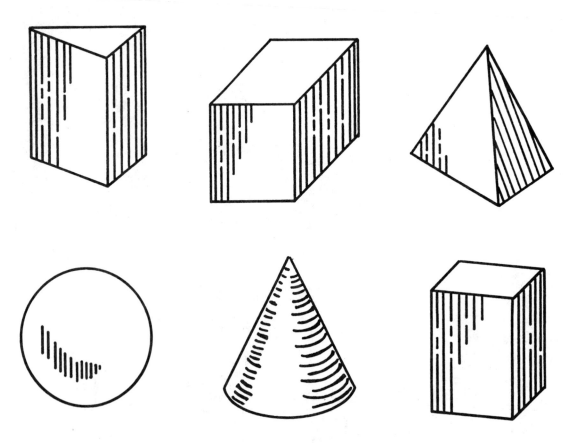

Figure 9–2   Geometric shapes can cause different types of feelings

Irregular forms are not members of the geometric family but are used to add beauty and form to a design. One must be careful not to overuse irregular forms because it is possible to render the design instability and discord through indiscriminate usage.

Another factor related to good form is weight. The actual weight is not as important as the implied weight which the design tends to communicate. As can be seen in Figure 9–3, the two identical shapes and mass do not look equal in terms of weight. The darker, more solid-looking rectangle *looks* heavier, even though we do not have any data that would indicate that such is true. Large bulky rectangles in solid colors do tend to look heavier than creative shapes in pastel colors, even though the volume and actual weight may be the same.

Size and form must also be considered against the background that the product is to be placed in. For example, a pyramid would lose much of its feeling of dominance and stability if it were moved from the desert to the foothills of the Rocky Mountain region. On the other hand, a modern skyscraper would look less stable in the middle of a desert because of its narrow and tall profile. In the middle of a city, however, it appears natural and strong.

Space

Space is difficult to define without the use of form and shape. In general, the area surrounding the form is known as the space. We are able to identify space through the use of

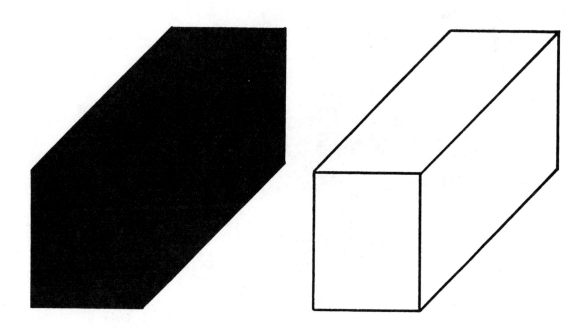

Figure 9–3   Some designs use mass color to suggest implied weight

three-dimensional objects much easier today because of the advancement in computer graphics. Using CAD (computer-aided design), the designer is able to use the full dimension of space in design (Figure 9–4). Each and every shape can be evolved and revolved in an unlimited number of positions in space. As a result of this innovation, designers are now making new headroads into the area of design through new use of space.

When one begins a design, whether a design draftsperson or a child, the first mark on the paper begins to define the space. Use of that space can provide feelings of openness, distance, and crowdedness. For example, tall ceilings can provide a feeling of openness, while lower ceilings may provide feelings of coziness and security.

In product design, space is important in terms of use and function. Large areas of unused space tend to make the consumer feel that the product is not well designed and, therefore, that it may not function well. Wise use of space is important for the overall design of a product.

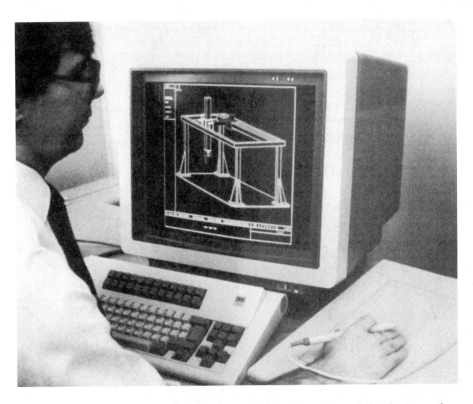

Figure 9–4   Advancements in computer graphics have allowed the designer the opportunity to use the full dimension of space *(From Jeffries,* AutoCAD for Architecture, *Copyright 1993 by Delmar Publishers, Inc. Used with permission)*

Line

Of all the elements in design, line is the only one which does not occur in the natural world. Instead, what we see is the edge of a surface, a division between forms or the change in the direction of a plane. Line is important to the designer because it defines shape and space (see Figure 9–5) and has the ability to create a different feeling. Horizontal lines tend to give the feeling of restfulness. Vertical lines are stable like the horizontal but are not as restful because they look like they take effort to remain vertical. Products which have diagonal or sloping lines appear to be active with the suggestion of movement. However, when the diagonals oppose each other, they tend to return to stability like the triangle.

Curved lines can be very useful for setting a mood. Indeed, the familiar yellow smiley face with its curved upward line makes us feel positive. If we were to turn the mouth line downward as in Figure 9–6, the feeling of sadness or unhappiness tends to occur. Combinations of curves and straight and diagonal lines can be used to combine various desired feelings. In addition, thick and solid versus thin and broken emphasis for lines can also affect the feeling that a design can communicate.

Figure 9–5   Line defines shapes in space

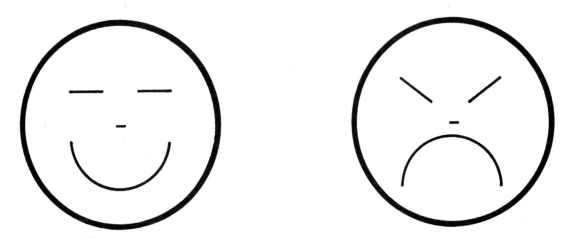

Figure 9–6 Lines can communicate feelings like the simple smiley faces

Texture

Texture refers to the surface of natural and human-produced objects. Texture then relates to the "feel" of an object both visually and actually. A surface may be soft, rough, smooth, dull, shiny, dark, or light. Whether you just see it or actually touch the item, a feeling of strong aesthetic response may be felt. Combinations or use of contrasts (see Figure 9–7) can be used to enhance or detract various feelings of texture in a product design. However, good design makes use of their added dimension and possible contribution.

Figure 9–7 Surface texture can convey special feelings

Color

As one might imagine, the world of color is very important when designing a product. As a design element, color can provide the necessary emotional and psychological influences necessary to make a product design a success. An experienced designer will select color for mood and intensify it by playing hue, value, and intensity against each other. These characteristics can enhance or decrease the value of color in design. As can be seen in Figure 9–8, the hue (color) is only one characteristic of a three-dimensional interaction.

Color in wood products is enhanced by the wood grain pattern. Heavy flame-grained woods such as oak or ash tend to convey a feeling of action and aggressiveness. Softer quarter-grained woods such as walnut and mahogany, on the other hand, tend to convey the feeling of warmth and smoothness. In both cases, the designer should let the true natural look of the wood come through in order to display the feeling of naturalness that is inherent in the use of wood as a material.

The use of color in the design solution can be one of the most important decisions made by the designer. Therefore, alternative options for color should be provided as part of the design solution. Such alternatives are obviously exciting and potentially unlimited.

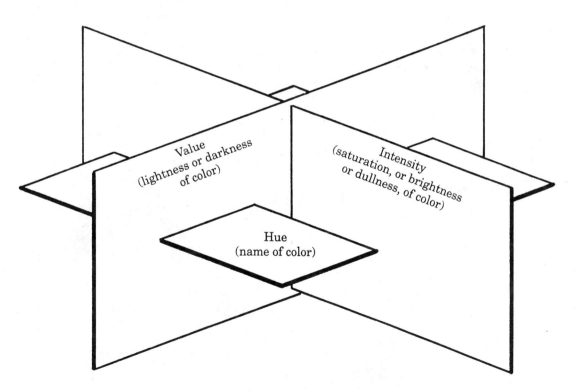

Figure 9–8   Color and design can affect mood

PRINCIPLES OF DESIGN

The elements of design can be arranged in an unlimited number of combinations. The procedure and choice for selecting the correct combinations of elements is called **following principles**.

Principles of design can directly affect the harmony or discord of a design. What we do with the elements can create harmony by itself; discord by itself; or, in some special cases, a combination of harmony and discord. Principles are classified under three major areas: balance, continuity, and emphasis.

Balance

The most common principle of design is symmetrical balance. Products which do not display symmetry (harmony) tend to show disorder (discord); that is, they do not appear to be pleasing or natural in organization but instead may demonstrate imbalance in arrangement. As can be seen in Figure 9–9, the design which does not demonstrate symmetry tends to look awkward. While the size, color, and mass may be the same in Figure 9–9, the shapes are not symmetrical (alike); and it seems awkward to the eye.

Sometimes a nonsymmetrical design is desired. When this occurs, the designer has usually balanced the design using another element such as texture, color, or the like. When a design demonstrates balance, we call it harmony. The Romans were very careful to use balance in their architecture (see Figure 9–10).

Combinations, regardless of whether they are symmetrical, can demonstrate total harmony or total discord, depending on the effect the designer is trying to accomplish in the design solution.

Figure 9–9 Design that does not demonstrate symmetry tends to look awkward

Figure 9–10 Roman style architecture demonstrates the use of balance and harmony *(Courtesy of Photri, Inc.)*

Continuity

Even a design which demonstrates elements in discord can be effectively balanced and create a feeling of harmony if it is repeated properly. Repetition tends to create a simple rhythm, as shown in Figure 9–11, and can provide a feeling of unity. Overuse of this technique can defeat its purpose and cause a feeling of monotony and dullness.

Alternating the repetition of a design is another way of creating the feeling of harmony. In Figure 9–12, the contrast of alternating the design actually enhances its feeling of unity. A checkerboard is a good example of alternating the design element of color, that is, light and dark (Figure 9–13). Continuity, then, is the principle that makes the elements fit together and provide the design solution with organization.

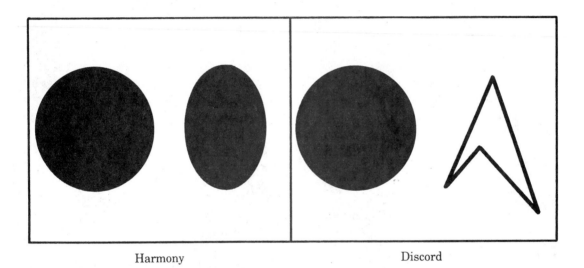

Harmony   Discord

Figure 9–11   Harmony and discord can be used for good design

Emphasis

When the designer selects the various elements which will become part of the design solution, attention should be given to what it is that the consumer should notice first. In other words, what feature of the product should the design emphasize? Like the photographer who composes a picture in the focal lens, the designer, too, must focus on the important feature of the product (Figure 9–14). By using various elements discussed earlier in this chapter, the designer is able to emphasize the important or dominant feature of the product design.

The ordering of dominant or subdominant emphasis is vital if the design is to display its strengths and capture the interest of the consumer.

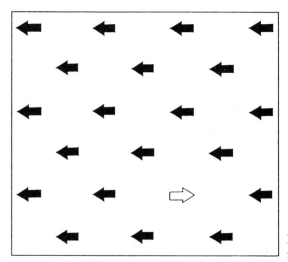

Figure 9–12 Harmony can be created in many different ways

Figure 9–13 A checkerboard shows alternate patterns of light and dark for harmony

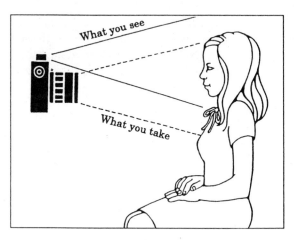

Figure 9–14 The photographer must select emphasis when composing a photograph

CREATING THE DESIGN SOLUTION

As mentioned earlier in this chapter, most designers agree that form should follow function. Not all designers follow this rule, but most designers of manufactured products consider it to be extremely important. If the design is based on what the product is able to do, it does not necessarily mean that the design cannot be pleasing to the eye. Indeed, that is the real challenge for the industrial designer.

Let us see if we can solve a design problem based upon what we have learned in this chapter. A good place to start is with the problem statement. Remember the problem statement at the beginning of this unit? We would like to have a desk organizer which will hold four pencils, a pad of notepaper, and fifty paper clips and provide a place to attach important memos.

Function

As one can see, the problem statement gives the designer a good idea about function of a product design. This is usually true because most consumers and market salespeople describe items by what they can do. In this case, the desk organizer is expected to do four things:

1. Hold four pencils
2. Hold a pad of paper
3. Hold fifty paper clips
4. Provide a place to attach memos

The function of this product should lead to several alternative design solutions. We can begin to express our early thoughts by brainstorming the various possibilities. To do this, we employ a technique called **"thumbnail sketching."** These are doodle-type sketches which help the designer refine his or her ideas. The designer is trying to solve functional, aesthetic, material, and production problems all at once.

In Figure 9–15, you see the basic thumbnail sketches which may lead to a design solution for the problem concerning the desk organizer. As you will notice in Figure 9–15, the size and shape can be varied while maintaining the function of holding four pencils. Each of the designs in Figure 9–15 *will* hold four pencils, but do they meet the functional require-

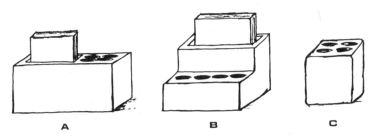

A   B   C

Figure 9–15   A thumbnail sketch of a desk organizer

ment of holding a pad of paper? Our choice is either to select the ones that do or to redesign those that cannot meet the requirement of holding a pad of paper. In this case, our desk organizer may look like the ones in Figure 9–16.

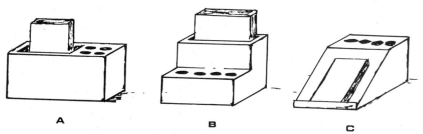

Figure 9–16 Desk organizers that can hold a pad of paper

Notice that the shape idea C in Figure 9–15 was eliminated as a possible solution. The reason it was eliminated was because it did not meet the functional requirement of holding a pad of paper. It could be redesigned to meet the requirement, so the designer created an alternative design to replace design C.

The next functional requirement is for the desk organizer to be able to hold fifty paper clips. Figure 9–17 shows the possibilities for designs. Notice the alternative design E has been dropped because it could not meet the functional criterion of holding fifty paper clips.

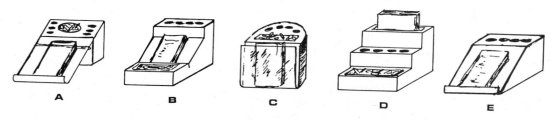

Figure 9–17 Desk organizers that can hold 50 paper clips

Our final criterion for the desk organizer was that it to be able to hold important messages. Reviewing alternative design solutions A, B, and C, it would appear that they will all be able to meet the requirement (see Figure 9–18).

Now we have three final solutions which have met the requirements of function. Our next task is to consider the form (i.e., design elements) and principles.

Solution C could be made out of multiple materials which would enhance its use of color and texture. It tends to look like a dull design primarily because of its curved-line approach.

Solution B also could take advantage of using multiple materials, but it tends to be very formal and almost monotonous.

Solution A is much more interesting in design and also makes use of multiple materials for effect. The contrast between wood, cork, and plexiglass is very attractive. With the addition of a

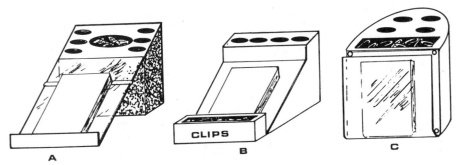

Figure 9–18   Three designs that will meet all criteria

tractive. With the addition of a brass band to hold the paper pad, an attractive desk organizer using good elements and principles has been designed and sketched. Our choice, then, is solution A.

Once the design solution has been selected based on the idea that form follows function, it is now time to begin to work out some of the details and problems. A refined drawing with measurements in proportion is appropriate at this time. Figure 9–19 shows the views that are used by design draftspersons to illustrate product ideas. Once the drawings are complete, it is necessary to begin the process of solving possible manufacturing or construction problems. This phase of product development is called the **design engineering process**.

DESIGN ENGINEERING PROCESS

So far in this chapter, we have identified the basic design principles and elements that are important to the product designer; and we briefly went through the process of designing a product to accomplish the needs of a design problem. The next task of the designer is to bring the design into reality by submitting it to several tests or checkpoints which will determine whether the product will be manufactured and marketed. In doing so, the proposed product will be prototyped and evaluated from an engineering and marketing perspective. This process, called the design engineering process, will determine the potential success of the product in the marketplace.

The design engineer is interested in producing a quality product that will meet a perceived need at a profitable level. In the review of the proposed product, the design engineer will use special criteria to determine the feasibility of the design. Design engineering criteria are focused on function, form, ergonomics, aesthetics, ease of manufacture, standardization, durability, and costs (see Figure 9–20). After evaluation of the design criteria, a numerical value will be assigned to each area for each product design and will be used to discriminate differences, weaknesses, and strengths for future decision making.

Function

The designer used function as a criterion when creating the alternative design solutions. The design engineer must now validate the original design by building a prototype and submitting it to operational testing. Not only must the product function as designed, but it must

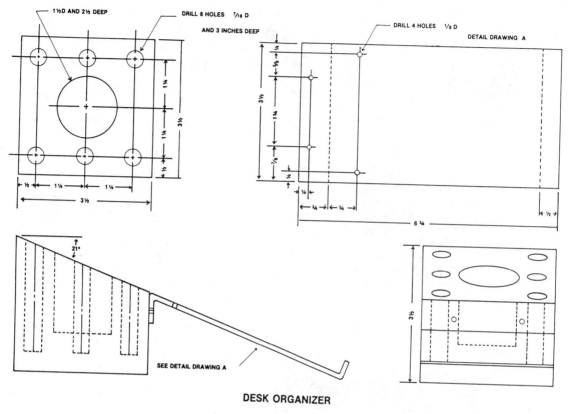

DESK ORGANIZER

Figure 9–19   A three-view drawing of product A

function in the environment for which it was designed. For example, if the product was a new seat design for an automobile, research must be done to to determine the space available in the automobile, the location of the driving controls, and the location of the seat-mounting hardware. Testing would involve not only the actual function of the seat (comfort, support, tilt, angle, etc.) but also structural and material durability and compatibility with the automobile.

Form

As you learned earlier in this chapter, form follows function. The design engineer is concerned about form because of the possible manufacturability. In most cases, the design engineer will be looking for ways to simplify the design so that standard parts, materials, and hardware can be used without compromising the designer's original ideas. The design engineer will provide testing on structural components in an effort to make sure that the design is safe and durable with a reasonable product life span. Unusual designs can raise manufacturing costs and provide unknown performance problems that may be undesirable. Market analysis is used to determine the cost–profit profile that will provide the window of flexibility in manufacturing techniques.

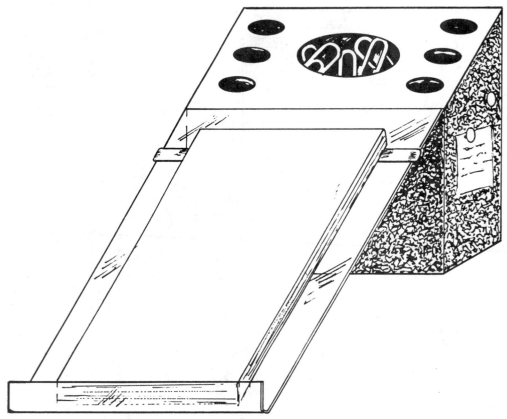

Figure 9–20 The final design meets all criteria for design

Ergonomics

Products are used by people and, as such, should be designed to function well in a variety of situations. People-friendly products range from comfortable seats, eye-level automobile instrumentation, and comfortable clothes to efficient homes and simple appliances. Much is known about the characteristics of human beings that can be used to design adaptive systems and products that are comfortable and easy to use. Figure 9–21 shows the many considerations that must be evaluated by the design engineer in assessing the ergonomic aspects of product design.

Human-factors engineering provides new data about the relationships between people and products. In addition to comfort, ergonomically designed products can provide much greater safety and product and human performance.

Aesthetics

Basic design principles and elements are employed by the product designer to achieve an aesthetic value that consumers will like. Products that are functional need not be

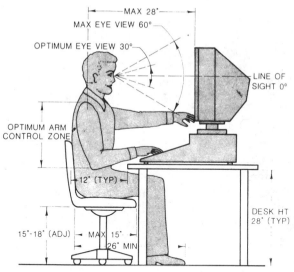

Figure 9–21 Ergonomic aspects of design are important factors for new products *(From Komacek, Lawson, and Horton,* Manufacturing Technology, *Copyright 1990 by Delmar Publishers, Inc. Used with permission)*

unaesthetically appealing. The design engineer must create the balance between function and form without losing the aesthetic value of the original design. While all product design solutions are a careful balance of compromise, those products that perform well and remain pleasing to the eye and human environment will be much more successful at the marketplace. We use the principles of design to create and our technological know-how to produce. Today's technological advancements provide new opportunities for designers and design engineers to blend the areas of function and form together for aesthetically pleasing products.

Ease of Manufacture

The ease of manufacture is important to the design engineer because he or she must recommend and help design the manufacturing system to produce the product at a profit. Whenever possible, the design engineer will recommend standard materials, machine tool fixturing, automated processes, fasteners, and finishes to cut costs of manufacture. Besides costs, the design engineer must plan for parts replacement, repair, and product liability over the life of the product.

Standardization

Standardization refers to the types of materials, parts, hardware, and equipment that can be purchased on a regular basis. The design engineer will refer to these items as "off-the-shelf" technology, meaning they have already been developed and tested and are available in ready form. Standardized parts and manufacturing fixturing and equipment save large amounts of research-and-development investments that must be made for custom-designed products and specialized equipment. In addition, standardized parts are available for service of the product and have better reliability potential than untested designs.

Sometimes manufacturers choose to develop products with special designs and new technologies (see Figure 9–22) to gain an edge in the market or to improve product performance. When this happens, they protect their investments by filing a patent on the process or procedure, which allows them exclusive rights for several years that their competitors would not have. When this happens, intensive testing and analysis for product safety and reliability must be performed by design engineers to ensure that the product will be accepted in the marketplace and function as advertised.

Durability

The life of a product is planned in the design process. The major factor is usually cost rather than capability. By selecting materials, parts, hardware, and finishes, the design engineer makes decisions on how long the life expectancy of a product will be. It would be defeating to produce a disposable cup using stainless steel as a material because the material will outlast the one-time function and send prices up so high that the product could not compete in the marketplace.

Design engineers try to match durability with function and costs in an effort to produce a quality product at a competitive price. The real challenge is to select materials and parts that will meet product needs and reduce manufacturing costs to capture the marketplace. Consumers are always trying to find that combination as well but sometimes end up purchasing a product because of a low price and a willingness to give up quality and durability. Others look for durable products with cost as a minor consideration because they intend to keep the product for high-frequency usage. Sears Roebuck Company provides options for consumers with its good, better, and best grades for its products. Good designers and design engineers are able to provide products for both purposes based upon the design objectives and goals.

Costs

Cost analysis is provided to estimate the cost of production, marketing, distribution, profit sharing, and servicing. The design engineer is able to gather data from various departments to put together the cost profile of the new product. The financial data are usually presented at meetings where product-development plans are developed (see Figure 9–23). The marketing department will have data on projected sales and price ranges; engineering will have a cost breakdown of capital investment, materials, parts, finishes, and subcontracted items; production will have estimated manufacturing timetables, labor estimates, and inventory requirements; and service will provide data on additional production costs for spare parts and service center needs.

All these factors must be clearly documented before the decision to produce the design is made. Timetables for production, marketing, and distribution must also be developed so that the product can be strategically placed on the market at the appropriate time. This is especially true for "season" products, such as clothes, or "yearly" products, such as automobiles.

The cost of a product must take into account all aspects of the process from design to service. It must also provide for unknowns and sometimes-difficult-to-predict overhead costs, including profit for the company.

Trade-offs and the Design Criteria

Designers may receive a set of design criteria for a certain product from management, or they may develop their own set of criteria. By examining any one product, you can probably identify the trade-off decisions that were made to design that product. As an example, consider the criteria used to design automobiles. The criteria and possible trade-off decisions are listed in Figure 1.

Design engineers must make trade-off decisions about the product design. Obviously, the trade-offs involved in designing cars give rise to a wide range of different automobiles. The first step in designing a quality product is to develop a list of specific design criteria.

CRITERIA	TRADE-OFF DECISIONS
Function	Speed, economy, passengers
Form	Compact, full-size, wagon
Ergonomics	Ease of entry and exit, rider comfort
Aesthetics	Styling, colors
Ease of Manufacture	Accessories, options available
Standardization	Standard engines, wheels, parts
Durability	Reliable, breaks down
Costs	Economy or luxury

Figure 1. Here are some design criteria and trade-off decisions for automobile manufacturers. *(Photo courtesy of Pontiac Division, General Motors)*

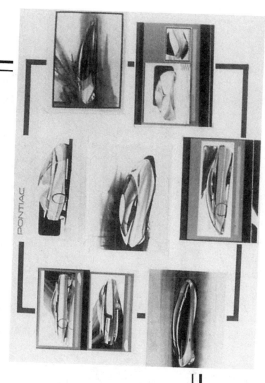

Figure 9–22 Special designs and new technologies can gain a company an edge in the market *(From Komacek, Lawson, and Horton, Manufacturing Technology, Copyright 1990 by Delmar Publishers, Inc. Used with permission)*

Figure 9–23   Meetings are held to discuss the process for bringing the product to market

SUMMARY

Product design has many aspects that can make the difference between success and failure. Good practice by designers using effective design principles and elements can create exciting new products for the marketplace. Design solutions can provide the design engineer many options and challenges as he or she refines the process using design criteria listed in this chapter. In some companies, the designer and design engineer may be the same person; in other companies, there may be several people involved. While the human resources may change, the procedures remain constant.

Good product design takes advantage of the aesthetics needed for appeal and the engineering needed for performance. Consumers are demanding new approaches that will provide quality products with good reliability at reasonable costs. The marketplace has expanded to a global economy with increased competition. These demands will provide new challenges for product design in the future. As manufacturing technology expands and new materials are developed, design options will change radically. New products will continue to improve, and our ability to produce them will be dramatically increased.

REVIEW QUESTIONS_____

1. What is product research? What does it tell you about the product?
2. Are all product profiles for new products? Explain.
3. List and describe the elements of design.
4. List and describe the principles of design.
5. Describe how a design solution is developed.
6. What does "form follows function" mean? How does it relate to good design?
7. Explain about and give examples of a thumbnail sketch.
8. What elements add formality to the design? Why?
9. What elements add motion to the design? Why?
10. Why should the designer also know about the materials-processing technology?

SUGGESTED ACTIVITIES _____

1. Conduct a brainstorming session to suggest products to be designed. Make up a class list of possible options.
2. Develop design solutions based on a written design problem.
3. Select a well-known product and have students suggest improvements and redesign it using the design solution process.
4. Show students actual industrial designs. Invite a designer to come in and talk to the class.
5. Have students design the actual product they will fabricate in the class. Once it is designed, have them convert it into working drawings and develop a bill of materials.

CHAPTER 10

Measurement and Layout Procedures

The need for a system of measurement has existed since the beginning of civilization. The need to identify methods of indicating ownership of land during early agrarian-based civilization and later developments in trade and commerce led to the development of concepts of mathematics and geometry which are the basis for measurement (Figure 10–1). Although many of the early standards of measurement were crude by today's standards, they provided a method of comparing quantities for the purpose of engineering canal and irrigation systems as well as defining boundaries for ownership of land.

The importance of standard measurement was not really apparent in the transformation of primary materials to usable products until the concepts of interchangeability and mass production were introduced. Prior to the introduction of these concepts, most products were custom made; that is, each item was unique, and the separate parts were made to fit one another in the assembly of each individual product. This meant that a part taken from one rifle would not necessarily fit another rifle made by the same craftsperson. Today, standardization of parts and measurement allows the mass production of products for mass consumption. Precision measurement is the key to interchangeability and, therefore, mass production.

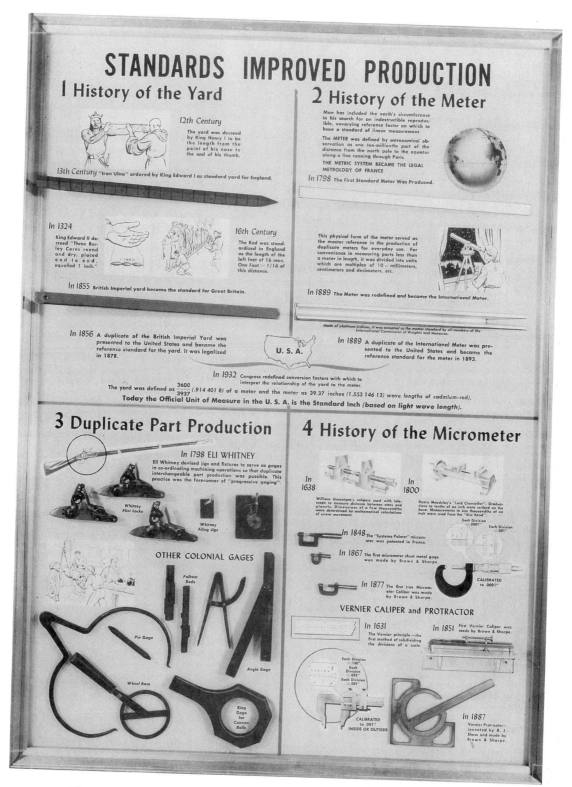

Figure 10–1 Early methods of measurement *(Courtesy of DoALL Company)*

MEASUREMENT

Measuring Units

The ability to produce interchangeable parts for mass production is based on standardized units of weight, volume, and length. While weight and volume are extremely important, the primary unit of measure used in manufacturing processes is linear measurement. Linear involves measurement in only one direction. The basic units of linear measurement are the foot and the meter. Depending on the application, each of these units are divided into smaller units or increased to larger ones.

The foot is divided into twelve equal units known as inches. Each inch can be further divided into fractional units usually no smaller than $1/64$ inch; decimal fractions of $1/10$, $1/100$, and $1/1,000$; or decimal equivalents of 0.1, 0.01, 0.001, and 0.0001. Today, decimal equivalents of one-millionth inch are no longer uncommon in precision industries. Some common fractions and their decimal equivalents are listed in Table 10–1.

Metric System

The meter is the basic unit of linear measure in the **metric system**. The metric system is the standard system of measurement used by the vast majority of the world today. This system was first developed in France after the French Revolution of 1789.

The metric system is a simple and very logical system of measure which is based on 10. The standard for the meter is two finely scribed lines on a special alloy bar which remains at a temperature of 0 degrees Celsius (32 degrees Fahrenheit) and is housed in Paris. Duplicates of this standard are in the possession of members of the International Bureau of Weights and Measures.

The meter can be divided into units which are divisible by ten to provide smaller parts or it can be multiplied by ten to provide larger units of measure. Prefixes have been given to these divisions and multiples of the basic unit. Figure 10–2 gives the prefixes and their equivalents for the commonly used multiples and divisions of the meter.

Divisions (Latin)		Multiples (Greek)	
Deci	(0.1)	Deka	10 times
Centi	(0.01)	Hecto	100 times
Milli	(0.001)	Kilo	1000 times

Figure 10–2   Metric prefixes and their values

DRAWINGS

Working Drawings

The manufacture of a product from primary materials could not be accomplished very well without a detailed representation of the product and all the dimensions and tolerances

Inches Dec.	mm	Inches Dec.	mm	Inches Frac.	Dec.	mm	Inches Frac.	Dec.	mm
0.01	0.2540	0.51	12.9540						
0.02	0.5080	0.52	13.2080	1/64	0.015625	0.3969	33/64	0.515625	13.0969
0.03	0.7620	0.53	13.4620	1/32	0.031250	0.7938	17/32	0.531250	13.4938
0.04	1.0160	0.54	13.7160						
0.05	1.2700	0.55	13.9700	3/64	0.046875	1.1906	35/64	0.531250	13.8906
0.06	1.5240	0.56	14.2240	1/16	0.062500	1.5875	9/16	0.562500	14.2875
0.07	1.7780	0.57	14.4780						
0.08	2.0320	0.58	14.7320	5/64	0.078125	1.9844	37/64	0.578125	14.6844
0.09	2.2860	0.59	14.9860	3/32	0.093750	2.3812	19/32	0.593750	15.0812
0.10	2.5400	0.60	15.2400						
0.11	2.7940	0.61	15.4940	7/64	0.109375	2.7781	39/64	0.609375	15.4781
0.12	3.0480	0.62	15.7480	1/8	0.125000	3.1750	5/8	0.626000	15.8750
0.13	3.3020	0.63	16.0020						
0.14	3.5560	0.64	16.2560	9/64	0.140625	3.5719	41/64	0.640625	16.2719
0.15	3.8100	0.65	16.5100	5/32	0.156250	3.9688	21/32	0.656250	16.6688
0.16	4.0640	0.66	16.7640						
0.17	4.3180	0.67	17.0180	11/64	0.171875	4.3656	43/64	0.671875	17.0656
0.18	4.5720	0.68	17.2720	3/16	0.187500	4.7625	11/16	0.687500	17.4625
0.19	4.8260	0.69	17.5260						
0.20	5.0800	0.70	17.7800	13/64	0.203125	5.1594	45/64	0.703125	17.8594
0.21	5.3340	0.71	18.0340	7/32	0.218750	5.5562	23/32	0.718750	18.2562
0.22	5.5880	0.72	18.2880						
0.23	5.8420	0.73	18.5420	15/64	0.234375	5.9531	47/64	0.734375	18.6531
0.24	6.0960	0.74	18.7960	1/4	0.250000	6.3500	3/4	0.750000	19.0500
0.25	6.3500	0.75	19.0500						
0.26	6.6040	0.76	19.3040	17/64	0.265625	6.7469	49/64	0.765625	19.4469
0.27	6.8580	0.77	19.5580	9/32	0.281250	7.1438	25/32	0.781250	19.8437
0.28	7.1120	0.78	19.8120						
0.29	7.3660	0.79	20.0660	19/64	0.296875	7.5406	51/64	0.796875	20.2406
0.30	7.6200	0.80	20.3200	5/16	0.312500	7.9375	13/16	0.812500	20.6375
0.31	7.8740	0.81	20.5740						
0.32	8.1280	0.82	20.8280	21/64	0.328125	8.3344	53/64	0.828125	21.0344
0.33	8.3820	0.83	21.0820	11/32	0.343750	8.7312	27/32	0.843750	21.4312
0.34	8.6360	0.84	21.3360						
0.35	8.8900	0.85	21.5900	23/64	0.359375	9.1281	55/64	0.859375	21.8281
0.36	9.1440	0.86	21.8440	3/8	0.375000	9.5250	7/8	0.875000	22.2250
0.37	9.3980	0.87	22.0980						
0.38	9.6520	0.88	22.3520	25/64	0.390625	9.9219	57/64	0.890625	22.6219
0.39	9.9060	0.89	22.6060	13/32	0.406250	10.3188	29/32	0.906250	23.0188
0.40	10.1600	0.90	22.8600						
0.41	10.4140	0.91	23.1140	27/64	0.421875	10.7156	59/64	0.921875	23.4156
0.42	10.6680	0.92	23.3680	7/16	0.437500	11.1125	15/16	0.937500	23.8125
0.43	10.9220	0.93	23.6220						
0.44	11.1760	0.94	23.8760	29/64	0.453125	11.5094	61/64	0.953125	24.2094
0.45	11.4300	0.95	24.1300	15/32	0.468750	11.9062	31/32	0.968750	24.6062
0.46	11.6840	0.96	24.3840						
0.47	11.9380	0.97	24.6380	31/64	0.484375	12.3031	63/64	0.984375	25.0031
0.48	12.1920	0.98	24.8920						
0.49	12.4460	0.99	25.1460						
0.50	12.7000	1.00	25.4000	1/2	0.500000	12.7000	1	1.000000	25.4000

For converting decimal inches in "thousandths," move decimal point in both columns to the left.

Table 10–1 English–Metric Conversion Table

added to that representation. The complete set of plans generally consists of an assembly drawing or exploded pictorial, the bill of materials, and the detailed orthographic projections of each part (Figure 10–3).

This representation, which allows the plan to be developed in one country and used in another to manufacture the product, is accomplished through a universal language known as **drafting**. It is a language of symbols which are universal to all languages and which are standardized throughout the world. All industries, whether manufacturing or construction related, use these symbols to develop product plans.

Assembly Drawings

The primary purpose of the **assembly drawing** is to show the completed product, whether an assembled machine or a part such as a pump assembly. The assembly drawing shows how the parts fit together. The parts of the assembly are generally numbered. This number is used to identify the part by name on an accompanying parts list.

The assembly drawing can be a multiview orthographic projection, such as Figure 10–4; a pictorial figure (Figure 10–5); or even an exploded pictorial, such as that illustrated in Figure 10–6. One advantage of the exploded pictorial is that it often eliminates the need for hidden lines, which can become confusing if the parts have a large amount of detail.

A variation of the assembly drawing is the **working assembly drawing**. In smaller, less-detailed assemblies, the detailed drawings and assembly drawing can be combined to form the working assembly. The working assembly not only shows the relationship of the parts to the whole assembly but also provides the detailed size and location dimensions all in one drawing. In addition, each part is numbered for identification and a parts list added (Figure 10–7). This drawing is used only when the assembly has a few parts without significant detail. The advantage is that it does not require a search of all the detailed drawings to find needed information. All information is contained on one or two sheets.

The assembly, exploded pictorial, and working assembly usually make extensive use of sectioning to eliminate the need for confusing hidden lines. Sectional views are accomplished through the use of imaginary cutting planes which slice through the part exposing hidden details beneath the surface. The cutting-plane lines represent the location of the removed section and *cross-hatching*. Figure 10–8 shows where the cutting plane passed through the sectioned part.

Detailed Drawings

The **detail drawings** are accurate scaled representations of the part or assembly with all the necessary information regarding size, shape, and location dimensions to produce the part or assembly. All information including notes for special operations are included in the detail drawings to produce the real object accurately from a two-dimensional representation of it.

In order to represent a three-dimensional part or assembly accurately on a two-dimensional surface (drafting paper or film), more than one view of the object is often required. The use of several two-dimensional views to represent a three-dimensional object is known

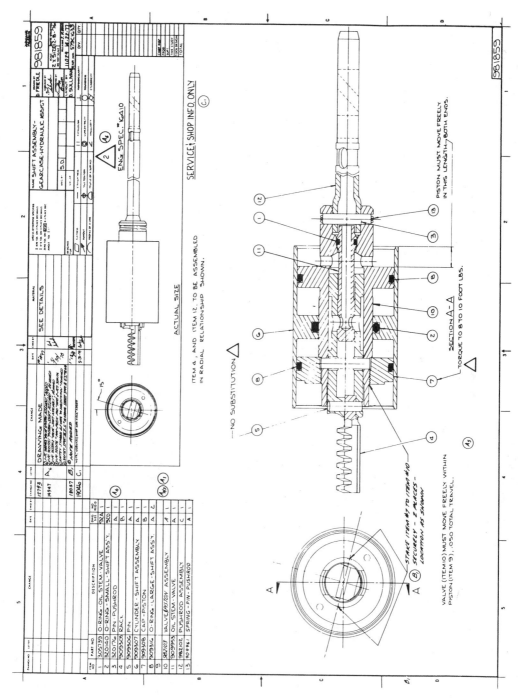

Figure 10–3 Working drawings *(From Pouler, Print Reading for the Machine Trades, 2nd edition, Copyright 1995 by Delmar Publishers, Inc. Used with permission)*

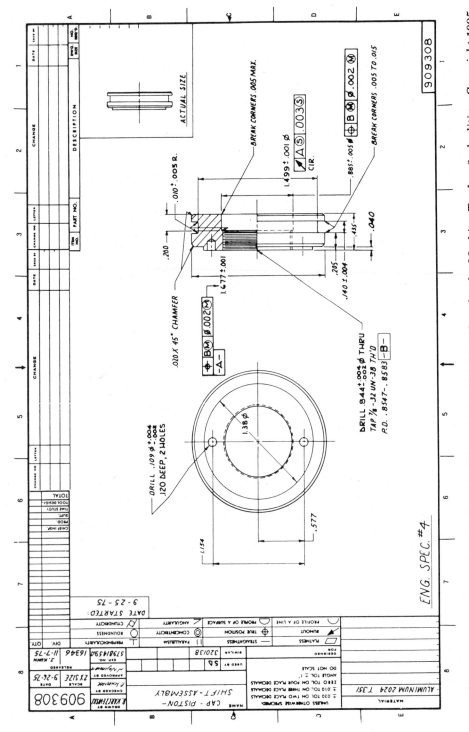

Figure 10–4 Multiview assembly drawing *(From Pouler, Print Reading for the Machine Trades, 2nd edition, Copyright 1995 by Delmar Publishers, Inc. Used with permission)*

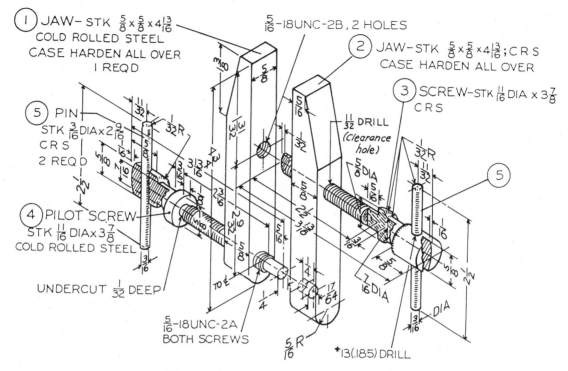

Figure 10–5 Pictorial assembly drawing

as **multiview drawing**. The development of these multiple views is based on a technique known as **orthographic projection** which means *straight* or *right-angle* projection.

Third-angle orthographic projection is used in this country to represent the detail of a part or assembly. The three views normally developed by third-angle projection are the front, top, and right side. Figure 10–9 shows the development of these three views from an object placed in space and cut by the horizontal and vertical profile planes. Figure 10–9 also shows the relationship of third-angle to **first-angle orthographic projection**, which is in common rise in Europe.

In most cases the front, top, and right side are sufficient to show all details. However, with parts that have much detail, additional views may be added for clarity. Figure 10–10 represents a detailed drawing with additional views.

GEOMETRIC DIMENSIONING AND TOLERANCING

American National Standard Y 14.5m established standards for **geometric dimensioning and tolerancing** in 1982. Geometric dimensioning and tolerancing provides a comprehensive system for accurate transmission of design specifications essential to modern manufac-

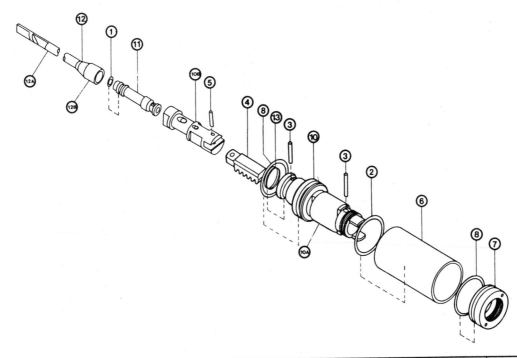

Ref. no.	Part no.	Name of part	QTY. Per assy.	Ref. no.	Part no.	Name of part	QTY. Per assy.
1	305739	O-Ring-Oil Stem Valve	1	10	982103	Valve and Piston Assy.	1
2	320140	O-Ring-Small-Shift Assy.	1	10A	909951	Piston-Shift Assy.	1
3	320176	Pin-Push Rod	2	10B	909952	Valve	1
4	909305	Rack	1	11	909953	Oil Stem-Valve	1
5	909306	Pin	1	12	982102	Push Rod Assy.	1
6	909307	Cylinder-Shift Assy.	1	12A	317812	Cam Blank-Push Rod	1
7	909308	Cap-Piston	1	12B	909954	Tube-Push Rod	1
8	909316	O-Ring-Large-Shift Assy.	2	13	909961	Spring-Push Rod	1
9	—	—	—				

Figure 10–6   Exploded pictorial drawing *(From Pouler,* Print Reading for the Machine Trades, *2nd edition, Copyright 1995 by Delmar Publishers, Inc. Used with permission)*

turing environments that emphasize modularity and interchangeability. Modularity and interchangeability rely on designs that must include clearly defined dimensions and allowable variations from basic dimensions (values used to describe, theoretically, exact size, shape, orientation, and location of a part or product's features).

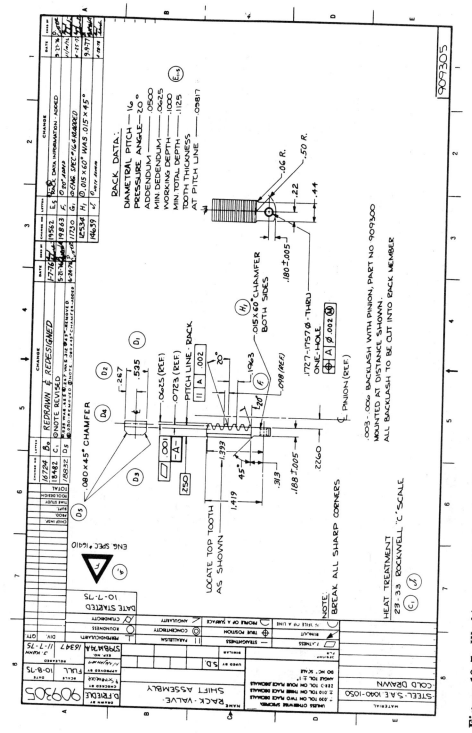

Figure 10-7 Working assembly drawing (*From Pouler, Print Reading for the Machine Trades, 2nd edition, Copyright 1995 by Delmar Publishers, Inc. Used with permission*)

185

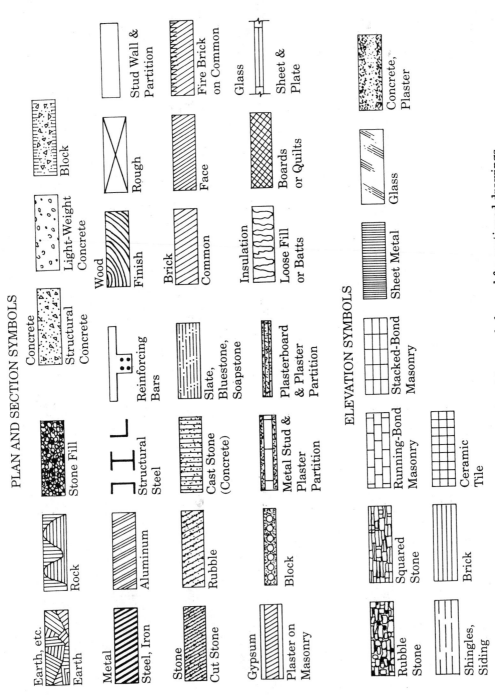

PLAN AND SECTION SYMBOLS

Earth, etc.
Earth

Rock

Metal
Steel, Iron

Aluminum

Concrete
Structural Concrete

Light-Weight Concrete

Block

Stone
Cut Stone

Stone Fill

Rubble

Reinforcing Bars

Structural Steel

Cast Stone (Concrete)

Slate, Bluestone, Soapstone

Wood
Finish

Rough

Stud Wall & Partition

Brick
Common

Face

Fire Brick on Common

Gypsum
Plaster on Masonry

Block

Metal Stud & Plaster Partition

Plasterboard & Plaster Partition

Insulation
Loose Fill or Batts

Boards or Quilts

Glass
Sheet & Plate

ELEVATION SYMBOLS

Rubble Stone

Squared Stone

Running-Bond Masonry

Stacked-Bond Masonry

Sheet Metal

Glass

Concrete, Plaster

Shingles, Siding

Brick

Ceramic Tile

Figure 10–8 Cross-hatching symbols used for sectioned drawings

186

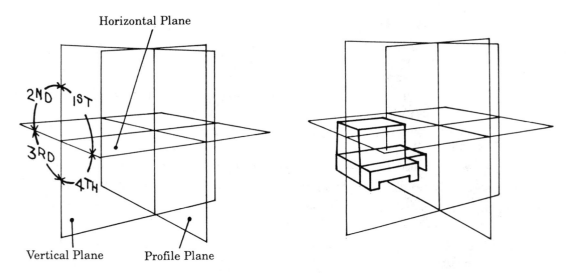

Figure 10-9   Third-angle projection

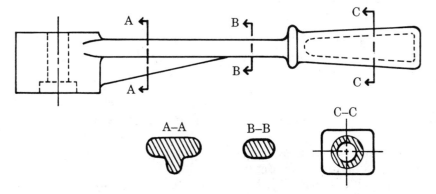

Figure 10-10   Details provided by additional views

American National Standard Y 14.5m helped eliminate confusion regarding methods used to express tolerances. For example, prior to Y 14.5m, corners were drawn without indicating whether any variation to the 90-degree angle was permitted. Dimensions were given from actual surfaces where datums were intended, and tolerances were given for which no precise interpretation existed.

Current standards of tolerancing use datums (theoretically exact points), which allow expression of permissible variations in a more precise manner. The following terms are commonly used in geometric dimensioning and tolerancing:

Actual size—the measured size.

Basic dimension—value used to the theoretically exact size, profile orientation, or location of a feature. It is the basis from which permissible variations are established by tolerances on other dimensions, in notes, or in feature control frames.

Datum—a theoretically exact point, axis, or plane derived from the true geometric counterpart of a specified datum feature. A datum is the origin from which the location of geometric characteristics of features of a part is established. *Datum feature*—an actual feature of a part that is used to establish a datum.

Datum target—a specified point, line, or area on a part used to establish a datum.

Feature—general term applied to a physical portion of a part (e.g., hole, surface, slot).

Geometric tolerance—the general term applied to the category of tolerances used to control form, profile, orientation, location, and runout.

Least material condition (LMC)—the condition in which a feature of size contains the least amount of material within the stated limits of size, for example, maximum hole diameter or minimum shaft diameter.

Maximum material condition (MMC)—the condition in which a feature of size contains the maximum amount of material within the stated limits of size, for example, minimum hole diameter or maximum shaft diameter.

Regardless of feature size (RFS)—the condition where a specified geometric tolerance must be met irrespective of where the feature lies within its size tolerance.

Tolerance—total permissible variation in size, shape, profile, or orientation.

Geometric dimensioning and tolerancing standards specify tolerances using feature-control symbols (Figure 10–11). The control symbol consists of a rectangular frame containing a symbol representing the geometric characteristic to be controlled and a second symbol specifying the required tolerance value. Figure 10-12 shows two examples of typical feature-control symbols used on drawings. Figure 10–12A indicates that each line element of the surface may vary from a straight line by the tolerance given in the block to the right (0.06 inch). In Figure 10–12B, symbols are stacked specifying requirements for straightness and circularity (roundness).

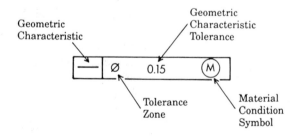

Figure 10–11   Feature-control symbols used for geometric dimensioning and tolerancing

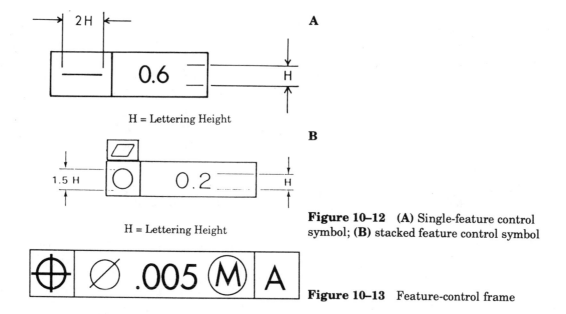

H = Lettering Height

H = Lettering Height

Figure 10–12   **(A)** Single-feature control symbol; **(B)** stacked feature control symbol

Figure 10–13   Feature-control frame

The feature-control frame also dictates the datum surfaces and order precedence for part setup, as well as assigning material condition modifiers to the tolerance and datum reference letters. The material condition modifiers, defined earlier, are maximum material condition (MMC), least material condition (LMC), and regardless of feature size (RFS). Figure 10–13 shows a control frame that includes a material condition. In Figure 10–13, the symbol M means the tolerance 0.005 applies when the feature being verified is produced at its MMC.

The preceding few paragraphs represent only a brief introduction to geometric dimensioning and tolerancing. It should, however, provide the reader with a basic understanding of its advantages over traditional methods found in earlier drafting textbooks. Many excellent textbooks and reference books have been written explaining, in detail, the conventions of geometric dimensioning and tolerancing.

BILL OF MATERIALS

The bill of materials is an extremely important part of the working drawing, since it is a complete list of all the materials, including any items such as fasteners, hinges, locks, and other commercial hardware not normally represented in the detailed drawings. The bill of materials can be compared to a recipe. The recipe for a cake includes all the ingredients needed to complete the product. In manufacturing, the bill of materials is the list of ingredients needed to complete the desired product or subassembly. None of the ingredients can be omitted from the recipe in order to have a complete product.

The typical parts list or bill of materials should include the following:

1. Part identification number
2. Part name or description
3. Number of particular parts required
4. Material specifications (type of material), including standard code designations indicating specific composition
5. Size of part (generally finished size)
6. Total amount of material required for each part in standard units (i.e., square feet, board feet, linear feet, etc.)
7. Special notes, unit cost and total cost may also be added

Figure 10–14 illustrates a conventional format used for the bill of materials. Some products and industries require extremely detailed material lists, while others are extremely simple with few details.

ITEM NO	PART NO	DESCRIPTION	DWG SIZE	NO REQD
1	305739	O-RING-OIL STEM VALVE	92A	1
2	320140	O-RING-SMALL-SHIFT ASS'Y.	92D	1
3	320176	PIN-PUSHROD A		1
4	909305	RACK B		1
5	909306	PIN A		1
6	909307	CYLINDER-SHIFT ASSEMBLY	A	1
7	909308	CAP-PISTON B		1
8	909316	O-RING-LARGE-SHIFT ASS'Y.	A	2
9				
10	982103	VALVE & PISTON ASSEMBLY	A	1
11	909953	OIL STEM-VALVE	A	1
12	982102	PUSHROD ASSEMBLY	C	1
13	909961	SPRING-PIN-PUSHROD	A	1

Figure 10–14   Conventional format used for bills of material

MEASUREMENT AND LAYOUT TOOLS

The understanding of precision measurement and its importance to manufacturing is critical to the idea of mass production. It is also important to understand the symbolic language of drafting, which is used to represent the three-dimensional object on two-dimensional surfaces. However, being able to read a scale to the nearest fraction of a foot or meter is not enough. To produce a finished product from the drawing requires the ability to use measuring and layout devices to locate dimensional information precisely from the drawing to the three-dimensional raw material. The following section is designed to familiarize the reader with some of the more commonly used measuring and layout devices.

Rules

Straight-line distances are generally located with a rule. This device could be a bench rule like the one illustrated in Figure 10–15A, which is typically used for measuring to the

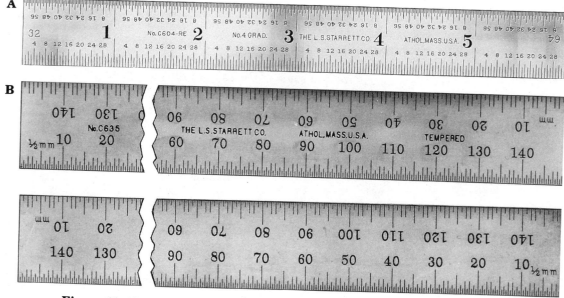

Figure 10–15 **(A)** Bench rule; **(B)** machinist's rule *(Courtesy of L.S. Starrett Co.)*

nearest centimeter or $1/16$ of an inch. It could also be a machinist's scale which is used to measure to the closest millimeter or $1/64$ of an inch (Figure 10–15B). Machinists' scales often divide the inch into units divisible by ten and often read to the nearest $1/100$ (0.01) of an inch.

Squares

Another device which can measure distances but also check squareness and layout lines perpendicular (at right angles) to an edge or line is the square. There are several types of squares available for use by the cabinetmaker, the carpenter, or the machinist. Figure 10–16 identifies a variety of squares used in various industries today.

The *combination square* Figure 10–16A is used for laying out a variety of materials. It has an adjustable head that slides along a metal blade and often contains a spirit level. The head and blade can be used to perform a number of measurement, inspection, and layout activities. The blade typically has a number of scales of inch or metric units (or both), and the head can be used to verify or layout angles of 45 degrees and 90 degrees.

The *machinist's precision steel square* (Figure 10–16B) and the *woodworker's try square* (Figure 10–16C) are similar in appearance, except that the try square generally has a scale on it and can be used for measurement as well as layout. This square consists of a head and a blade which is at a right angle to the head. Unlike the combination square, the head is fastened rigidly to the blade. A variation of the standard try square is the try and meter square illustrated in Figure 10–16D. This square allows both right angles and 45 degree angles to be verified or laid out.

The *framing* or *carpenter's square* is generally larger than the squares mentioned previously and contains special tables for cutting roof rafters and braces. The framing square and rafter tables allow the carpenter to cut the correct angles and lengths for any roof slope. The framing square can also be used to lay out a number of geometric shapes. Figure 10–16E illustrates the framing square and some of the tables typically found on it.

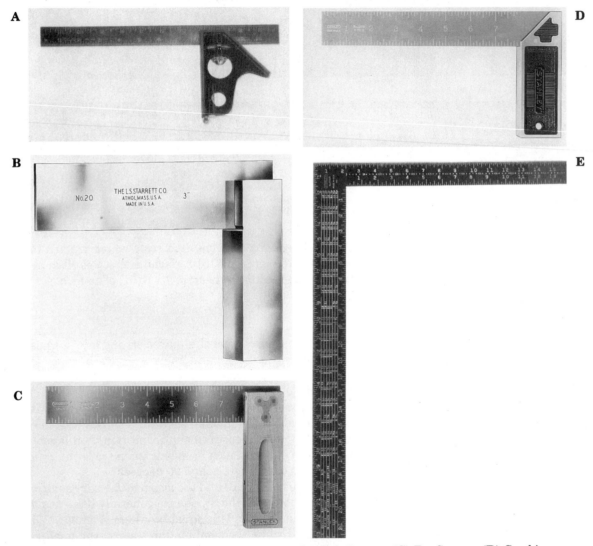

Figure 10–16 **(A)** Combination Square, **(B)** Machinist's Square, **(C)** Try Square, **(D)** Combination Try and Miter, **(E)** Framing Square *(A, C, D, & E courtesy of Stanley Tool Company; B courtesy of L.S. Starrett Co.)*

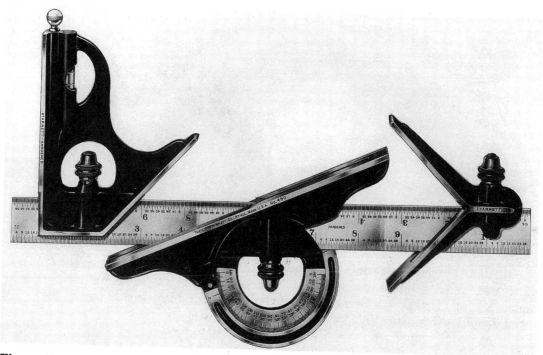

Figure 10–17   Sliding T-bevel and protractor used to set desired angles *(Courtesy of L.S. Starrett Co.)*

Protractor Head and Sliding T-Bevel

Some applications require instruments that are capable of angles other than 45 degrees or 90 degrees. The sliding T-bevel and the protractor head on the combination square are normally used when laying out or verifying various angles (Figure 10–17). The protractor head is generally used with the combination blade and has a scale on it which reads to the closest degree. Although generally used in the metals area, it could be used with any material. The sliding T-bevels are primarily used to check or transfer angles on to wood. Although they have no scale of their own, any angle between 0 and 180 degrees can be set with the aid of a protractor or square (Figure 10–18).

Micrometers and Verniers

In many cases, precision measurement to the nearest $1/64$ of an inch or closest millimeter is sufficient. However, many industries require the machining of parts to tolerances that are much more precise. Required measurements are typically to the nearest $1/1,000$ of an inch or 0.05 of a millimeter. In some cases, the required tolerance is even smaller. The device that has been developed for measuring such small units is the micrometer caliper. There are a

Figure 10–18 Sliding T-bevel and framing square used to lay out angles

number of types of micrometer calipers on the market today. The selection normally depends on its application. Figure 10–19 illustrates some of the commonly used micrometers and their applications.

Micrometers can be obtained in various sizes from 1 inch (25 mm) to 12 inches (300 mm) or greater. Usually the micrometer will have a 0- to 1-inch or 0- to 25-mm range. For example, a 1- to 2-inch (25- to 50-mm) micrometer will be used to measure objects between 1 inch (25 mm) and 2 inches (50 mm).

Today, many companies are selling micrometers with electronic or mechanical digital readouts. Little skill in reading the micrometer is required with this type of instrument. Figure 10–20 illustrates a digital-readout micrometer.

A   **B**   **C**

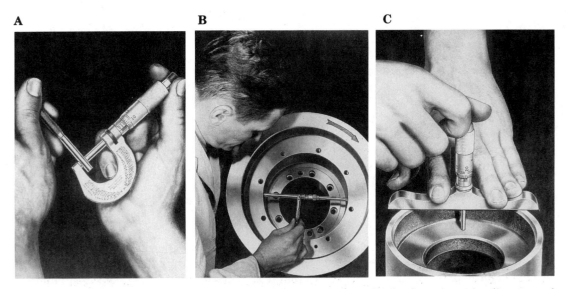

Figure 10–19 **(A)** Outside micrometer, **(B)** inside micrometer, **(C)** depth micrometer *(Courtesy of L.S. Starrett Co.)*

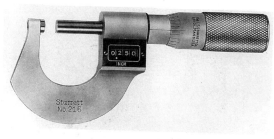

Figure 10–20   Digital readout micrometer
(Courtesy of L.S. Starrett Co.)

Reading the Micrometer

Micrometers in use in the United States today employ either the millimeter or the inch as their basic unit of measure. Typical outside micrometers, whether metric or inch, look very much alike; and the terms that describe their parts are the same. Figure 10–21 shows a typical outside micrometer and identifies the various parts.

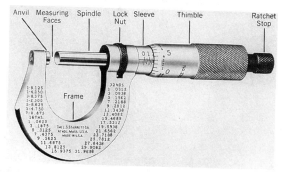

Figure 10–21   Parts nomenclature of standard outside micrometer *(Courtesy of L.S. Starrett Co.)*

Inch Micrometers

Inch micrometers are designed to measure to the nearest $1/1,000$ of an inch. The spindle of the inch micrometer has forty threads per inch, and each revolution or the thimble is $1/40$ or 0.025 (1 divided by 40 equals 0.025) of an inch. Each time the thimble rotates one complete turn, the spindle moves 0.025 closer or further from the anvil. The sleeve consists of markings in tenths above the horizontal axis, and each tenth is divided into three spaces of 0.025 below the horizontal axis (Figure 10–22). Each line on the thimble itself represents 0.025. Figure 10–22 illustrates a typical micrometer reading to the nearest $1/1,000$ of an inch.

Some micrometers have a vernier scale which permits a measurement to the nearest $1/10,000$ of an inch. Figure 10–23 gives an example of a vernier micrometer and typical reading to the nearest $1/10,000$ of an inch. The fourth place to the right of the decimal is taken from the vernier scale located on the sleeve of the micrometer caliper.

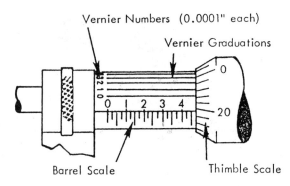

Figure 10–22 Inch micrometer graduations *(Courtesy of L.S. Starrett Co.)*

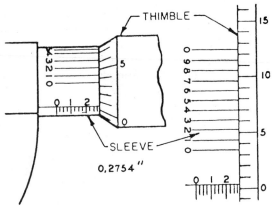

Figure 10–23 Inch micrometer with vernier scale *(Courtesy of L.S. Starrett Co.)*

Metric Micrometer

Many metric micrometers are designed to measure to the nearest $1/100$ of a millimeter. The operation of the metric micrometer and its appearance are very similar to the inch micrometer. The spindle of the metric micrometer has 50 threads for each 25 mm along its axis. In this case, the thimble must be rotated twice in order to adjust the spindle 1 mm closer or farther away from the anvil. Each mark above the horizontal axis line on the sleeve represents a millimeter, with each 5 mm being identified (i.e., 5, 10, 15, 20).

The lines below the horizontal axis line each represent the half millimeter marks (0.5 mm). Figure 10–24 illustrates a typical reading to the closest $1/100$ of a millimeter.

Vernier scales are also added to metric micrometers for greater precision. Figure 10–25 illustrates a vernier metric micrometer reading to $1/1,000$ of a millimeter.

PATTERNS, JIGS, AND FIXTURES

The ability to produce interchangeable parts begins with the availability of well-prepared working drawings and the ability to read and interpret those drawings. The second step in the process requires accurate measuring devices and appropriate layout tools. With these,

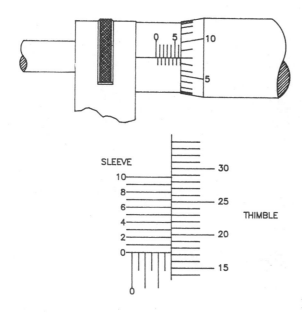

SLEEVE

THIMBLE

Reading 6.585 mm

Figure 10–24 Metric micrometer graduations; accuracy $1/100$ of a millimeter

the individual could interpret the two-dimensional drawing and transfer that information
to the three-dimensional material using the layout equipment discussed earlier. The problem is that to lay out each individual piece would be extremely time consuming and ineffi-
cient in the manufacturing industries.

To compensate for this, the layout of cuts to be made or holes to be drilled along with
other operations is built into a device that eliminates the layout of each part but ensures
standardization and repeatability. In order to provide for such standardization and permit
mass production—while, at the same time, ensuring quality, economic efficiency, and repeatability—special patterns, jigs, and fixtures are designed. These devices are usually designed
to meet two requirements:

1. The device can be easily adapted to the existing equipment.
2. It provides the most economical method of ensuring repeatability of operations.

The Pattern

Product design consists of a number of different geometric shapes that give form to the
object. In many instances, these geometric shapes consist of curved lines, arcs, and angles
which are difficult to machine without the use of a pattern for layout purposes. Straight lines
and simple angles can easily be produced on most equipment with the standard accessories provided with it. Irregular shapes require additional planning and preparation for processing.

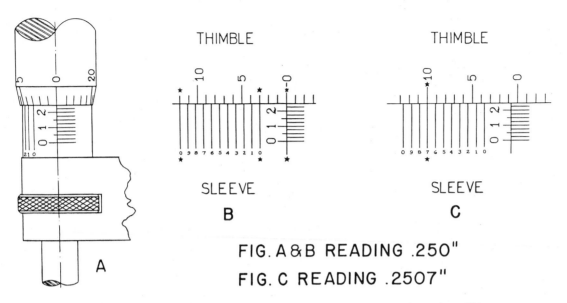

Figure 10–25   Metric micrometer with vernier scale; accuracy $1/1{,}000$ of a millimeter

The materials used for the development of patterns can be something as permanent as metal, which would withstand repeated use and maintain necessary accuracy, or as simple as a paper pattern for limited production. The decision to use one material over another for the making of a pattern would most likely be based on the number of units to be produced. A material that is often used because of its availability, strength, and ease of machining is hardboard. Hardboard is strong enough to last through repeated use for mass production runs. It also can be shaped with simple hand and power tools and equipment.

Whether the material for the pattern is as permanent as metal, the development of the pattern begins with the basic layout on paper. Once the layout is complete, it is then transferred to the more permanent material to be used in the manufacturing process. These layouts consist of a series of straight lines and curved lines that are connected to produce a pleasing design.

Jigs and Fixtures

The layout process for the custom production of one product is often accomplished by placing the desired shaped dimensions or location dimensions directly on the parts that will eventually be assembled to form a completed product. Most production today does not involve the custom making of individual products but rather the production of multiple products for a mass market. The ability to mass produce low-cost quality products relies on efficient, effective methods of ensuring that each process that involves changing the shape of the material can be repeated many times.

This system of mass production relies heavily on being able to produce standardized parts as efficiently as possible. This need for effective, efficient operations generally does not allow for the time it would take to lay out each piece by hand, using conventional layout techniques. Instead, devices must be designed which can be attached to conventional manufacturing equipment that ensure the standardization that is needed.

These devices are generally referred to as **jigs** and **fixtures**. They can be as simple as the stops you see on the saw in Figure 10–26 or as complicated as the one in Figure 10–27. A *jig* is basically a device designed and constructed to hold and locate the work or to guide the tool while the operation is being performed. The jig is generally not fixed to the machine but can be repositioned. Figure 10–28 shows a simple drill jig used to locate and guide a twist drill. The *fixture* is a device that is designed and constructed to support and locate the part while the operation is being completed. Unlike the jig, which can be repositioned, the fixture is fixed to the machine performing the operation. Figure 10–29 illustrates some fixture designs for various operations.

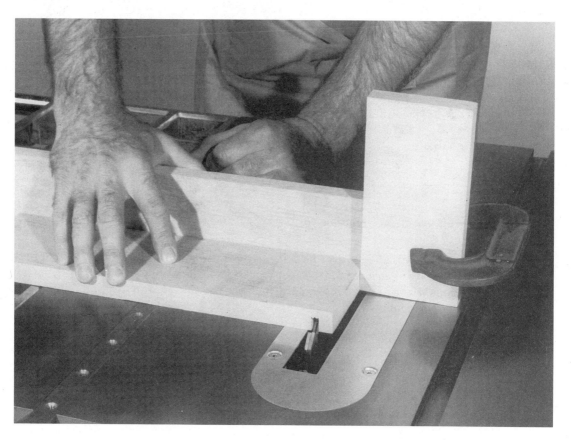

Figure 10–26 Simple stop used to cut parts to length

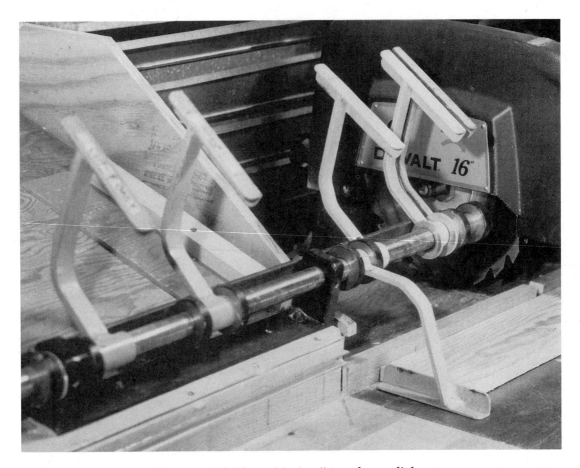

Figure 10–27   Multiple positioning jig used on radial arm saw

It is important that jigs and fixtures be extremely accurate and remain so over the length of the run. Jig and fixture accuracy are important to ensure product accuracy and repeatability of operations. Some general considerations for the design and development of jigs and fixtures include the following:

1. The device should be designed to be easily adapted to conventional manufacturing equipment.

2. The design takes into account the safe operation of the equipment and provides for required tool clearances.

3. The material selected for the device is dimensionally stable so that repeated tolerances can be achieved.

4. The devices are designed to hold the part securely during the operation and provide for safe efficient loading and unloading of the machine.

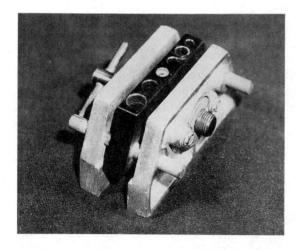

Figure 10–28   Drill jig used to locate and guide twist drill

Figure 10–29   Examples of fixtures used for positioning or guiding tools *(Courtesy of Rockwell Tools Inc.)*

These and other factors should be considered when planning for a mass production. The jig or fixture must meet the close tolerance specifications of the product design and, at the same time, ensure safe efficient operations.

This section on jigs and fixtures has been included in this chapter because the purpose of layout and measurement is to ensure dimensional accuracy of the product shape and size. This accuracy can be achieved by reading or interpreting the working drawing and then using the layout tools to transfer that information to the part. In mass production efforts, the accurate layout (location of parts to be cut or drilled) relies on the accurate design and development of a fixture or jig that ensures standardization and interchangeability.

SUMMARY

Mass production of usable goods from primary materials has provided people with more and less expensive quality products. This concept of manufacturing relies on the standardization and interchangeability of parts, which is based on an accurate system of measurement and layout.

Continuing developments in lasers and other precision devices have enabled tolerances to the closest $1/1,000,000$ of an inch a reality.

The sophisticated, miniaturized, electronic products of today make this kind of accuracy not just a reality but a necessity. The layout of parts to such tolerances is extremely important and extremely difficult. Such layout often relies on the pinpoint capability of machines with computer numerical control to produce such accuracy.

Whether the accuracy is built into the machine tool itself or designed into the fixture or jig attached to its table, the basis for either is still a precise standardized system of measurement and graphic language.

? REVIEW QUESTIONS _____

1. Describe the early development of systems of measurement.
2. Compute decimal equivalents for the following fractions: $1/10$, $1/8$, $1/4$, $3/8$, $1/2$, $5/8$, $3/4$, $7/8$.
3. Convert the following from meters to centimeters and then millimeters: 1 m, 2.2 m, 30 m, 100 m.
4. What is the purpose of sectional views?
5. What information is generally included on the bill of materials?
6. Name two instruments that can be used to lay out and verify angles other than 40 degrees and 90 degrees.
7. The vernier scale on measuring devices allow accurate measurement to _____ (decimal equivalent).
8. What two requirements must patterns, jigs, and fixtures meet?
9. Describe the difference between a jig and fixture and give examples of how each is used.

SUGGESTED ACTIVITIES

1. Select an item from home that has some intricate detail and sketch an orthographic detailed sketch with hidden lines, sectional views, and dimensions.
2. Develop a set of working drawings and bill of materials for a product to be produced in class.
3. Using the carpenter's framing square and rafter table, compute the correct length of common rafter for a 4 to 12 slope and a run of 10 feet. Using a short piece of rafter material and the carpenter's framing square, lay out the ridge cut and bird's mouth for a 4 to 12 slope.
4. Select several pieces of round stock and pipe and determine the outside and inside diameters with metric and standard inside and outside micrometers.
5. Select five objects from the home or classroom and measure their length in feet and inches to the closest ¼ of an inch. Record each measurement and convert it to an appropriate metric measurement (meter, centimeter, millimeter). Record the metric measurement in a column next to the English measurement.

CHAPTER 11

Product Research, Development, and Analysis

Modern industry depends on getting new product ideas from several sources. We have already discussed the important role that the industrial designer plays in the development of new product ideas. Working along with the engineers and designers are the people in research and development (R & D), who conduct industrial research, experiment with working models, and develop experimental products and manufacturing processes. To make sure that design ideas are practical and within the capabilities of the company, the R & D team is active in discovering or refining the materials, processes, and technology to accomplish the task (Figure 11–1).

The R & D department conducts many activities which can be broken down into three basic areas: industrial research, product analysis, and process development.

INDUSTRIAL RESEARCH

There are several types of industrial research (both pure and applied) that the R & D department employs during the process of creating new products. **Pure research** is the process of gaining knowledge and facts which are discovered during the process of basic research and have no intended purpose. Universities and special research firms often conduct research in this manner. **Applied research** is the process of gaining knowledge and facts which are discovered while conducting a research project to solve a stated problem. Most R & D departments conduct this type of research as an ongoing process. The government funds both

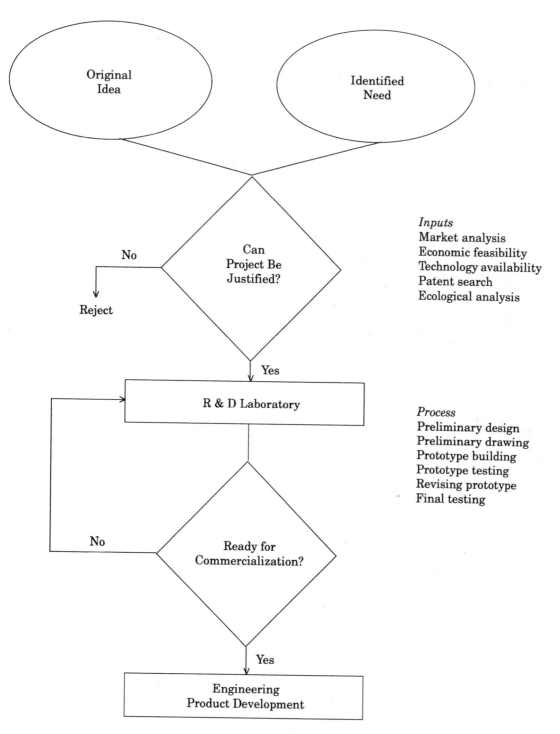

Figure 11–1 Flow diagram of R & D process

types of research through such agencies as the National Aeronautics and Space Administration (NASA), Department of Transportation (DOT), Environmental Protection Agency (EPA), National Science Foundation (NSF), Department of Energy (DOE), and National Institute of Standards and Technology (NIST). Some knowledge is gained from the successful completion of the research project, and some knowledge is gained from the ancillary spinoff technology that was discovered inadvertently during the research process. In recent years, much of our synthetic materials in technology have been discovered in this manner (Figure 11–2).

A balanced R & D program can serve the company well and keep it in the market for new and creative products. Let us discuss two of the most common approaches that R & D departments use to explore and develop new product ideas.

Fundamental Research

Some things exist as phenomena (unaccountable occurrences) in nature. R & D programs are constantly trying to uncover the mystery concerning these unusual events. The purpose of fundamental research is to seek the answers to "what is" and, therefore, gain an advantage over known phenomena as well as the unknown types of materials, processes, and technology.

Fundamental research also makes use of re-examining existing materials, processes, and technology. By constantly re-evaluating procedures and known technology, a company can improve efficiency and thereby cut costs and increase profits. An example of this type of research can be seen by comparing the differences between processing materials in space versus processing them on the planet earth (Figure 11–3). By researching the advantages of weightless manufacturing, new processes or materials can be discovered and developed in the laboratory. Many chemical companies are engaged in this type of fundamental research. While some R & D programs may not conduct a great deal of fundamental research themselves, most of them monitor this type of research with great interest. A simple spinoff bit of technology can mean new products and millions of dollars for the "early bird" company that is able to develop the technology first.

Exploratory Research

Creating or discovering new technologies, gaining new information about materials and processes, and applying this knowledge toward the development of a new product is called **exploratory research**. It is research designed to come up with a functioning product that can be marketed for a profit. This type of research is the primary goal of the R & D department.

Some companies call it "scouting research" because it explores new territories that other companies have not yet entered. Many times, the first company into a new territory can enjoy advantages such as patents, copyrights, licenses, and distributorships that can provide large economic successes. For instance, the computer game market of the 1980s was pioneered by Atari, who captured a large portion of the market. Those companies who entered the computer game territory at a later date had to play a catch-up game in both hardware (machine) and software (program) technology. In the meantime, Atari was able to make its original investment a worthwhile venture.

Figure 11–2 Space shuttle launching *(Courtesy of NASA)*

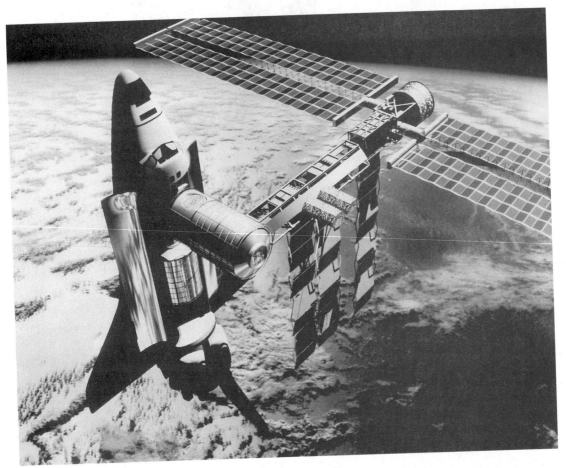

Figure 11-3A   A manufacturing laboratory in outer space *(Courtesy of NASA)*

Some companies choose not to enter a new territory until it is proven profitable by another company. Not all exploratory research is worthwhile. If the manufacturing process is too complicated or expensive and the marketing possibilities questionable, the product could be shelved (not used or developed) for years. Such was the case of unleaded gasoline and gasohol. Until the price of oil increased worldwide, production of these worthwhile low-pollution fuels had to wait for the market to develop.

While a growing number of companies maintain their own R & D programs to conduct exploratory research, some are too small and must depend on external agencies for assistance. A company that manufactures fine furniture may make use of the information and research conducted by the Forest Products Laboratory, a government-sponsored institution; or a manufacturer of housing may make use of a university-sponsored solar collector or pas-

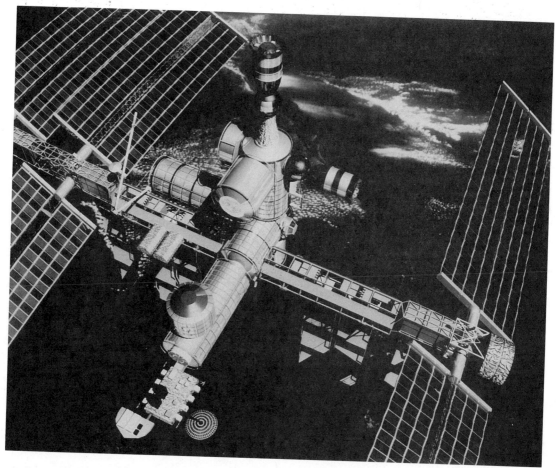

Figure 11–3B A manufacturing laboratory in outer space *(Courtesy of NASA)*

sive-design research project. Even the basic trade associations have practical programs that can help the smaller manufacturer conduct exploratory research.

PRODUCT DEVELOPMENT

Product Analysis

Consumers tend to take for granted the steady parade of new products offered by industry each year. However, what appears on the market each year is only a part of the many ideas developed through new product research and design. Many apparently useful products never go to the production stage because an analysis of the market reveals little or no

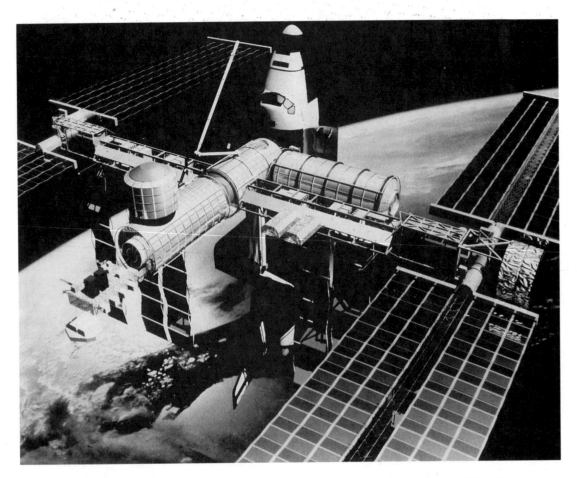

Figure 11–3C   A manufacturing laboratory in outer space *(Courtesy of NASA)*

profit margin. Regardless of its apparent worth, sufficient demand for an item must be identified before a company will invest the revenue to begin production of a new product.

Profit is the basic goal for any manufacturing company and the ability to make a profit usually requires a market and efficient conversion of natural materials into finished products. **Product analysis** is the identification of the most economical and appropriate method of manufacturing a new product to meet required specifications and regulations.

R & D departments' primary function is to engage in product research *and* product development. During the process of discovering and developing new product ideas, several common techniques are used. These techniques can be classified under three types of product development.

1. Imitation
2. Adaptation
3. Innovation

Imitation

Look-alike products and products that have similar features or functions are designed to get a company into a profitable market quickly. They also legally avoid patent constraints which tend to prevent duplication of technology for a given period of time.

Smaller companies with limited R & D facilities but flexible manufacturing capabilities can and often do find a profitable market in look-alike products. Even larger companies will play the imitation game by producing a product under a brand name and subleasing the design or even manufacturing the product under another name. In such cases, minor design changes are made in order not to duplicate the original product exactly.

Examples of products in industry which have used this technique include automobiles, tires, shoes, appliances, and others. Major national retail stores sell products that are look-alikes (imitations) of famous brand names at lower prices under their own name. Sometimes these products are manufactured by the original brand name company, and sometimes they are manufactured by a smaller company or by a multinational company with facilities in Japan or Taiwan. In either case, the imitation product–development business is big and profitable.

Adaptation

Changing a product to make improvements or to provide features that will make the product perform or look more attractive is another responsibility of R & D. Ford's "a better idea" slogan sums up the process of improving product lines and designs in an effort to convince the consumer that their product is better than their competitors. Indeed, the microcomputer industry has produced one product after another in the adaptation game for product improvement and capability. Each generation (adaptation) of microcomputer, disk system, memory, or printer has added features that have improved the product and advanced the technology. R & D departments are constantly monitoring competitors' and their own products in an attempt to make improvements or add features that will impress the consumer enough to purchase the product. Product adaptation benefits each of us as consumers because it tends to increase product quality and decrease product price.

Innovation

The most difficult method for developing new products is the process known as innovation, yet it is also the most rewarding process for both the individual and company. Innovative products are unique in the sense that they are usually based on the use of new technology or materials and are able to set the trend in that particular product line. When the R & D department has such projects in progress, the facility is usually closed to the public for tours and photographs. Keeping an R & D research project a secret is very important to some industries. For example, the automobile industry has many new designs that they are working on several years in advance. Even some manufacturing processes are kept "off limits" in order to maintain the technological edge over competitors.

Often, when an innovative product is discovered, new manufacturing facilities must be set up (tooled) and commercial markets explored and developed. Sometimes the product is released too soon, and the consumer rejects the innovation. This was a problem with some of the earlier automobile designs by Studebaker and the Ford Edsel. Sometimes the product is released too late, and another company is able to capture the market first. Timing the release of an innovation is very important and can make or break its acceptance by the consumer.

Refining the Product

Once the product design has been accepted by management, the R & D department begins the process of refining the product. The design drawings are converted into mock-up models. Traditionally, these models have been made made of clay, plastic, or wood and represent what the finished product will look like. However, a relatively new technique called stereo lithography is being used to replace the conventional processes. Stereo lithography creates a three-dimensional plastic model from images created by computer-aided design (CAD), computer-aided manufacturing (CAM), and computer-assisted engineering (CAE). During the process, CAD/CAM/CAE data are used to yield a series of cross-sections of the part or product. These cross-sections are then used to create a solid prototype of the part or product by forming polymer layers identical to each computer-generated layer. Successive cross-sections are laid one on top of the other to form an accurate three-dimensional model of the computer-generated design.

The next step is to develop a working model called a prototype. Usually, it is full sized and represents what the finished product will look and function like.

Prototypes are tested, revised, redesigned, and reconstructed until all the appearance and performance bugs are worked out. Prototypes are then turned over to the manufacturing specialists for use during process development. During this phase, manufacturing decisions will be made.

PROCESS DEVELOPMENT

Once the prototype has been fully tested, the R & D department will begin to work on manufacturing problems. A team of professionals from the R & D department—along with an engineer, industrial technologist, and production foreman—will begin to develop an experimental production line in the R & D facility or, in the case of a smaller company, somewhere in the plant. The purpose of this experimental line is to develop the manufacturing process and work out the bugs and problems of production. This process is often accomplished using computers and simulation software.

After the decision to produce is made, questions related to the actual production are considered. In this case, the *process analyst*, sometimes referred to as a *methods analyst*, or a team of process analysts will carefully study the product and provide technical information regarding the most economical and appropriate methods of manufacturing. The major questions to be considered during this analysis are the following:

1. What materials will be used in the manufacture of this product?
2. What machines and processes are most economical and efficient in the manufacture of this product?

The raw materials selected must meet specified standards for strength, formability, and environmental effects. A careful analysis of sample raw materials will provide the analysts with the most appropriate materials. In getting the plant ready for the manufacture of a new product, they will make up a new plant layout or design or order new tooling; set up a transportation system; design production and quality controls; and provide for storage of the finished product.

A decision regarding the most appropriate processes could be based on a careful analysis of the following criteria established by the Society of Manufacturing Engineers:

1. Nature of parts, including materials, tolerances, desired finishes, and operations required
2. History of fabrication, including machinery or assembling of similar parts or components
3. Limitations of facilities, including the plant and equipment available
4. Possibility of product design changes to facilitate manufacturing or cost reduction
5. In-plant and out-of-plant material-handling systems
6. Inherent processes to produce specified shapes, surfaces, finishes, or mechanical properties
7. Available qualified workers to be used for the production

After a careful analysis of materials, processes, and environmental factors, the process analyst or team determines the most efficient method of combining the flow of raw materials with required operations. An analysis of the sequencing of operations with specifications for tools and equipment should result in an efficient, effective system of manufacturing.

The basic procedure that follows in the process analysis is as follows:

1. Analyze specifications and drawings
2. Study operations
3. Sequence operations
4. Combine basic operations
5. Specify tools and equipment

The product analysis would not be complete without a careful study of environmental impact of the use of the raw materials and processes selected for manufacturing. Problems related to the neutralization of waste products leaving the production site, recycling of materials, and on-site environment quality are given careful consideration. Consideration must be given to regulations regarding waste disposal and air quality established by the EPA. Raw materials which may be less expensive initially and may form more easily than others may require expensive equipment to ensure minimal pollution of air and land as regulated by the EPA.

Process development is also used to improve existing product-manufacturing process. The R & D team has many support functions, of which efficiency studies, troubleshooting, and technical consulting are a few. In fact, as a support function, the R & D team may work with every faction of the company from engineering to marketing and with both management and labor.

The economic, technical, and environmental analysis can be combined in a final analysis for planning purposes. One decision model that can be used is known as the break-even analysis chart (Figure 11–4). Three factors are used in this decision-making model:

1. *Fixed costs*—costs that do not vary as a function of the output rates. These costs include the initial cost of the equipment, building, research, and other costs related to the design of the process.
2. *Variable costs*—these costs do vary as a function of the output rates. They include the cost of materials, labor, and energy.
3. *Revenue*—this is a product of selling price and the number of units made and sold.

Since the fixed costs are not variable, they are represented by a horizontal line. Total cost is a function of both fixed costs and variable costs. The total cost line in this chart will vary with increase in production. The revenue line, which is a product of selling price and volume sold, is also variable. This chart is used to determine the level of production needed to break even. The **break-even point** is the point at which cost of production and revenue are equal—that is, no profit, no loss. This point is represented on the chart by B—the point at which the total cost line and the revenue line cross.

A careful analysis of economic, technical, and environmental factors associated with producing a new item is critical to the final decision of whether to go ahead with the project. Whether building a new nuclear energy plant or manufacturing a new toy product, analysis is a vital part of the decision-making process.

SUMMARY

The development of new products and processes is a function of the research and development department. The R & D team must understand the process of engineering, material processing, design, and economics to function effectively. New ideas are difficult to find, and the task of the R & D team is most challenging. Today, more than $20 billion is spent annually on this important function of industry. Over one million people are employed in some function of the R & D process.

Only when a new product achieves continued success and acceptance by the consumer at a satisfactory profit for the company, can R & D say that their professional efforts have been successful.

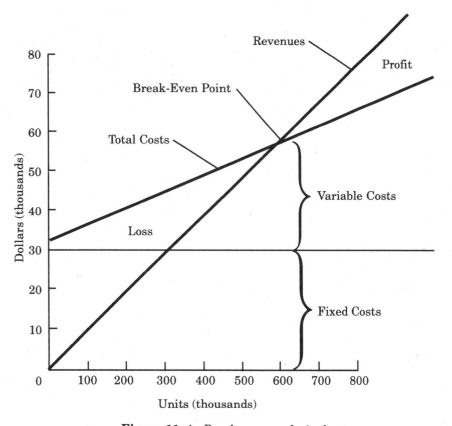

Figure 11–4 Break-even analysis chart

REVIEW QUESTIONS

1. List the three basic areas in which R & D departments conduct their activities.
2. List two agencies that fund pure and applied research efforts.
3. What circumstances would help a company from manufacturing a product that was a result of new exploratory research findings?
4. Define *product analysis*.
5. What is the difference between product and process analysis?
6. List the various functions served by the prototype.
7. What major questions are considered during process analysis?

8. List the criteria established by the Society of Manufacturing Engineers for identifying appropriate processes for new product manufacturing.
9. What role does the environment play in decisions made by the process analysts?
10. What model was used in this chapter to examine the economic, technical, and environmental impact of manufacturing a new product?

SUGGESTED ACTIVITIES _____

1. Research and develop a functional product that can be made from recycled materials. Prepare refined sketches and drawings.
2. Follow the procedure listed below in completing a process analysis of the product developed in Activity 1:
   a. Analyze specifications and drawings
   b. Study required operations
   c. Sequence operations
   d. Combine basic operations
   e. Specify tools and equipment
3. Write to the Forest Products Laboratory, Forest Service, U.S. Department of Agriculture, One Gifford Pinchot Drive, Madison, Wisconsin 53705-2398, and request a listing of research bulletins published. Request research bulletins of interest to you. Try repeating an experiment and record your findings. Are they the same?
4. Write to NASA, Educational Program Officer, NASA, Langley Research Center, Hampton, VA 23665, and request information on Skylab Experiments. Try repeating one of these experiments.
5. Write or phone industries in your area and request information regarding their R & D departments.

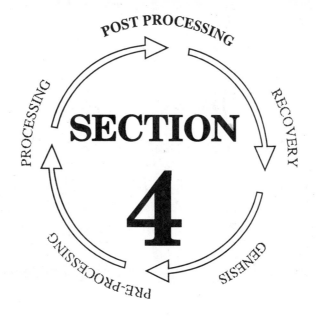

SECTION

4

POST PROCESSING

RECOVERY

GENESIS

PRE-PROCESSING

PROCESSING

MATERIAL-
TRANSFORMATION
PROCESSES

CHAPTER 12

Preprocessing Materials

Once the product design has been selected and the proper working drawings have been completed, the next step in material processing is to mark and cut each part of the product to proper size. Many of the marking and cutting procedures are similar even though the material may change. Most materials are purchased in standard sizes which have been produced by the manufacturer of the material. In this chapter, we discuss the preprocessing techniques used to convert standard sizes into the various parts of a product.

STANDARD SHAPES AND SIZES OF MATERIALS

In Chapter 10, you learned how to use layout tools to develop a working drawing. In this chapter, you will use many of the same types of tools to mark the material. First, you must know what shapes and sizes are produced and sold by the manufacturer of different materials.

All materials are sold in standard sizes and shapes (Figure 12–1). Wood is usually sold in boards and sheets. Boards are sold by thickness, width, and length. In addition, they are sold by grades, which indicate the quality of the material and the way in which it was cut from the tree. As can be seen in Figure 12–2, boards can be slash-sawn or quarter-sawn and are usually graded by the number of and size of knots.

Wood sold in boards is measured and priced by the board foot method; that is, you are charged for each 1- × 12- × 12-inch piece in the board. Figure 12–3 is an example how to figure the board feet in a standard board for the purpose of determining its value.

Wood is also sold by the sheet. The most common sheets are called plywood because they are made up of more than one sheet glued together. Some plywood has a special sheet that

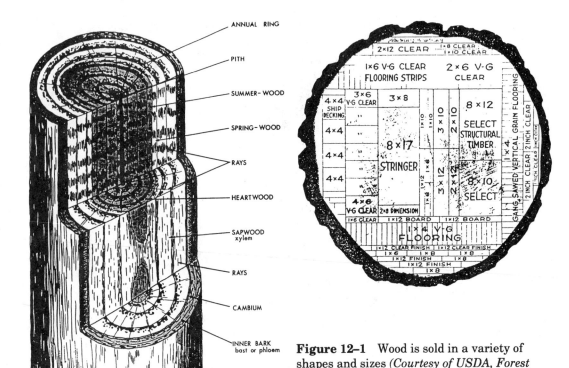

Figure 12–1   Wood is sold in a variety of shapes and sizes *(Courtesy of USDA, Forest Products Laboratory)*

is made out of a more expensive wood that has a better appearance. This procedure is called **veneering** and is a special process for initiating large sheets of solid woods. In Figure 12–4, a special rotary cutting method is used to cut a very thin sheet of hardwood which will be used to give plywood a special appearance.

Wood sheets are sold by size, thickness, and grade. They are usually 4 × 8 feet or 4 × 12 feet in size and a variety of thicknesses which range from $\frac{1}{8}$ inch to $1\frac{1}{4}$ inches. Grades range from CDX (used for construction) to fine, special grain patterns used for paneling or fine furniture.

Metal is sold in shapes, sheets, plates, and bulk ingots. Ingots are used by the foundry industry to melt down and recast into a desired shape. The most common metals used by foundries are brass, aluminum, cast iron, and cast steel. Figure 12–5 shows how melted ingots are used to make parts for a product. This process is called **casting** and will be described in further detail in Chapter 14.

Metal is also sold in sheets and plates (see Figure 12–6). They are similar in size to plywood except that they are usually classified by their thicknesses. Usually if the metal is less than $\frac{3}{8}$ inch thick, it is called a sheet. Sheets are measured by gauges and represent graduations by the thousandths of an inch. For example, a 12 gauge reading is equal to 0.125 inch. Figure 12–7 shows a chart that is used to compare gauge sizes to numerical sizes. Plates

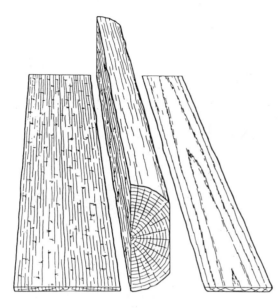

Figure 12–2 Boards can be slash-sawn or quarter-sawn *(Courtesy of USDA, Forest Products Laboratory)*

Some advantages of plainsawed and quartersawed lumber

Plainsawed	Quartersawed
Figure patterns resulting from the annual rings and some other types of figures are brought out more conspicuously by plainsawing.	Quartersawed lumber shrinks and swells less in width
	It twists and cups less.
Round or oval knots that may occur in plainsawed boards affect the surface appearance less than spike knots that may occur in quartersawed boards. Also, a board with a round or oval knot is not as weak as a board with a spike knot.	It surface-checks and splits less in seasoning and in use.
	It wears more evenly.
	Types of figure due to pronounced rays, interlocked grain, and wavy grain are brought out more conspicuously.
Shakes and pitch pockets, when present, extend through fewer boards.	It does not allow liquids to pass into or through it so readily in some species.
It is less susceptible to collapse in drying.	It holds paint better in some species.
It shrinks and swells less in thickness.	The sapwood appearing in boards is at the edges and its width is limited according to the width of the sapwood in the log.
It may cost less because it is easier to obtain.	

are measured by inches and range from $3/8$ inch to more than 6 inches in thickness. In addition, sheets and plates are graded by the way in which they were produced. Hot-rolled metal is usually used for common fabrication, and its thickness is variable with flakes of carbon which will rust and fall off if exposed to weather. Low-carbon steel is the most common hot-rolled metal in use today. Cold-rolled metal is more even in thickness and is usually used

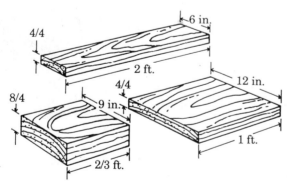

Figure 12–3 Wood is sold by the board foot

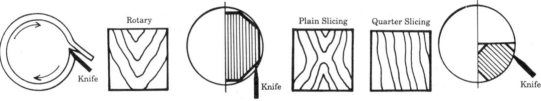

Rotary

The log is mounted centrally in the lathe and turned against a razor sharp blade, like unwinding a roll of paper. Since this cut follows the log's annular growth rings a bold, vaiegated grain marking is produced. Rotary cut veneer is exceptionally wide.

Plain Slicing (or Flat Slicing)

The half log, or flitch, is mounted with the heart side flat against the guide plade of the slicer and the slicing is done parallel to a line through trhe center of the log. This produces a variegated figure.

Quarter Slicing

The quarter log, or flitch, is mounted on the guide plate so that the growth rings of the log strike the knife at approximately right angles, producing a series of stripes, straight in some woods, varied in others.

Figure 12–4   Expensive hardwood can be sliced into veneer sheets for covering plywood and hardboard

for better grades of metal which require a smooth finish. Some alloy metals require other special techniques for shaping, which include a variety of hot- and cold-forming processes.

Most metal sheets and plates are priced by weight and can be special ordered in non-standard sizes at extra cost. Standard metal handbooks by the various companies will list the various choices, shapes, and weight per foot. By looking up the weight per foot, one can figure the cost without actually weighing the material.

Metal is also sold in a variety of shapes. The most common shapes are flats, angles, channels, I and H beams, rounds, and half rounds. Usually they are sold in 16-foot or 20-foot lengths and are priced by the length or weight. Shorter or longer lengths are available at extra cost.

Plastic is sold in sheets, pellets, and shapes. Extruded standard shapes include flats, rounds, half rounds, squares, and special forms. An example of a special form is the plastic handle of a screwdriver. Most plastic shapes are extruded in a hot process, but some shapes such as blanks can be cast or rolled into form. Special shapes for plastic are usually sold in

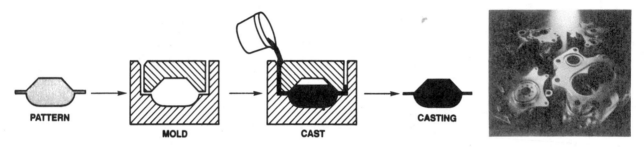

Figure 12–5   Basic casting processes *(From Komacek, Lawson, and Horton,* Manufacturing Technology, *Copyright 1990 by Delmar Publishers, Inc. Used with permission)*

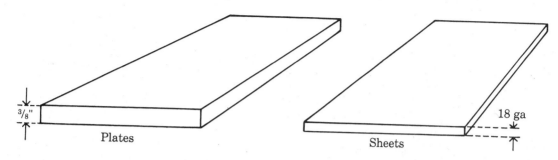

Figure 12–6 Sheets and plates are determined by thickness

6-foot or 8-foot lengths and priced by the foot. The price is based on the type of plastic material and the process used to shape it.

Plastic sheets are sold in all sizes. Common sizes are 3 × 5 feet, 4 × 4 feet, and 4 × 8 feet. They vary in thickness using the same measurement system as for metal sheets. Some plastic sheets are clear and look like a sheet of glass. Special protective coverings are used on these sheets to prevent them from being scratched during shipping and processing.

Pellets are sold by bulk for injection molding processes. They are used in a variety of sizes, shapes, and colors. In a similar method to metal casting, they are melted and shaped into a usable part. This process will be further explained in Chapter 14. For the most part, plastic pellets are used in production processes by machines for multiple part production. Some plastics may also be purchased in liquid form. Casting techniques may be used by mixing the liquid resin with a hardener to cause a solid plastic material. In this case, it is sold by the pint, quart, gallon, or 5-gallon lot.

Sheet-metal gauges and thicknesses

Gauge	Ferrous		Nonferrous	
	Inch	mm*	Inch	mm*
16	0.0598	1.60	0.050	1.20
18	0.0478	1.20	0.040[a]	1.00
20	0.0359	0.90	0.032[b]	0.80
22	0.0299	0.80	0.025[c]	0.65
24	0.0239	0.65	0.020[d]	0.50
26	0.0179	0.45	0.015	0.40
28	0.0149	0.40	0.012	0.30
30	0.0120	0.35	0.010	0.25
32	0.0097	0.22	0.007	0.18

[a]32-ounce   [b]24-ounce   [c]20-ounce   [d]16-ounce
*Metric replacement sizes are based on ANSI B32.3-1974.

Figure 12–7 Gauge sizes have numerical equivalents

Ceramic materials are sold in bulk as a clay or slurry material. They may also be purchased in sheet form and are referred to as tiles. Tiles are ceramic pieces which have been formed and fired (processed in a furnace-like oven) to hardness. They may also have a shiny finish, which is called a glaze. Bathroom ceramic tiles are a good example of this type of material. Concrete blocks and clay-based bricks are also ceramic materials which can be purchased for product processing. The best ceramic materials are used in flatware (china) for use in plates, bowls, and other food containers.

One of the most common ceramic materials is glass. It is most commonly sold in sheets which are available in a variety of sizes and thicknesses. Glass is graded by its process and special shapes. A variety of ceramic materials are available for use as a finished product or part of a constructed or manufactured product.

Bulk ceramic materials are sold by weight; tiles, blocks, and bricks by the hundred; plates by the piece; and glass by the square foot. Prices vary based on quality, quantity, and location.

Material Sizing

Once you have selected the material and purchased the standard sizes that are available, it is necessary to begin the preprocessing activity known as material sizing. Simply stated, this procedure is the expansion or reduction of material sized to meet the requirements for laying out the various parts.

Wood materials are purchased in finished or rough form. Sheets of plywood, veneers, particleboard, and masonite are usually finished, squared, and ready to use as is or cut into smaller pieces. Structural wood used for building construction is also sized, finished and ready for use in a variety of shapes and lengths.

EXPANDING SIZES

Both metal and wood can be made into larger pieces by welding or gluing processes. Usually, this is not necessary for ceramic or plastic parts.

Metal

Metal sheets and structural shapes may be joined together by welding. The most common methods of welding are discussed in Chapter 18 and, therefore, will not be explained here. However, the procedure for expanding metal by this process does have some unique techniques that will help you.

Structural shapes can be joined together by using some simple jigs or fixtures to make sure the alignment is correct. The following is the procedure for setting up structured shapes to be welded:

1. Cut both ends to be joined so that they are flat and square.
2. For material over ¼ inch thick, it is recommended that a bevel similar to that in Figure 12–8 be used.

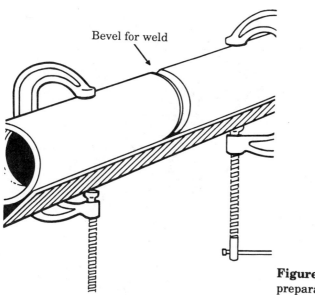

Bevel for weld

Figure 12–8 Welding requires special joint preparation

3. Angle iron can be used for aligning pipe; and straight pieces of steel can be used to align angles, plates, and structural shapes.
4. Leave a space (one-half the thickness of the piece to be joined) between the two pieces for the weld.
5. Tack the joint (short welds), remove the alignment fixtures, and check for accuracy (Figure 12–9).
6. Weld the joint (Figure 12–10). (Check Chapter 18 on how to avoid heat warping.)

Metal plates can be joined by butting edge to edge or overlapping the pieces. When the pieces are lined up, C-clamps are used with a supporting large-angle iron (see Figure 12–11) along the bottom edge. As with the structural shapes, a bevel should be used for pieces over ½ inch thick; and a space should be left at the butt joint for the weld.

Wood

Wood planks and boards can be ordered finished or rough in a variety of grades. Because of cost, most boards and planks are purchased in rough form and must be sized and finished. This procedure is known as planing and squaring stock. As can be seen in Figure 12–12, a board has two faces, two edges, and two ends. Both hand and machine tools can be used to plane and square stock. The following procedure applies to both techniques:

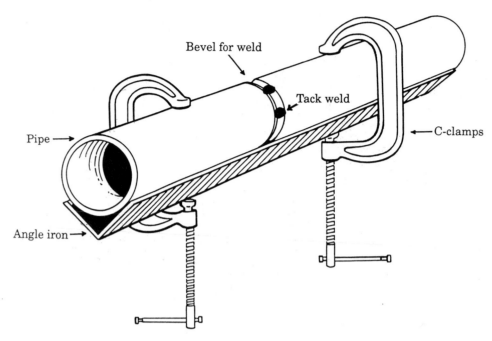

Figure 12–9   Pieces to be welded must have a close fit

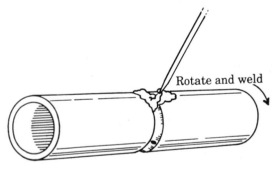

Figure 12–10   The weld makes the metal one piece

1. Plane one face flat and smooth using a hand plane or power jointer.
2. Joint the adjacent edge flat and square to the planed edge using a hand plane or jointer. (If a jointer is used, make sure that the flat face is placed against the fence. If chipping occurs, reverse the board and joint in the opposite direction.)
3. Plane the second face with a hand plane or wood surfacer, being careful to make it parallel and the proper thickness.
4. Joint the final edge with a hand plane or power jointer. (If more than ¼ inch is desired to be removed, it is recommended that the board be cut to width with a rip saw or power table saw prior to final jointing.)

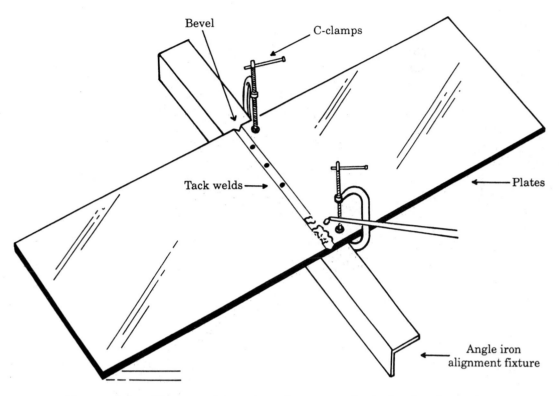

Figure 12–11 Flat materials such as plates must also be lined up for welding

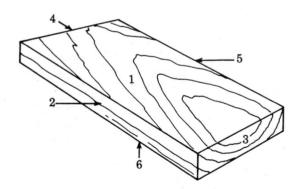

Numbered surfaces of board:
1 and 6—faces
2 and 5—edges
3 and 4—ends

Figure 12–12 A board has two faces, two edges, and two ends

After the wood has been planed and squared, it can be expanded by gluing the boards together. The following is a procedure for gluing up:

1. Lay the planed boards side by side until the correct size plus 1 inch is obtained.
2. Reverse the grains (by looking at the edge) of every other board in the series (see Figure 12–13).
3. Select grain patterns so that the overall surface will be coordinated and attractive.
4. By using a hand plane or lowering the rear table of the jointer, cut a spring joint as shown in Figure 12–14.
5. Cut a spline using a table saw or drill holes using a dowel jig to provide additional strength for the joint as shown in Figure 12–15.
6. Glue edges and clamp tight with furniture clamps as shown in Figure 12–16 and let dry for 24 hours.
7. Remove clamps, scrape excess harden glue, and surface piece to correct thickness.

Once the metal or wood has been expanded to the proper size, it is ready for layout procedures.

Material Layout

The actual marking of the material is necessary to make up the various parts. Similar to the drafting procedure, you will now need to use the various layout tools to create shapes for future processing.

The following layout procedures are applicable to all materials once they have been sized.

1. **Determining Length** The first procedure for layout is to determine the length of each piece. Using a squared end, the length is usually measured using a measuring tape or ruler. A simple check mark is made, which will indicate where the cutoff line

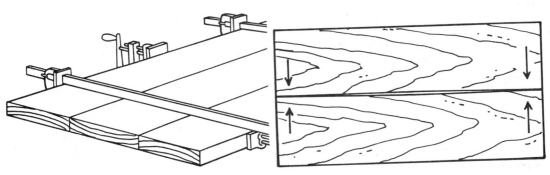

Figure 12–13 Boards are reversed by grain pattern for stability

Figure 12–14 A spring joint is used to assure tight joints

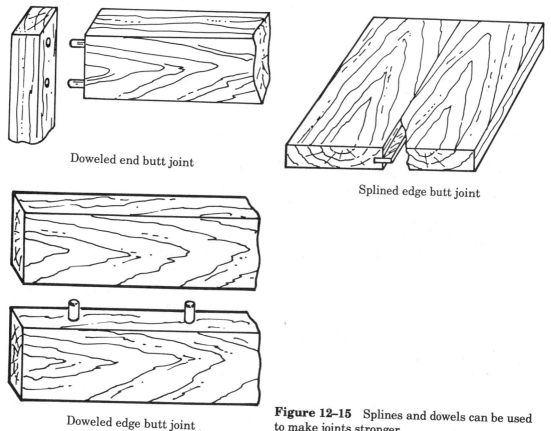

Doweled end butt joint

Splined edge butt joint

Doweled edge butt joint

Figure 12–15   Splines and dowels can be used to make joints stronger

should be drawn. Using a square against the straight edge, a line is drawn to indicate where the cut will be made. In general, pencils are used to mark wood, plastic, and ceramic materials. Soapstone and scribes are used to mark metal. If the metal is shiny, a bluing fluid may be applied first to help the scribe marks show better. Never use a pen or magic marker on wood or plastic because it will stain the material. If multiple cuts are to be made on a longer piece of material, care should be given to allow for the cut. Saws need more allowance than shears, but it is a good idea to rough cut all the parts first and then finish cut each part separately.

2. **Determining the Width**   Once the length has been determined, the width of each piece should be measured and marked using the same methods as previously described. It is best to measure from the same straight edge if possible. Squares or straight edges can be used to connect the points and lines. Not all parts are in the shape of a square or rectangle, as can be seen in Figure 12–17. Therefore, several marks and a variety of angles or circles may be used to lay out each piece. If the longer

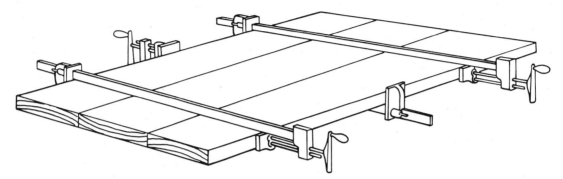

Figure 12–16   Glued edges are clamped with pressure and left to dry overnight

and wider piece has several smaller parts in it, the widest and longest piece will determine the width and cutting procedure. Remember, careful planning during the layout procedure can make maximum use of basic materials.

3. **Determining Thickness** Most materials are purchased in standard thicknesses and may be used without further modification. However, solid woods, some plastics, and a variety of metals may require a reduction in thickness for the entire part or subsections of a part. If possible, it is better to mark the piece from a common face or reference point. Measuring and marking procedures are similar to the aforementioned techniques with careful attention to cutting requirements. Most materials are reduced in thickness by resawing or machine tool-cutting techniques. Care must be given to allow for the thickness of each cut.

SPECIAL LAYOUT PROCEDURES

Once the thickness, width, and length of each part have been marked, special layout procedures may be required for location of holes, bevels, fasteners, or joints. Layout tools such as dividers, compasses, marking gauges, thickness gauges, and French curves may be used to help locate and mark future processing requirements. Figure 12–18 shows a variety of special layout tools that can be used to measure and mark wood, metal, plastic, and ceramics.

Because each part has been identified and marked for size and shape, it is now possible for one to transfer special layout needs directly from the drawing to each piece. For some metals that will be cut with an oxyacetylene process, it may be necessary to also use a prick punch to score the layout lines. This will allow the operator to see the lines during the cutting operation.

Some machine tool metal-cutting procedures require very close tolerances. It is common practice to use a special layout table known as a **precision block** (see Figure 12–19). The block is used for a point of reference with an angle plate to locate special points for cutting, bor-

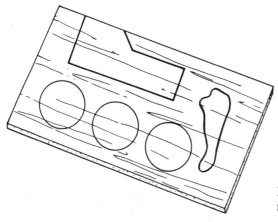

Figure 12–17   Not all shapes are squares or rectangles

ing, or fastening procedures. A variety of gauges and markers is used to locate points which are either scribed into the metal or prick punched for future use during machining operations. Plastics and ceramics that will be machined also use the same marking procedures as metal.

Curves and other irregular shapes may be transferred to the workpiece by first making the design on a grid sheet. The grids (usually ½-inch or 1-inch squares) provide points of reference as the curves are laid out with French curves. Once the design has been completed, it is cut out and traced on the workpiece. If the pattern is to be used more than once, it may be cut out on a piece of sheet metal or masonite and then transferred to the workpiece. Figure 12–20 shows how the grid transfer process can be used.

If a special cutter is to be used to make a special shape, location of the cutting area may be all that is necessary for layout. For example, holes or dovetails need only a center point or guideline to help the operator locate the special process. Drawing the circle for the hole or marking each finger of the dovetail would be a waste of time during the layout process because it is not helpful or necessary.

Once the part of the product and special processing requirements have been measured and marked, it is time to begin the material-separation process.

MATERIAL SEPARATION

There is a variety of methods used to separate (cut) materials. Most of these processes can be applied to metals, woods, plastics, and ceramics using the same major cutting tools. Conceptually, the major types of material-separation processes can be classified under one of four groups:

1. Chip removal
2. Shearing
3. Flame cutting
4. Special process cutting

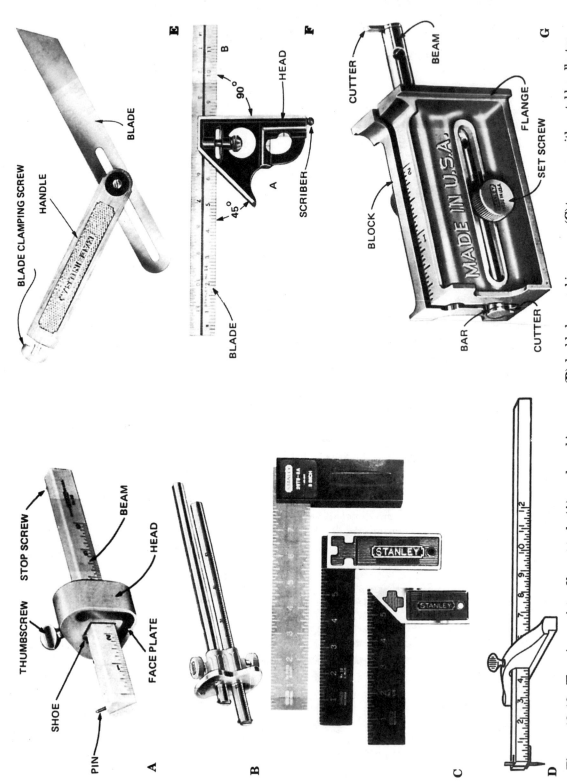

Figure 12-18 There is a variety of layout tools: (**A**) wood-marking gauge; (**B**) double-bar marking gauge; (**C**) try square with metal handle, try square with wood handle, try and mitre square with metal handle; (**D**) panel gauge; (**E**) t-bevel; (**F**) combination square; (**G**) butt gauge (*From McDonnell and Kaumeheiwa, The Use of Hand Woodworking Tools, Copyright 1978 by Delmar Publishers, Inc. Used with permission*)

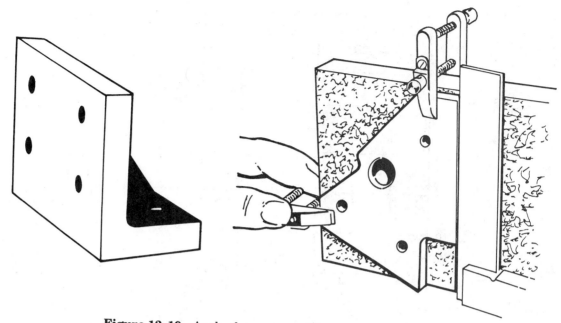

Figure 12–19 Angle plates are used for precise measurements

Chip Removal

The most common material-separation process is chip removal. This process works on the principal of using a wedge-type tooth blade or cutting tool to remove a thin section of the material. The removed section of material is usually called a **kerf**, and the most common process for this method is sawing. Some chip-removal cutting processes use multiple-wedge teeth like a band saw, and some use a single tooth like the shaper or lathe.

Multiple-Tooth Cutting

Saw blades may be of the straight or circular design. For example, the table saw uses a round blade. One exception is the band saw, which uses the straight-blade design but revolves in circular motion. In most cases, the multiple-tooth blade is made up of teeth that are slightly angled to each side in order to provide clearance for the blade. This idea of bending the teeth is called **"set"** and is shown in Figure 12–21. As can be seen, the most common forms of set are raker, straight, and wave.

Saw blades are also classified by their size, number of teeth, and speed of cutting. The following terms are used to describe a saw blade:

1. *Composition.* Most saw blades are made from carbon or high-speed alloy materials. Sometimes the teeth are made from a material called tungsten carbide and welded

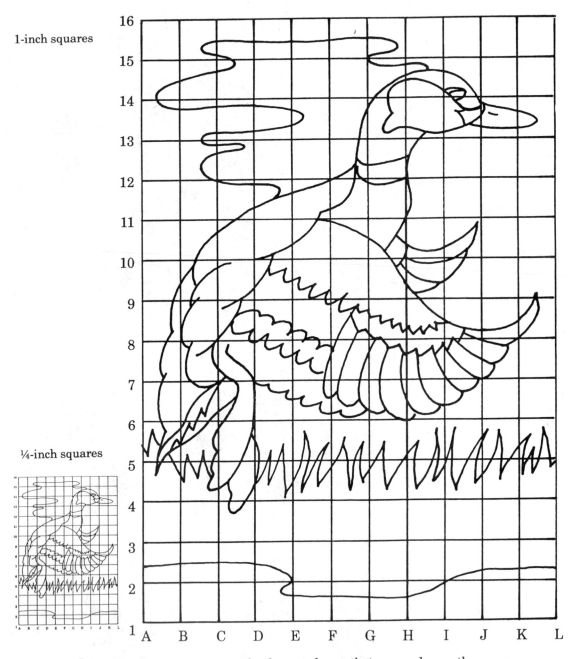

1-inch squares

¼-inch squares

16
15
14
13
12
11
10
9
8
7
6
5
4
3
2
1

A B C D E F G H I J K L

Figure 12–20 Patterns are used to lay out shapes that are used more than once

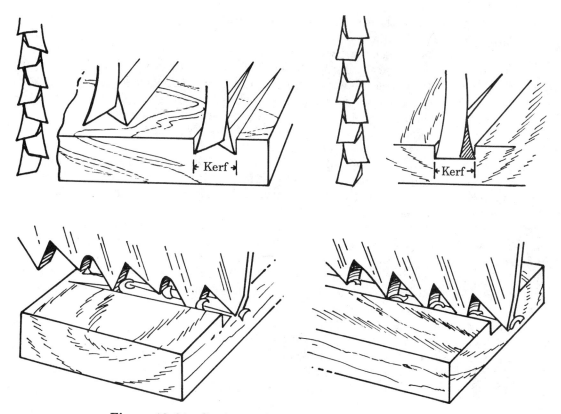

Figure 12–21   Cutting teeth must have set to remove waste

to the blade. Carbide-tipped blades are able to cut tougher materials and hold their sharpness longer.

2. *Blade width.* The width of a saw blade is the distance from the tip of the tooth to the back of the blade.

3. *Blade gauge.* This refers to the thickness of the blade teeth. They can range from 0.025 inch to 0.250 inch, depending on the type of material being removed.

4. *Blade pitch.* The pitch of a saw blade refers to the number of teeth per inch. A 12 pitch blade has twelve teeth in each linear inch. In any sawing operation, at least three teeth should be in contact at all times. Therefore, it is necessary to increase or decrease the pitch of a saw based upon both the hardness and thickness of the material being removed.

5. *Lip clearance.* The lip clearance is the relief given the cutting edge so that it may enter the material to be cut. The angle may range from 6 degrees for metal to 35 degrees for woodworking tools.

6. *Rake angle.* This is the angle at which the face of the cutting edge enters the work. A large rake angle requires a great force to drive the tool. Too small a rake angle results in a thin cutting edge. Saw manufacturers usually design the rake angle based upon the material and sawing operation. However, machine tool operators who grind their own cutting tools will need to follow the recommended chart in Figure 12–22 in order to obtain the proper rake and clearance angles.

7. *Crosscut teeth.* This type of blade cuts both as a knife and chisel. Figure 12–23 shows the various tooth angles used for the crosscut blade. Generally, it is used to cut across the grain in wood and to prevent excessive splintering.

8. *Ripsaw teeth.* This type of blade has chisel-shaped teeth which can cut metal, wood, and plastic materials. The teeth are angled to each side (set) but aggressively remove material in one direction. Generally, this saw style is popular for soft metal and cutting wood with (along) the grain.

When possible, the piece to be cut should be set up so that the largest area of surface is in contact with the blade. Some metals need a cutting fluid to help the cutting process and maintain longer blade life. Steels use cutting oil, cast irons use a water and oil cutting fluid mixture, and aluminum uses a kerosene fluid base. The decision to use a coolant is based on how much metal is being removed per minute and the speed of the blade. Wood and plastic materials depend upon the choice of blade pitch to control blade clogging and heat buildup. Plastic materials, in particular, require course (skip tooth) blades to prevent chip welding (material which attaches itself to the rake side of a tooth).

A cutting speed chart allows you to select the necessary speed for the material with which you are working. Ceramic materials are not listed on the chart. That is because special cutting blades are used for ceramic materials, which are discussed under the special process section of this chapter.

Single-Tooth Cutting

The most common use of single-tooth cutting occurs in the metals industry using machine tools. The horizontal milling machine uses the saw multiple-tooth method described earlier, and the vertical mill can separate materials using a multiple-tooth cutting tool. However, the engine lathe and metal shaper make exclusive use of the single-tooth cutting tool. Figure 12–24 shows how a single-tooth cutting tool is used on the lathe and shaper. The major disadvantage of single-tooth cutting is that the larger kerf (caused by the necessity of the thicker cutting tool for strength purposes) causes more waste in material. Its advantage lies in the ability to cut material without changing the machining setup.

Shearing

Shearing, unlike chip removal, does not produce a waste material as a result of the separation of the material. Much like a pair of scissors are used to cut paper, shears are able to cut wood, plastic, metal, and ceramic materials.

1. Front clearance

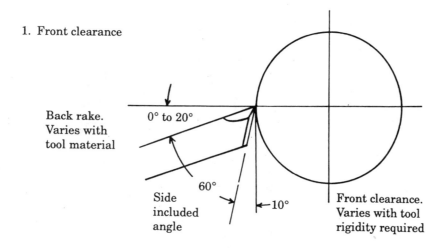

Back rake.
Varies with
tool material

0° to 20°

60°

Side
included
angle

10°

Front clearance.
Varies with tool
rigidity required

2. Side clearance

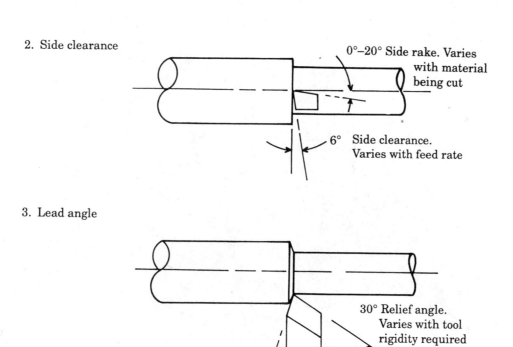

0°–20° Side rake. Varies
with material
being cut

6° Side clearance.
Varies with feed rate

3. Lead angle

30° Relief angle.
Varies with tool
rigidity required

75° included angle

15° lead angle

Figure 12–22 A metal-cutting tool is shaped by the operator according to specifications

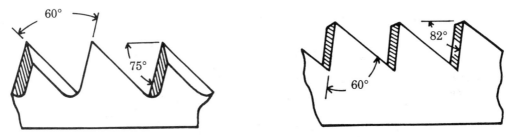

Figure 12–23   Crosscut teeth are used to cut across the grain of wood

The most common form of shearing is found in the sheet-metal industry, where a variety of hand-operated shears are used to cut and shape metal. They are operated like a pair of scissors and can cut a variety of shapes, circles, and angles. Straight cuts are usually performed on the squaring shear which is foot operated for metal which is 18 gauge or thinner. Punches also operate on the principle of shearing and are used to make holes in the material.

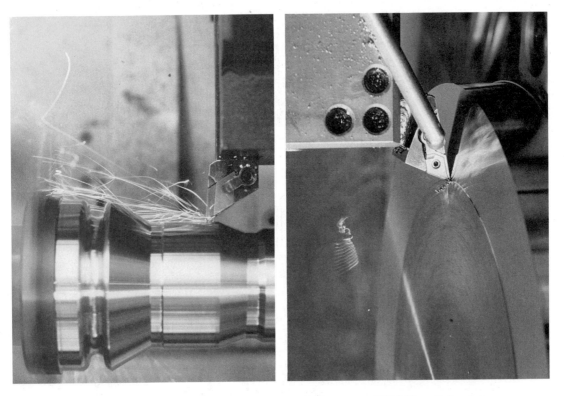

Figure 12–24   A single-tooth cutting tool *(Courtesy of Giddings & Lewis)*

Thicker metals can also be sheared using a heavy-duty iron worker. Plastic materials can be sheared using metal shears, paper scissors, and standard paper cutters. Approximately the same rules that exist for metal thickness also apply to plastic.

Ceramic materials, because of their difficulty to work after being fired, are usually cut in their green condition. When ceramic materials are in their green condition, standard metal-shearing tools are used to cut and shape the material along with special wire-cutting devices.

Last, wood products can be sheared with special razor-sharp guillotine-style shears. Some veneers can be cut with standard paper scissors and paper cutters. Special mitre shears are also used to trim joints for precision joints in fine woodworking.

Shearing is fast and clean and leaves little waste. One of the drawbacks to shearing (particularly in thicker materials) is the deformation of the material because of stress. Bending and warping of materials that have been sheared will require further work to return them to their original shape.

Flame Cutting

Unique to only the carbon-based steel family is the ability to cut metal by oxidizing the carbon particles. Oxygen–acetylene torches can be used to separate a number of steels. However, nonferrous metals and cast steels and irons do not respond well to this type of material separation.

Figure 12–25 shows how the oxygen–acetylene torch can be used to separate steel. Like the chip-removal process, the flame-cutting process will require additional material for the cut which will result in waste. Today, computer numerical control (CNC) and optical-scan gang flame cutters are used in industry. Several pieces can be cut at one time using this method. CNC-controlled oxygen–acetylene cutters are programmed on the X and Y axes. For complex shapes, preprogrammed shapes such as circles, angles, and rectangles may be selected.

Optical-scan oxygen–acetylene cutters are designed to follow the lines on a drawing placed on a table. Complex shapes can be cut and repeated by reusing the pattern or committing the commands to the CNC memory disc.

Arc cutting is used primarily for cutting and beveling cast steels and iron. A special cutting electrode is used with an arc welder to produce heat and chemical reactions to oxidize the metal, causing separation. This process is used primarily by the metals repair industry and field repair application. The major disadvantage to arc cutting is excessive material waste, uneven cuts, and slag residue which must be removed for further processing (welding, fitting, etc.).

Other materials such as woods, plastics, and ceramics do not respond to flame-cutting techniques and, therefore, do not apply for this type of material-separation process.

SPECIAL PROCESS CUTTING

Special techniques are available to separate materials that do not respond to standard cutting procedures. The three most common special techniques are abrasive cutting, laser cutting, and hot-wire cutting.

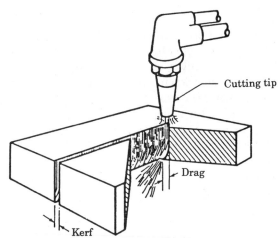

Cutting tip

Drag

Kerf

Figure 12–25   Metal can be cut with an oxyacetylene torch

Abrasive Cutting

Abrasive cutting can be used to cut metallic and nonmetallic materials. In addition to the metals industry, the abrasive saw is popular for cutting glass, brick, stone, concrete, and fired-ceramic materials. Each abrasive acts as a small tooth and actually cuts a small bit of material. Because of their durability and high speed (10,000 to 15,000 feet per minute), abrasive cutoff saws are very fast cutting. Like other separation processes, there is waste in the cutting operation; and this should be kept in mind for layout purposes in order to provide excess material for the waste area.

Laser Cutting

One of the most recently developed cutting methods is laser. The heat from a light beam is so intense that is will produce a very small hole in the hardest of materials. By releasing a tremendously great concentration of energy as a pulse which is produced by the use of a ruby and controlled by the use of a lens, the laser can systematically cut any type of material known to humankind.

Energy of the laser is derived from the monochromatic light characteristics of extreme intensity and coherence emitted from a charged ruby. Once directed onto a material, the beam will penetrate with great accuracy. As can be seen in Figure 12–26, special applications for laser cutting are restricted to processes that cannot be performed by conventional cutting methods.

Plasma Arc Cutting

Another special cutting process is plasma arc cutting. This process is very popular for cutting stainless steel which is difficult to cut by conventional methods and nearly impossible to cut when its thickness is increased. It also is used to cut aluminum and other noniron powder–based alloys.

Using temperatures in excess of 10,000 degrees Fahrenheit, the cutting of all metals is merely a matter of directional control of the plasma. The plasma is formed by the presence of some type of gas (usually a mixture of hydrogen and nitrogen) and a high-frequency electric arc. The results of this type of cutting process are similar to the laser, producing a very narrow waste area with unusual edge smoothness.

Electron Beam Cutting

Like the laser and plasma arc cutting processes, the electron beam cutting process will also sever most materials and require very little waste for the cut. By controlling the diameter path of electrons in a vacuum, they move from the tungsten electrode to a hole in the anode and continue on until they collide with the workpiece. This collision causes the material to vaporize, thus making the separation.

In Figure 12–27 a very high-voltage (60 kV to 150 kV) concentrated beam of electrons penetrates the workpiece to cause a separation. Because of the possible radiation effects, a shield is used to protect the operator. Also a limiting factor is the box like device needed to provide a vacuum environment. This limits the process to smaller pieces and raises the cost of the process. Therefore, the electron beam cutting process has a limited production use and is usually considered for special applications.

Hot-Wire Cutting

The process for cutting styrofoam plastics is known as a hot-wire process. A wire is stretched between two points and a voltage is passed through the wire to heat it. Like the principle on which a band saw or jigsaw works, the material is pushed into the wire and can cut a straight or curved path. The process is fast and leaves a reasonably smooth edge.

SUMMARY

This chapter has focused on the preprocessing procedures and techniques of material sizing, layout, and separation. Once the material has been purchased in a standard shape or size, a modification may be necessary to expand or reduce the size of the material. This modification is a preprocessing technique called material sizing. It may be as simple as cutting a smaller piece from a larger piece or as complicated as gluing several smaller pieces

Figure 12–26 Lasers cut very accurately. *(From Jeffus,* Welding Principles and Applications, *3rd edition, Copyright 1993 by Delmar Publishers, Inc. Used with permission)*

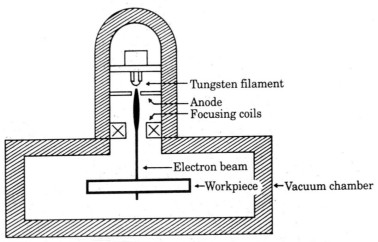

Figure 12–27 Shields are used to protect the operator from possible radiation in electron beam cutting

together to make a large piece. The purpose of the material-sizing process is to get the basic material ready for layout procedures.

The process of material layout is similar to the drafting procedures studied in Chapter 10 except that the markings are made on the actual material. The markings can be made by measurement or they can be transferred by special patterns made of paper, metal, or masonite. Care must be taken to make efficient use of the material, provide the most attractive aspect of the material, and leave enough space for waste.

Once the material has been divided and marked into its various components, the proper material-separation process must be selected. Different materials require different cutting processes, although they may share a number of commonalities. Select the most efficient and cost-effective method available to do the job. Remember, the special cutting processes may be much more costly than conventional methods because of their complex technology. Yet, for very difficult materials like ceramics and special alloy metals, one of the special processes may be the only alternative.

When the preprocessing of material has been completed, the various parts may be complete or require further processing. Chapters 13 to 19 will provide more detailed information and procedures on transforming raw and primary materials into finished products.

 REVIEW QUESTIONS _____

1. Woods are sold by the various grades and the way they were cut from a tree. Describe the terms *slash-sawn* and *quarter-sawn*.

2. Special rotary cutting machines are used to cut veneers. Why are veneers used in the woodworking industry?

3. What is the difference between a metal sheet and a metal plate?

4. List the materials that are sold in sheets.

5. How is ceramic material sold?

6. Material sizing is a process used to _____ or _____ standard-sized materials.

7. How is a line "special marked" for the oxygen/acetylene cutting process?

8. What are the most common material-separation processes? Give at least three examples.

9. Which cutting process is used to cut very thick pieces of stainless steel?

10. Which cutting processes can be used to cut fired-ceramic materials?

SUGGESTED ACTIVITIES

1. Using basic layout tools, have students lay out the parts of their project on the material.

2. Have students glue up several pieces of wood, being careful to follow correct procedures.

3. Using scrap material, have students cut shapes in metal using the oxygen/acetylene cutting torch.

4. Have students select a special material-separation process and write a technical paper based on library research.

5. Take the class on a field trip to an industrial plant that must use preprocessing techniques.

6. Have students use several of the cutting processes as they fabricate or produce their class project.

7. Using a display panel, have groups of students show examples of the various preprocessing techniques.

8. Using a test and measurement approach, have students evaluate and demonstrate appropriate and nonappropriate material-separation processes.

CHAPTER 13

Molding Processes

Molding processes can be defined simply as the forming of a desired part or product in a mold. Various types of equipment and operations are employed in the molding of metals, woods, plastics, and ceramics. The following is a list of operations that can be classified as molding processes:

Compression	Transfer	Blow	Injection	Extrusion
Die casting	Powdered metal	Forging	Rolling	Wood flour

The basic elements of any molding operation include the following:

1. Temperature required
2. Force required
3. Type of mold
4. Physical state of the material before, during, and at the end of the process

These elements will be used to describe the molding processes listed above.

MOLDING

Compression Molding

Compression molding is one of the primary manufacturing processes used to shape thermoset plastics. Figure 13–1 illustrates the compression molding operation. This process requires a two-piece matched mold capable of withstanding considerable force.

The correct amount of the thermoset plastic is measured and placed into the open mold which is generally heated to temperatures between 250 and 410 degrees Fahrenheit. The

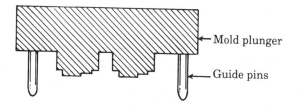

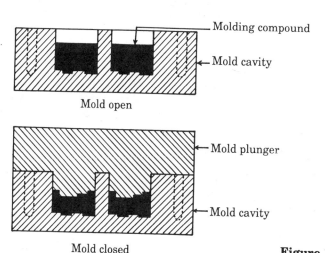

Figure 13–1 Compression molding

thermoset plastic is in granular or powder form prior to being molded. The top half of the mold is then carefully placed on the half containing the plastic, and the complete mold is placed between the platens of a hydraulic press (see Figure 13–1). The function of the press is to apply both heat through the platens and the necessary pressure to cause the plastic first to plasticize the material and then cure or polymerize the plastic. This process causes the plastic to retain the shape of the mold cavity.

The presses range from small benchtop hand-operated machines to large, highly automated high-production models. The typical force required can range from as little as 1,000 psi to pressures which exceed 15,000 psi.

The most commonly used plastic for compression molding is phenolic plastic. Products manufactured using this process include such things as distributor caps, electrical resistors, gears, toys, and ashtrays.

Transfer Molding

The transfer molding process is essentially the same as compression molding except that the design of the matched mold is changed to allow for a gating system. The gating system (Figure 13–2) is added to reduce the direct pressure on the part. This process is used to manufacture plastic parts whose design incorporates intricate detail and thin sections.

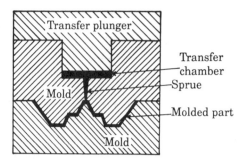

Figure 13–2   Transfer molding

The gating system is very similar in appearance to the ones found in a green sand mold used in the metal casting area. The material used is typically a phenolic plastic which is solid rather than granular or powder and is known as a **preform**. The preform has been preprocessed to the approximate shape and size of the mold cavity. This facilitates handling and, therefore, makes the processing more efficient.

The disadvantage of compression and transfer molding is that any waste produced as a result of the processing cannot be reused. This is particularly true of the transfer molding operation, which utilizes the gating system. Unlike thermoplastics, thermosets cannot be granulated and recycled. The required temperatures, pressures, equipment, and procedures used in transfer molding are the same as those used for compression molding.

Powder Metallurgy

A process similar in principle and process to the compression molding of plastic is the forming of metal parts and products from metal powders. The process known as powder metallurgy (P/M) is used to make a wide range of machine parts, gears, filters, and self-lubricating bearings. Figure 13–3 illustrates the wide range of applications for this process.

The modern industrial use of powder metallurgy as a shaping or forming process came into its own during World War II and has grown steadily since that time. The process lends itself to high production rates and the shapes that can be produced with this process are virtually without limit. Production rates range from several hundred to several thousand per hour. However, the weight of parts produced in conventional equipment generally does not exceed thirty-five pounds.

Three basic steps are used to produce parts using this process. These steps involve mixing the powders, compacting the powders, and sintering (bonding metal particles with heat) the compact.

Mixing

The first step involves selecting and mixing the correct metal powders and additives such as lubricants. This blend of metal alloys and additives is either done by the supplier of the metal powders or by the manufacturer of the product.

Figure 13–3 Parts made by the powder metallurgy process *(Courtesy of Metal Powder Industries Federation)*

Shaping

The method of forming the powdered metal mixture into the desired shape or form is very similar to the process described earlier in the section on compression molding. Perhaps the most popular shaping method is *compacting*.

Compacting consolidates the loose powder in a hardened steel or carbide die to the desired size and shape. The consolidated powders in the form of the desired part is known as a *green compact;* this has adequate strength to be handled and transported for further processing. During the shaping process, a controlled amount of the blended powders is fed into a die and compacted at pressures that range from 10 to 60 tons per square inch. The presses employed for compacting operations are generally mechanical, hydraulic, or a combination of both. Many of the presses are specially designed for high production rates and, therefore, include such options as rotary tables with a series of closed dies at the top and bottom for mass production. Although special presses and optional equipment have been designed, many commercial presses which are used for other materials and processes can be used to demonstrate powdered metallurgy. Figure 13–4 illustrates a typical flow used in the P/M process.

Sintering

Unlike the compression molding of plastics, the compacting of metal powders typically requires a postprocessing step known as *sintering*. After leaving the compacting press, the green compact is conveyed to a controlled-atmosphere furnace where it is heated to a temperature below the melting point of the base metal. The part is held at this temperature for a specified period of time and then cooled.

This application of heat in a controlled furnace is a solid-state process that results in bonding the metal particles together, thus providing such functional properties as strength. The powdered metal part is often ready for use after sintering. However, secondary operations are often necessary to provide additional qualities such as finish, size, and shape. These include heat treating, machining, re-pressing, grinding, tumbling, or milling. These and other processes are discussed in later chapters.

Wood Flour Molding

A forming technique similar to compression, transfer, and powdered-metal molding is wood flour molding (see Figure 13–5). Wood flour molding is a process that uses fine particles of wood to which dry adhesives and colorants are added and then formed in a closed die with heat and pressure.

The process is a two-step process. It involves (1) mixing the wood flour with the adhesives and colorant in correct proportions and (2) pressing the mixture in a closed die to desired shape and size. This is a process that can be done on smaller laboratory presses. It is used in industry to produce such commercial products as toys, coasters, handles, and decorative furniture parts.

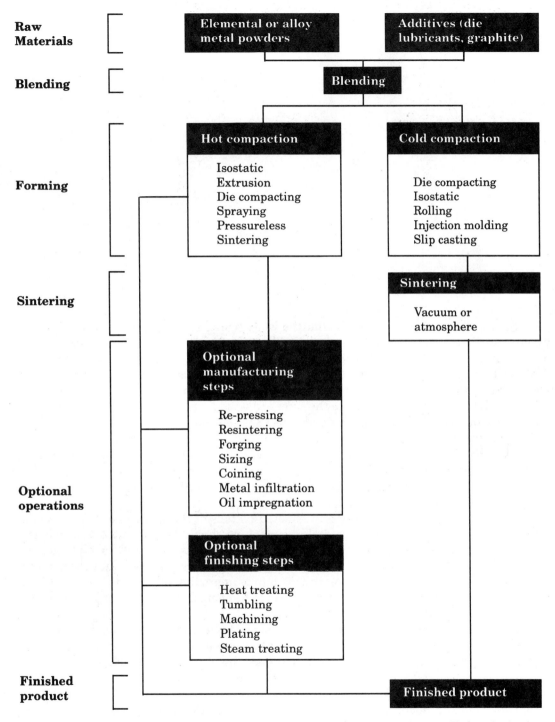

Figure 13–4 P/M processing steps *(Courtesy of Metal Powder Industries Federation)*

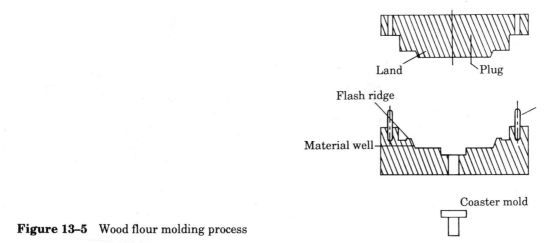

Figure 13–5 Wood flour molding process

One advantage of this process is that the wood flour used in the process is often wood waste that is generally burned or discarded. The wood waste is collected from sawmills and other woodworking factories and sized by mechanical or air screening methods. The wood flour particles are typically sifted through a screen to produce uniform particle size. The mesh, sized according to its openings per square inch, can range from 8 to 140.

Dry adhesives powder is added to the screened wood flour and mixed thoroughly. Adhesive resins such as *melamine, urea,* or *phenolic* are typically used in this process and account for as much as 30–40% of the volume of the mix.

Powdered pigments can also be added to the mixture to enhance the color of the molded items. The wood flour, adhesives, and pigments are blended thoroughly in a mixing container before being placed into the mold. The mold is generally coated with a mold release prior to being filled with the mix.

The filled mold is placed between the heated platens of the compression molding machine being used. The platens of the molding machine should be preheated to a temperature of 250 to 300 degrees Fahrenheit (120 to 150 degrees Celsius). After the platens reach the prescribed temperature, the mold is carefully placed between them and the required pressure is applied. The total molding time will take approximately two minutes. Depending on the type of resin used, the mold may need to be opened for a brief period of time during the process to allow gas to escape.

At the end of the processing time, the pressure is released and the die removed from between the platens. Extreme care should be taken in handling the hot die. Heavy gloves are recommended for handling the die. The two-piece die is then separated, and the molded part is removed. Ejection pins are commonly included in the design of the die to facilitate the removal of the molded part.

Today, many products once made of wood flour using the process described are being made of plastic. However, the molding process and equipment are very similar to those used to wood flour molding; and its use as a manufacturing process is still very prevalent.

INJECTION MOLDING

Injection molding is a high-speed, high-volume forming process that has found its greatest application in the plastics industry. However, this process was first developed for the ceramics industry during the 1930s and has recently been revitalized by that industry to mass produced industrial ceramic objects that have complicated shapes.

The process involves forcing a material through a cylinder and into a closed mold or die. Injection molding is a principal method of forming thermoplastics and can be used for themosets with some modification to existing injection machines.

The plastic is typically heated in the cylinder to temperatures ranging from 300 to 650 degrees Fahrenheit, and forces between 5,000 and 40,000 psi are used to force the heated plastic into the closed die. The thermoplastic material cools upon entering the die and solidifies within a few seconds.

Although the products made using this process are too numerous to mention, they include toys, telephone cases, laundry baskets, automobile dashboards, handles, knobs, and many other parts for the automobile and appliance industries.

Equipment

There is a variety of kinds and sizes of molding machines used for injection molding. Injection molding machines are either the screw-ram or plunger type (Figure 13–6) and are rated according to the amount of material injected during a cycle. Many machines of this type have a range of capabilities from less than an ounce to several hundred ounces of plastic per shot.

Molding machines consist of two functional units: one for melting and injecting the plastics, and one for opening and closing the mold. The majority of the production machines are

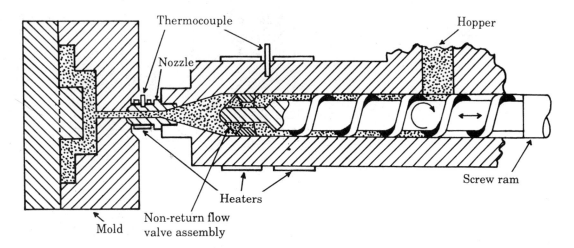

Figure 13–6 Cross-section of screw-ram and plunger type injection molding machine

hydraulically controlled and are of the screw-ram type. However, smaller, pneumatically controlled plunger-type molding machines are often used in smaller shops and school laboratories.

Molds

The molds used in injection molding are typically made of hardened steel or metal alloys and are designed with slight draft or taper in order to facilitate part removal. They are a two-piece design and are matched for accuracy and detail. In high-production operations, the molds are water cooled between cycles. Figure 13–7 illustrates a typical cross-section of a matched mold. The channels running through the edge of the mold are used to carry the water that cools the plastic, thereby making it rigid.

Many of the terms used to describe a matched mold are also used in die casting and metal casting molding operations. Listed here are the parts of the matched mold with explanation (see Figure 13–7):

Sprue—inlet to mold cavity from die head; also used to describe the plastic appendage to the part

Runners—grooves running alongside the mold cavity from the sprue; the runners connect the sprue and the gates

Gates—inlet from runners to mold cavity

Vents—small grooves running to edge of mold which permit gas to escape

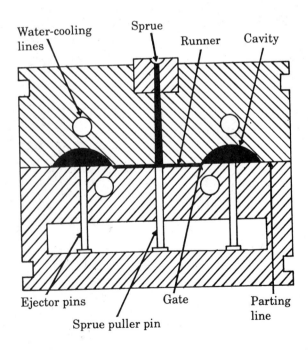

Figure 13–7   Cross-section of water-cooled matched mold

Ejection pins—small-diameter cylinders extending through mold body used to force plastic part from mold cavity

Procedure

A variety of thermoplastics can be formed by injection molding. Polyethylene is perhaps the most commonly used for this process, closely followed by polystyrene, acrylics, and cellulosics. The following is a step-by-step general procedure for injection molding:

1. Preheat molding machine to required temperature (300 to 650 degrees Fahrenheit).
2. Load hopper with plastic pellets or granules.
3. Close matched mold halves.
4. Move screw or ram forward, forcing melted plastic through die head and into the matched mold.
5. Cool mold for specified time (approximately 5 to 15 seconds) and open mold halves.
6. Eject part from mold and remove all flash, gate runners, and sprue.

As stated earlier, injection molding is now being used for ceramics as well as plastics. The advantage for ceramics is that little if any postprocessing is required. This is important because ceramics are extremely hard materials and difficult and expensive to machine.

The process used for injection of ceramics begins with submicron-sized particles which are immobilized in a thermoplastic binder and injected into a die cavity. The die is then heated; and pressure (as much as 100,000 psi) is applied, causing the ceramics powders to flow into the desired shape. The piece is then removed from the die and fired twice—once to evaporate the binder and again to fuse the powders. The smaller particles (50 to 100 microns) used in this process increase strength, density, and permit a more intricately detailed component to be shaped.

Injection molding is a primary molding method that has provided many inexpensive products and parts for use in today's society. It is a process that has been used extensively because of the following advantages:

- High speed, high volume
- Inexpensive
- Finished parts require little or no machining
- Permits wide range of size and shapes
- Allows for intricate detail
- Very little waste

BLOW MOLDING

Blow molding is a high-production process typically used to form thin-walled, hollow plastic containers of various sizes. Bottles which hold a few ounces to 55-gallon drums are easily formed using the blow molding process. Recently, this has become a very popular production method whose products are replacing those formerly made of glass and metal. The process uses thermoplastic materials to form a variety of plastic products, including shampoo bottles, milk containers, plastic drums, and a variety of other containers.

Process

Blow molding requires the use of compressed air to inflate a heated hollow tube known as a **parison** against the walls of a two-piece closed mold. Figure 13–8 shows a typical setup for blow molding operations.

During this process, the heated thermoplastic material is forced through a heated cylinder and into a die head where it is formed into a hollow tube of moldable plastic. The hot plasticized tube (parison) is injected between the mold halves and pinched at the bottom (see Figure 13–8).

Air pressure (20 to 40 psi) is then forced through the die head into the hollow hot plastic tube. The result is that the tube inflates forcing the plastic against the mold cavity walls and conforms to the shape of the mold. After the parison is inflated, the air pressure is maintained for a few seconds to cool the plastic, causing it to solidify and maintain the desired shape.

The mold is then opened and the part is removed from the molding machine. In general, postprocessing consists of removing the flash which occurs at the parting line and pinch-off area. This scrap, along with any defective products, can be chopped up and used again.

Equipment

The equipment used for blow molding operates, in principle, much like that used in injection molding. The molding machines are either hydraulically and pneumatically controlled or strictly pneumatically operated.

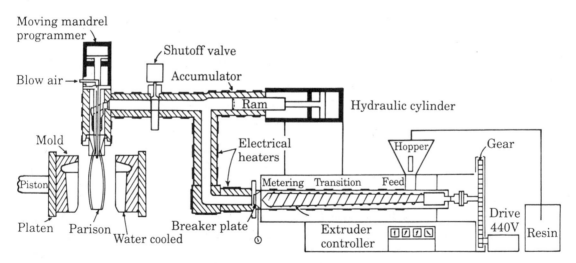

Figure 13–8   Extrusion process

Like injection molding machines, the blow molding equipment is designed to perform two functions. The first function is designed to open and close the matched molds and relies on a pneumatically or hydraulically controlled cylinder for its operation. The second function is designed to melt the plastic and force it through the die head and into the mold cavity. This function is also hydraulically or pneumatically controlled and uses a ram- or screw-type cylinder to force the plastic through the heated chamber. The temperatures required for this process vary according to the type of thermoplastic used but typically range between 300 and 500 degrees Fahrenheit.

Mold

Molds for this process generally have deeper cavities than those for injection molding and vary slightly in design. Figure 13–9 shows a typical mold used for this process. One minor difference in design is the pinch-off area required on the blow forming mold. This area closes off the hollow parison, allowing it to inflate. These molds are made of machined aluminum.

DIE CASTING

A process that is very similar to injection molding is die casting. This process uses both equipment and molds like those used in the injection molding of plastics to force liquid metal into two-piece dies or molds. These two processes are so similar that some books refer to them as one in the same.

Like injection molding, die casting is a high-speed, high-volume process which can produce as many as 800 pieces per hour. Nonferrous metals such as aluminum, brass, zinc, tin,

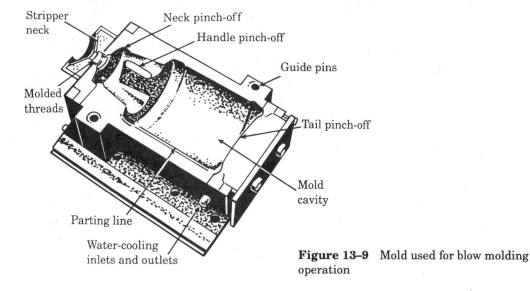

Figure 13–9 Mold used for blow molding operation

lead, and magnesium are used in the die casting process to produce a variety of parts and products weighing from one ounce to well over 90 pounds. Small engine blocks, toys, handles, carburetor parts, and pump housings are only a few examples of the products made using this process.

Advantages of this process include

- Extreme accuracy
- Excellent surface finish (little machining)
- High-volume production

In addition, the scrap rate is extremely low, since the materials used can be remelted easily and reprocessed.

Equipment

The equipment used for die casting is either hydraulically or pneumatically controlled and is classified as either a cold- or hot-chamber molding machine (Figure 13–10). Hot-chamber die casting machines are designed to be used with low-melting alloys of zinc, tin, and lead. It is referred to as hot-chamber die casting because the plunger system is immersed in molten metal, thus providing a constant supply of heat and, therefore, increased production rate. This system operates at pressures from approximately 80 to 600 psi.

The cold-chamber machines are used to die cast higher-temperature nonferrous metals like aluminum, brass, and magnesium. These metals require greater pressures (5,600 to 22,000 psi) than those used in the hot-chamber machines. In the cold-chamber process, the molten metal is ladled into the cylinder just ahead of the plunger. The plunger is then moved forward, thus forcing the metal into the die where it solidifies under pressure and is ejected when the die is open. This process eliminates the continuous contact of melting pot and plunger to molten metal, which would reduce the life of the equipment with the high-temperature metals.

The molds used in die casting are made from steel and are similar in design to those used in injection molding. The molds are highly polished, which results in extremely good surface finish on the completed part, thus reducing the need for machining. The life of the mold or die is a function of the type of metal being used and ranges from 10,000 fillings per mold to several million.

EXTRUSION

Extrusion is a manufacturing process that is used to produce parts and products of a specific cross-sectional design. The manufacturing principles used in the process are similar to those discussed previously with respect to injection and blow molding operations. Extrusion is a process used to manufacture plastic, metal, and ceramic parts and products.

The variety of plastic, metal, and ceramic products made by the extrusion process is extensive. Plastic rods, tubes, monofilament line, cable coatings, and moldings are made by

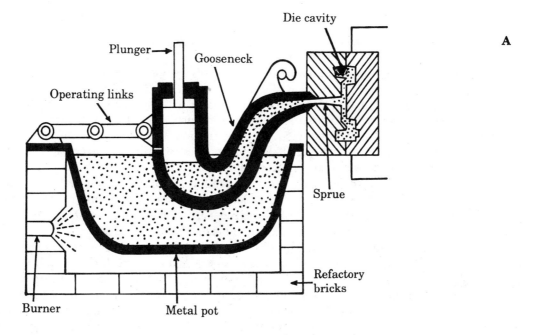

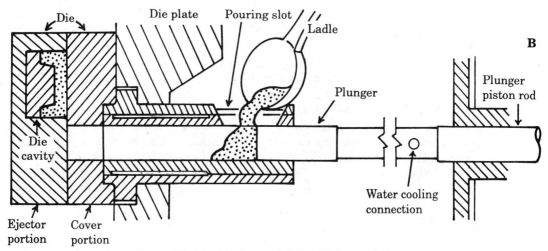

Figure 13–10 **(A)** Hot and **(B)** cold die casting process

extrusion. Seamless tubing, moldings, windows, doors, ladders, and appliance trim are among the more than two billion pounds of metal extrusions made each year. Ceramic extrusions include brick, hollow tile, tubes, and rods.

Extrusion was developed more than 100 years ago but did not become widely used until approximately seventy-five years ago with the development of the hydraulic extrusion press.

Substantial compressive force is required to force the materials through a die or opening of a desired shape and size.

Metal extrusions are used extensively in the building and construction industry. More than half of all metal extrusions are used in that industry in the form of windows, door frames, and other building components. Thirty percent of all metal extrusions are used in the transportation industry. They form the structural framework for airplanes, ships, and land vehicles.

Although steel can be extruded, most metal extrusion is done with nonferrous metals, including aluminum, copper, and lead alloys. The most commonly used extrusion method is the *hot-extrusion* method. During this process, a heated billet is placed in a cylindrical chamber and compressed by a hydraulically operated ram. At the opposite end of the chamber is a steel die having an opening of a desired shape and size. The die opening represents the path of least resistance; and the billet, under force, is squeezed out through the opening as a continuous bar or tube of the desired shape and size (Figure 13–11).

Plastic extrusions are made much the same way injection molded products are made. Pellets or granules of plastic are fed into a heated cylinder from a hopper attached to the extruder. The plastic, primarily thermoplastics such as polyethylene, polyvinyl chloride, and nylon, is fluidized and homogenized in the heated cylinder before being forced through a die and cooled at the other end. Figure 13–12 shows a typical ram-type extrusion machine used for plastics.

Large quantities of ceramic parts and products including brick and drainage tile are made using the extrusion process. The state of the material at the time of processing is plastic. The various clays and additives are mixed with water to form a thick plastic mass. This mass is then forced through a die and is formed into a long continuous length of the required cross-sectional shape.

The extruded shape moves from the extruder to a cutting machine where it is trimmed and cut to a specified length. Once cut to length, the product is then transported to a drying area where it is further treated and then baked or fired in kilns (large ovens) for extended periods of time (40 to 150 hours) at temperatures exceeding 1,500 degrees Fahrenheit.

Plastic, ceramic, and metal extrusions are used extensively in the manufacturing, construction, and transportation industry as well as in the home. The extrusion process and equipment is very similar for each of the three aforementioned materials. The differences are related to the state of the material before and during the process and the pressures needed to force the material through the die.

THERMOFORMING

Thermoforming is a molding process that uses mechanical or air pressure to force a heated plastic sheet around a prepared form. It is used to form a variety of items, both large and small, including such products as automobile tops and complete auto bodies.

Thermoforming is used with a variety of plastics, and the selection of the type is dependent on the required specifications. Mechanical and thermal properties as well as appearance are among the considerations used in determining the type of material to use. The methods used for thermoforming plastic sheet stock are classified in the following manner:

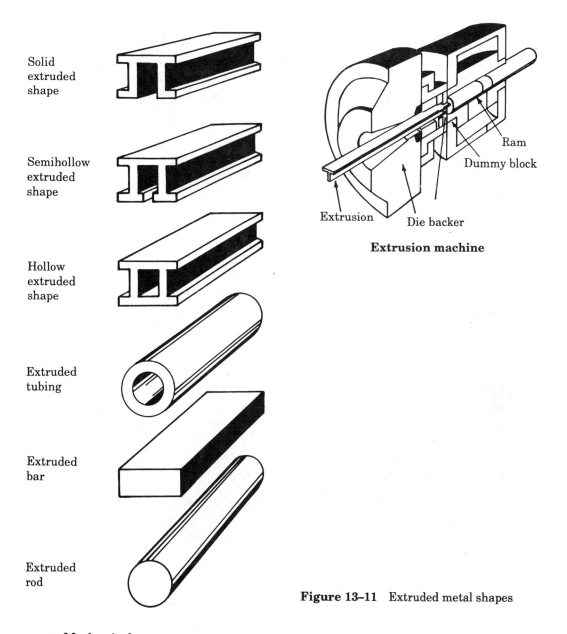

Solid extruded shape

Semihollow extruded shape

Hollow extruded shape

Extruded tubing

Extruded bar

Extruded rod

Ram

Dummy block

Extrusion

Die backer

Extrusion machine

Figure 13–11 Extruded metal shapes

- Mechanical
- Vacuum
- Blow forming

The *mechanical* method (Figure 13–13) requires a set of matched molds that are forced together with a heated sheet of plastic between the two halves of the mold. The sheet or film

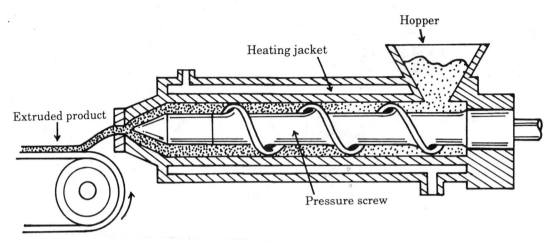

Figure 13–12 Cross-section of typical ram-type extrusion press

is stretched tightly between the mold halves and heated to a temperature below its melting point (275 to 400 degrees Fahrenheit) to increase its elasticity. When the plastic is ready, the mold halves are forced together and held for a short period of time until the plastic becomes rigid again. This process is often used to form objects with deep pockets or recessed areas.

Vacuum forming relies on air pressure differentials to form the part or products. A one-piece mold is typically used in the process. The one-piece mold is placed between the plastic sheet and the source of the vacuum. Small holes are drilled around the surface of the pattern to allow the vacuum to evacuate the air from between the plastic sheet and the pattern surface.

The plastic sheet is heated until it becomes pliable, and the sheet is brought in contact with the pattern to form a seal between the two. The vacuum is turned on and the air between the two evacuated. Once the air is evacuated between the heated plastic sheet and the pattern surface, the outside air pressure forces the plastic around the form of the pattern. The plastic cools while the vacuum is held for 10 to 15 seconds, and the part is removed (Figure 13–14).

Blow forming uses air pressure to force the heated plastic sheet against the prepared mold. This process is used to form bubble-type products such as windows for buildings or canopies for airplanes.

The heated plastic is securely clamped to a frame with a hole in the middle. Compressed air is then forced through the hole and against the pliable plastic. The shape of the blow-

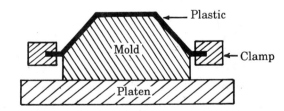

Figure 13–13 Mechanical thermoforming

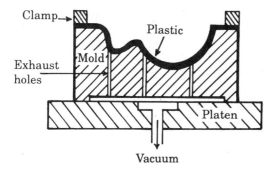

Figure 13–14   Vacuum thermoforming

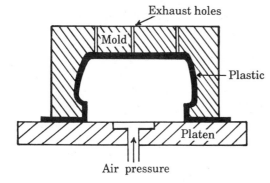

Figure 13–15   Blow thermoforming

formed product is controlled by the air pressure and heat used in the process. Figure 13–15 illustrates the process used in forming a small bubble-type product.

CALENDERING

Calendering is a process used in the plastics industry to form film and sheet stock. The forming mechanism used in this process is a series of rolls used to form the plastic resin into a uniform thickness and width. The rolls can be adjusted to control the thickness of the finished sheet or film stock.

The process begins by placing plasticized thermoplastic into a material feed or hopper unit. This unit feeds the resin between two rolls that are spinning in opposing directions. The plastic is then pulled through a series of rolls that are progressively closer together. The final rolls are designed to cool or set the plastic. Figure 13–16 shows the different stages of the calendering operation.

The heated rubbery mass that is fed through the calendering machine is prepared by mixing the desired thermoplastic with colorants, lubricants, stabilizers, and other additives that give the finished product the desired properties and appearance. Calendering is a pro-

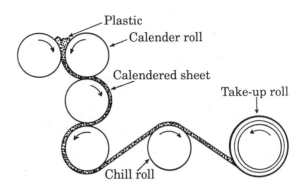

Figure 13–16   Calendering operation

cess used to produce numerous products for home and industry. Floor coverings, shower curtains, inflatable toys, and car seat covers are just a few examples of the many products manufactured this way.

ROLL FORMING

Roll forming is a process used to form thin metal into various cross-sectional shapes. The process is somewhat similar to the calendering operation in that a series of rolls is used to control both the shape and thickness of the finished product. Figure 13–17 shows several of the shapes formed using the roll-forming process. The transformation from flat sheet to finished shape is completed as the metal passes through a series of rolls that gradually change the cross-section to the desired form.

The primary advantage of roll forming over other processes is speed and economy. Rolled welded tubing, for example, can be formed much faster and at a substantially lower cost than seamless tubing formed from solid billets. Typical production speeds for roll-formed products are between 50 and 300 feet per minute. Although used primarily for thin metals—up to $5/32$ inch thick and 16 inches wide—special machines can be designed for heavier and wider sheets.

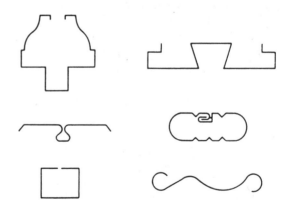

Figure 13–17   Roll forming *(From Goetsch,* Modern Manufacturing Processes, *Copyright 1991 by Delmar Publishers, Inc. Used with permission)*

FORGING

Forging is a method of forming metal by hammer blow or pressing. The cost of forging is higher than other manufacturing processes but is used on parts and products that require great strength because of high stress loads. Forging improves the quality of the metal, refines the grain structure, and increases the strength. Forging is used to manufacture items such as aircraft landing struts, tools, crankshafts, gears, and automobile axles. The range of shapes and sizes varies significantly. Forgings can weigh less than a pound or several tons.

The forging process is accomplished using one of two basic types of machines. The *drop-* or *hammer*-type machine uses the force of a sudden impact to form the metal blank between die halves to a desired shape. Capacities for these machines range from 100 to 10,000 tons of force. *Press*-type machines use a squeezing force to form the product between die halves. The capacity for the press-type machine ranges from 100 to 50,000 tons of force.

Forging can be classified as *drop, press, roll,* or *cold* forging. While the first three refer to the method used, the fourth refers to the state of the material during the forming process. Heating the metal prior to forging increases its plasticity and reduces its resistance to compressive forces. However, heating can also alter physical properties and appearance. Cold forging increases strength and improves surface finish.

Drop Forging

Heated metal is placed between die halves that are attached to the hammer and anvil portions of the machine. The upper die is dropped several times on the hot metal blank, causing it to flow into the die cavities. This is typically done in stages beginning with what is referred to as a "fuller" die. The intermediate die, called a "blocker," brings the forging closer to its finished form. The "finisher" die forms the forging into its specified shape and size (Figure 13–18).

Press Forging

Slow squeezing action is used rather than sudden impact like that used in drop forging. It is primarily used for nonferrous metals, especially aluminum and magnesium, and produces extremely smooth castings.

Roll Forging

This process involves metal bars rolled between two revolving dies. It is used for tools, axles, levers, and other parts requiring high strength.

Cold Forging

This involves a hammering or squeezing of unheated metal. This process is used to form gears and screw threads as well as bolts, nails, rivets, and screws.

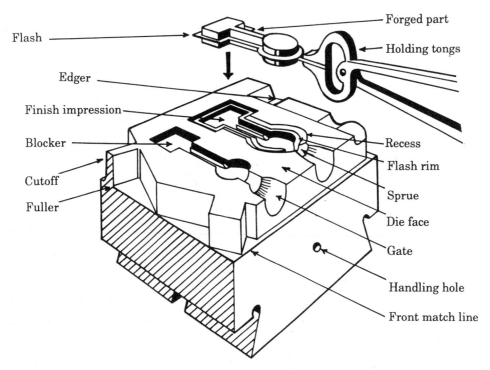

Flash
Edger
Finish impression
Blocker
Cutoff
Fuller

Forged part
Holding tongs
Recess
Flash rim
Sprue
Die face
Gate
Handling hole
Front match line

Figure 13–18   Combination die

SUMMARY

The forming techniques described in this chapter are used to shape a wide variety of parts and products out of metals, plastics, and ceramics. The range of products vary in size from a small plastic toy to a landing strut for a 747 jet. The products formed from molding operations are used in homes, offices, businesses, transportation systems, amusement parks, and industries around the world.

The processes in this chapter, unlike the processes described in subsequent chapters, change the shape of the material without addition or subtraction of material. There is no volume change during the process of shaping or forming the product through the molding processes described. These processes rely on compressive forces applied by machines that are hydraulically or pneumatically operated and the elastic and plastic deformation properties of the materials being formed.

Elasticity is the property that allows the material to be stretched or deformed temporarily and recover when the force is released. If a sufficiently large force is applied, plastic deformation takes place. This means the shape will change and remain changed rather than returning to its original form.

Table 13–1 illustrates the characteristics of the forces and states of the materials before and during the various processes used to mold plastics, metals, woods, and ceramics.

Process	Force Required (psi)	Forming Device	Temperature Required (°F)
PLASTICS			
Compression	1,000–10,000	Matched Molds	280–400
Transfer	6,000–12,000	Closed Matched Mold	280–400
Injection	5,000–20,000	Closed Mold	300–650
Blow	40–100 (air)	Hollow Cavity Mold	300–500
Extrusion	500–6000	Die-Opening	250–600
Calendering	Varies with Material and Finished Size	Series of Rolls	300–400
Thermoforming	Varies with Type Process	Matched Molds, Single Molds, Single Mold, or Free Formed	10–200
METAL			
Powder Metallurgy	20,000–120,000	Matched Die	Room Temperature
Die Casting	600–22,000	Closed Die	Melting Temperature of Metal Used
Extrusion	100,000–200,000	Die with Desired Shape	Varies According to Metal and Process
Forging	100–50,000 tons	Matched Die	Room Temperature
WOOD			
Wood Flour Molding	1000–10,000	Closed Die	250–300
CERAMICS			
Extrusion	500	Opening of Desired Shape	Room Temperature
Injection	5,000	Closed Die	300

Table 13–1   Summary of molding processes

 REVIEW QUESTIONS_____

1. What primary differences exist between compression molding and transfer molding operations?
2. Compare and contrast powder metallurgy forming with compression and transfer molding.
3. Describe the three-step process used in forming parts using powder metallurgy.
4. What process used to form various molded products from wood is similar to compression molding?
5. Describe the two functional units of an injection molding machine.

6. Describe the function of the following parts of a mold used in injection molding:

   sprue   runners   gates   vents   ejection pins

7. List three or four advantages of injection molding.

8. Compare extrusion processes for metals, ceramics, and plastics in terms of temperatures and pressures used for each material.

9. Contrast blow molding and injection molding. List three or four products made using the blow molding operation.

10. What two classifications are used for die casting equipment?

11. What is the principal forming method used for plastic sheet or film products? What similar process is used in forming metal products?

12. Describe three different techniques used to forge metal products. What is the advantage to forging over other processes?

SUGGESTED ACTIVITIES

1. Create a chart illustrating the molding processes discussed and example products made from each process.

2. With available equipment, complete as many of the processes described as possible. Record temperature, force, and equipment requirements for each process.

CHAPTER 14

Casting Processes

Casting is one of the oldest manufacturing processes still being used today. Bronze castings have been found that date to 2000 B.C. The various processes classified under casting are neither additive nor subtractive processes. Like the processes described in Chapter 13, the volume is unchanged. The state of the material is changed during the process, but material is neither added nor subtracted in doing so.

Casting, a process usually associated with metal, is being used extensively to shape rubber, plastic, and ceramic products as well. There are a number of different casting methods; and although they differ substantially in technique and equipment, each requires the following systems for a successful product:

1. Pattern
2. Gating systems
3. Mold or refractory

Chapter 14 focuses on the most popular casting techniques used to shape metals, ceramics, plastics, and rubber. The casting processes included in this section are as follows:

| Green sand casting | Investment casting | Centrifugal | Rotational | Slush |
| Slip | Foamed | Shell | Vacuum process | Full mold |

In many instances, a particular process used for casting one material is very similar if not identical to a process used for another. For example, slush casting is a process used when a hollow area or cavity is required in the finished product. It involves filling a mold with a material that is in a liquid state, allowing the material against the mold walls to solidify, and then pouring off the excess to form the desired cavity. This casting principle is used to form metal, ceramic, and plastic parts and products.

GREEN SAND CASTING

Green sand casting is one of the oldest and most commonly used casting processes. It is used to cast various metals into a wide variety of parts and products. Castings made by this process range in size from the smallest of trivets to enormous turbine parts that are cast in huge pits and transported on railroad flat cars.

The term *green sand casting* refers to the moisture that is contained in the sand mixture when molding takes place. Usually, the moisture content range is between 2% and 8%; and other additives such as clay produce the plasticity needed to mold the sand without it breaking. An analogy can be drawn between green wood and green sand. A branch taken from a young, living tree is high in moisture content and will bend nearly in half without fracturing, while a dry dead branch will snap in half quickly. Dry sand is very brittle—much like the dry branch—and, therefore, will not mold easily and hold its shape around a pattern.

Since green sand casting is the most popular of the casting techniques, the process, tools, and equipment used are discussed in greater detail. The discussion focuses on the following components of a green sand casting system:

Patterns   Sands (refractory)   Gating system   Risering   Cores and coremaking

Patterns

Pattern making is a special craft that requires highly developed skills in woodworking and, at the same time, a comprehensive knowledge of the principles of metal casting. The pattern maker must be aware of metallurgical concepts related to the nature of solidification and shrinkage rates for various metals. This information is used to manufacture a pattern that will produce a casting of specified shape and size.

The first or original pattern is referred to as the master pattern. The master pattern is made from white pine, mahogany, or cherry and is constructed with double-shrinkage allowance. **Shrinkage** is the amount of contraction that takes place as the metal cools and solidifies in the mold. Table 14–1 lists the shrinkage rates for various metals.

The master pattern, with double-shrinkage allowance, is used to cast the patterns from which the molds will be made. These patterns will have the required single-shrinkage allowance required to produce a casting within specified size tolerance.

There are a number of different types of patterns depending on the requirements for the finished casting. The more commonly used patterns are classified in the following manner:

Single   Split   Matchplate   Loose piece pattern   Irregular parting line

Material	Contraction (in/ft)
Gray Cast Iron	$\frac{1}{2} - \frac{1}{8}$
Cast Steel	$\frac{1}{8} - \frac{1}{4}$
Aluminum	$\frac{3}{16} - \frac{1}{4}$
Magnesium	$\frac{1}{8} - \frac{5}{32}$
Brass	$\frac{3}{16} - \frac{7}{32}$
Bronze	$\frac{1}{8} - \frac{1}{4}$

Table 14–1   Shrinkage rates

A slight taper is added to the sides of the patterns to allow it to be drawn from the mold without causing damage to the face of the mold cavity. The amount of taper varies with the type of pattern and design requirements. In general, the smallest amount of taper is about $\frac{1}{16}$ inch per foot of **drawface** (drawface is the area of the pattern that is drawn parallel to the sides of the mold). The normal practice is to allow $\frac{1}{64}$ inch per inch of drawface.

Sand for Green Sand Casting

The term *refractory* means to hold back or resist heat. In green sand casting, natural or synthetic sand is the material used to form the mold cavity into which the molten metal is poured. There is a variety of sands and additives that can be selected depending on the type of metal and desired results, including finish quality of casting. The following is a brief description of different sands and additives used for green sand casting.

Natural Bonded Sands

Naturally bonded sands come from natural deposits around the country. This sand is readily available and usable as it comes from the ground after being properly tempered (correct moisture content added). These sands are most abundant in the eastern Great Lakes region and parts of the South. Albany and Tennessee Gold Seal are among the more popular brands of the naturally bonded sands.

These sands generally contain between 15% and 25% clay and impurities, including organic matter and other elements typically found in the earth. Although these impurities present some problems, such as reduction of bond strength, these sands are popular because of their availability, lower initial cost, and wider working range of moisture content.

Synthetic Molding Sands

The use of the word *synthetic* to mean artificial does not apply in this case. The use of the term *synthetic,* when referring to molding sands, means that additives have been mixed with the sand to produce desired properties. The following is a list of some of the more popular synthetic molding sands with typical proportions of various additives:

Olivine sand—a water-based synthetic sand. The greenish color makes it easily recognizable.
  Typical Mixture
  Olivine sand of appropriate AFS (American Foundrymen's Society) fineness
  2.5–3.5% water
  5–6% Bentonite clay (Southern Bentonite or half Southern and half Western)
  0.75% cereal (optional)

Petro bond or oil-based sand—popular sand for educational laboratories. This sand is tempered by oil; therefore, it is not necessary to condition it before each molding session.
  Typical Mixture
  Clay-free silica sand of desired AFS fineness
  5% clay binder (red flour–type material)
  3% light-weight (no. 10) oil:
  Texaco Ursa P-40
  Gulf Endurance #81
  Gulf #926
  Shell Corena 72-67116
  1 ounce per 100 lbs of P-1 catalyst ($\frac{1}{8}$ lb of Methyl alcohol can be used as a substitute)

Synthetic sand mixtures have advantages as well as disadvantages:

Advantages

1. Greater strength
2. Better detail and surface finish
3. Lower moisture contents in water-based synthetics reduce chances of defects related to moisture
4. Can be rammed hard
5. Better thermal-expansion qualities
6. Excellent dimensional accuracy
7. Longer life than natural bonds

Disadvantages

1. Higher initial cost to purchase sand and additives
2. Oil-based sand produces fumes and smoke

Other Additives

A variety of additives are used in foundries to improve certain qualities of the sand system. Basically, there are two reasons for additives:

1. Reduce potential for casting defects
2. Improve or control physical properties

The following is a list of the categories of additives and their uses as they relate to the two reasons for using additives listed above:

1. *Carbonaceous:* sea coal, pitch, fuel-oil graphite
   Uses: retards expansion, reduces contraction, improves green strength and permeability
2. *Cellulose (cushioning additives):* wood flour, cereal hulls
   Uses: minimize expansion defects, increase green strength, flowability, and collapsibility
3. *Fines:* silica substance, flours, iron oxide
   Uses: reduces metal penetration, increases hot strength
4. *Cereal additives:* cereal flour, dextrin, sugars
   Uses: generally improves all strengths and collapsibility
5. *Chemical additives:* boric acid, ammonium compounds, sulfur, sodium salts, diethylene glycol
   Uses: improve casting surface, control sand acidity

PHYSICAL PROPERTIES OF FOUNDRY SAND

The following is a partial list of the physical properties that are used when describing the characteristics of molding sands used for metal casting:

Grain fineness: Standards for sand grain sizes have been established by American Foundrymen's Society. Grain sizes range from AFS 6 (3,360 microns) to 270 (53 microns). The AFS number refers to an average grain size. An AFS fineness between 100 and 180 is typically used for aluminum castings.

Green strength: The ability of the tempered (moisture added) sand to withstand the pressure imposed on it by the liquid metal.

Thermal stability: Resistance to drastic changes in temperature. If sand fails as a result of expansion and contraction due to temperature change, the result is defective castings.

Permeability: Expressed as the volume of air in cubic centimeters that will pass per minute under a pressure of one gram per square centimeter through a specimen one square centimeter in cross-sectional area and one inch high. Typical permeability range is 40 to 150. The higher the number, the more permeable the sand. Permeability is a measure of the ability of the sand to exhaust gases that are produced when the hot metal is poured into the mold cavity.

Collapsibility: The ability of the sand to contract with the metal. If the metal contracts and the sand does not collapse against it or contract with it, tears can form in the casting.

Flowability: Ease with which the sand flows around the pattern. The ability to compact or ram the sand to conform to the pattern shape.

PROCESS

Gating System

The gating system refers to passageways that are formed to permit the flow of molten metal into the mold cavity. The gating system is perhaps the most important element of metal casting. An improperly designed gating system will increase the probability for defects in castings. Some of the defects that can result from poorly designed gating systems include the following:

1. *Expansion defects:* defects caused by inadequate flow of metal into mold cavity
2. *Misrun:* an incomplete casting due to failure of metal to fill mold cavity
3. *Cold shut:* a definite discontinuity due to imperfect fusion. This defect may have the appearance of a crack or a seam.
4. *Gas defects:* cavities caused by gas pressure. Improper gating system will produce turbulence which increases gas pressure.
5. *Cuts and washes:* voids and rough surfaces or erosion of the mold when the velocity of the metal flow is too great
6. *Shrinks:* a depression in the casting due to improper solidification of the metal in the mold cavity

The American Foundrymen's Society has identified the following objectives for a properly designed gating system:

1. The molten metal should enter the mold cavity quietly (as low in velocity and as free of turbulence as possible).
2. The system should be designed to ensure that the first metal into the gating system, which is generally damaged, does not enter the mold cavity.
3. Metal entering the gating system flows into the mold cavity in a manner that will promote proper solidification.
4. The gating system should be economical.

Figure 14–1A illustrates a properly designed gating system, and Figure 14–1B shows one that is typically found in textbooks. The properly designed system will meet the objectives described previously, while the other one will cause casting defects. A brief description of each element of the properly designed gating system will serve to explain the importance of proper gating.

The pouring basin (Figure 14–2) should be located close to the edge of the flask and be large enough to keep a continuous flow of metal feeding into the sprue. The bottom of the basin should be convex and a raised lip should be cut between the sprue and the basin reservoir.

Proper design reduces the turbulence, helps prevent damaged metal from entering the sprue, and helps slow the velocity. With the basin located close to the edge of the flask, the lip of the crucible can be lowered close to the basin, thereby reducing turbulence. Some of

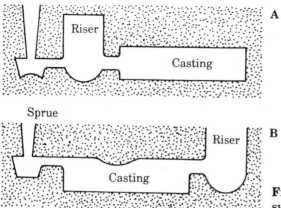

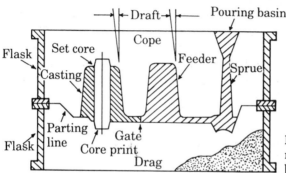

Figure 14–1   (A) Properly designed gating system; **(B)** poorly designed gating system

Figure 14–2   Properly designed pouring basin reduces turbulence and traps damaged metal before it can enter the mold cavity

the damaged metal will be trapped in the basin and prevented from flowing into the gating system. The elevated lip before the sprue helps slow the metal down before dropping down the sprue and into the sprue base.

The sprue should be tapered; and its cross-section should be round or, better yet, rectangular. Although some books illustrate the use of a straight sprue hole, this in practice can cause air to be trapped in the gating system and forced into the mold cavity area. When this happens, defective castings can result. As a liquid material falls, its cross-sectional area becomes smaller. If the sprue does not conform to the falling metal, air spaces form on the sides and can become trapped in the gating system. The *choke* or bottom of the sprue should be approximately one-half the cross-sectional area of the top; ½ inch is typical for the diameter of the choke.

The *sprue base* (see Figure 14–2) serves to reduce the velocity and turbulence of the metal before entering the mold cavity and also helps trap damaged metal. The molten metal picks up velocity as it falls down the sprue; and, if allowed to continue into the mold cavity at that rate, it could cause cuts and washes or gas blows in the castings. When the metal hits the larger reservoir formed by the sprue base, its rate of flow is reduced before entering the runner or gate leading to the mold cavity.

The bottom of the sprue base should be convex rather than concave. This is to help reduce turbulence. A simple example of a commonly occurring incident can be used to illustrate the reason for the convex shape. If the water is turned on at a sink and a bowl is beneath the spigot, the water will splash out. If the bowl had a convex shape at the bottom, the tendency to splash would be substantially less than it is with the concave shape.

The sprue base should be five times the cross-sectional area of the base of the sprue. If the cross-sectional area of the sprue choke is 0.20, then the cross-sectional area of the sprue base should be 5 × 0.20, or 1.0 inch. The depth should be two times the depth of the runner.

Figure 14–3 illustrates some typical gating systems with runners and gates attached. When runners are used, they are generally placed in the bottom or *drag* side of the mold and the gates in the top or cope side of the mold. The runners are placed in the mold cavity so that the metal from the sprue base will rise and flow into the runner and eventually into the gates leading to the mold cavity. The runners should be wider than they are deep, and the cross-sectional area should be four times that of the sprue choke.

Runner extensions or sumps (see Figure 14–3) are placed beyond the last gate to help trap damaged metal. If the flask is sufficiently large enough, the runner extension should extend 6 inches beyond the last gate. If the size of the flask will not permit a 6-inch runner extension, then increase the depth of the extension to accommodate the same volume of metal. For example, if the runner extension can extend only 3 inches beyond the last gate, then increase the depth to twice that originally designed.

The cross-sectional area of the gates should be four times the cross-sectional area of the sprue choke. If two gates are required, then each gate should be two times the choke's cross-sectional area. Gates, like runners, should be wider than they are deep and should be wedge shaped, that is, shallower as they enter the mold cavity.

Risering

A riser is used to feed the casting as it begins to cool and contract. Risers are not always needed and should not be used to determine whether the mold is full. Quite often, the riser is placed in the mold in a location that actually promotes shrinks in the casting. Figure 14–1B illustrates the placement of a riser that is located beyond the mold cavity and is exposed to the air. The placement, shape, and the fact that it has been cut to the top of the mold, contribute to its inability to be effective.

The purpose of a riser is to feed thick sections of a casting that contract during solidification and leave depressions called shrinks. The riser should be the last metal to solidify so that as the casting begins to cool and harden it can draw metal from the riser. By placing the riser between the sprue and the mold cavity and not exposing it to the air, it will remain molten longer than the casting and feed the casting.

The shape of the riser should also be designed to promote solidification. The surface area/volume ratio should be lower than that of the casting. This can generally be accomplished by designing a riser that is shorter with a larger diameter. The riser height should not exceed 1½ times the diameter. Theoretically, the sphere would be the best shape for a riser because of its low surface area/volume ratio. However, because it is easier to form, the cylinder is commonly used for risers.

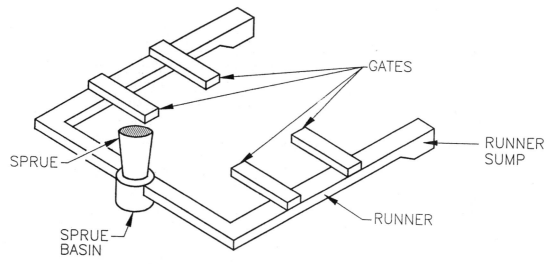

Figure 14–3 Properly designed gating systems

Cores and Coremaking

Cores are shapes that are used to form contours, primarily internal cavities, that are not molded through the use of the pattern. Cores for green sand casting are generally made from sand to which binders are added to give give them the strength needed to hold up under the pressure of the molten metal. Figure 14–4 illustrates the placement of a core in the mold cavity area to produce an internal cavity in the finished casting.

A number of processes have been developed to cure the core sand mixture to provide the strength needed for metal casting. Baked cores that use oil as a binder are quite popular in industry as well as education. Other processes include the following:

1. Shell core process
2. CO_2 sodium silicate process
3. Air set
4. Cold box
5. Hot box

Molding Operations

The process of molding the green sand around the pattern varies with the type of pattern and equipment used by the foundry. Larger foundries have automated molding machines (Figure 14–5) that have almost eliminated any hand work. The flasks are moved from one position to another by conveyors, sand is thrown into molds by sand slingers or overhead hoppers, and molds are rotated from drag to cope side by rollover machines.

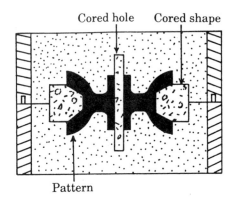

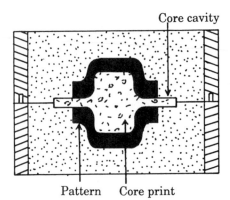

Figure 14–4   Core placed carefully in mold to produce internal cavity

Molding operations in smaller foundries and educational laboratories are accomplished through hand-molding operations or a combination of hand and machine operations. The jolt–squeeze machine is commonly used in smaller foundry operations as well as some school foundries. The purpose of the jolt–squeeze machine is to eliminate hand-ramming and thereby increase productivity. The name of this machine is derived from the action it performs in molding the green sand around the pattern.

The hand-ramming procedure is still commonly used in school settings and occasionally in smaller commercial foundries. The following procedure will vary slightly with the type of pattern being used in the molding operation.

Procedure—Hand-Molding Operations

1. Temper sand (add moisture) until a squeezed handful will support itself when held near one end.
2. Place pattern board on molding bench.

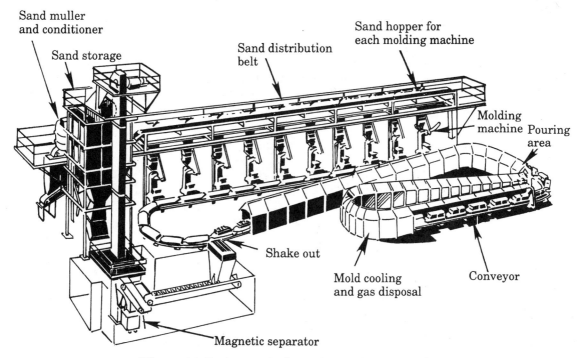

Figure 14–5 Automatic Green Sand Molding Machine

3. Position the pattern on the pattern board so as to allow sufficient space for gates and risers.
4. Position the half of the flask, with the pins down over the pattern and pattern board. Lightly dust the pattern with parting compound.
5. Riddle sand over the pattern until all surfaces of the pattern are covered with approximately 1 inch of sand.
6. Fill the remaining area of the flask half with heap sand. Using a hand ram, peen ram against the flask edges and butt ram over pattern area.
7. Using a straight edge or strike-off bar, remove excess sand from flask. Place a bottom board on top of the flask and roll the flask over. The pins should be facing up with a bottom or pattern board covering the top and bottom surfaces of the flask half.
8. Remove the bottoming board, exposing the pattern. Place the cope (top) portion of the flask on top of the drag (bottom) half and lightly dust the sand and pattern in the drag portion of the flask. *Note:* If the pattern is a split pattern, the cope half of the pattern would be placed in position on the drag half before parting compound was added.

9. Using a piece of chalk, mark for the location of the sprue by placing a chalk line on the side and end of the cope portion of the flask. Remember, the pouring basin should be close to the flask edge; therefore, the sprue should be located accordingly.

10. Riddle a layer of facing sand over the exposed sand and pattern. Fill the rest of the cope with heap sand and ram as described in Step 6.

11. Strike off excess sand and cut sprue where marked, making sure sprue hole is cut through the depth of cope portion of flask.

12. Cut pouring basin using slicks and molder's tools. Use design described in the section on gating for pouring basin.

13. Carefully lift cope half of flask from drag portion and set on edge or pouring basin surface.

14. Cut sprue base and runners in drag portion of flask. *Note:* Runners are not always needed.

15. With the cope on its edge or pouring basin surface, cut gates and risers as described in the section on gating.

16. Carefully pull pattern from the sand with draw spikes or screws. Tapping the pattern before drawing it from the sand will loosen it so that it draws more easily.

17. If necessary, place vents in cope side of mold using a thin wire.

18. Remove all loose sand from mold cavity, gates, runners, sprue, and sprue base. A hand bellows can be used for this purpose.

19. If cores are needed, place them in the mold at this point.

20. Carefully place the cope portion of the mold in position on the drag portion. Make sure the cope was not rotated during the molding process. *Caution:* Make sure the mold halves fit together tightly. Loose sand between mold halves can cause the molten metal to run out between the cope and drag.

21. Set the mold in pouring area.

Melting and Pouring

In all casting operations, the physical state of the material at the time it is cast is liquid. Many plastics are in a liquid state prior to pouring and become solid through the addition of a catalyst or by placing them in a heat source. Metals, ferrous as well as nonferrous, are solid and must be melted before being poured into the mold cavity. Table 14–2 provides melt and pouring temperatures for commonly used metals.

There is a variety of furnaces used for melting metals typically used in the casting process. The selection of melting equipment is generally based on a number of factors, including

- Temperature requirements
- Size of castings
- Cost

Crucible Furnaces

The crucible furnace is perhaps the most popular of melting equipment for educational laboratories and small foundries specializing in nonferrous castings. These furnaces are gas-

or oil-fired burners and are limited in size because the metal is contained in a crucible that must be lifted from the furnace before pouring.

Investment (Lost Wax) Casting

Accuracy and finished detail are among the advantages and reasons for using the investment casting process. Investment casting, often referred to as "lost wax" casting, is nearly 6,000 years old and was first used by the Chinese. This process has been modernized as the result of new materials and machinery and is used to produce a variety of products, including jewelry, art objects, and complex parts for the aerospace industry.

The pattern used in this process is made from wax and is disposable. The mold is formed by surrounding the wax pattern with a slurry of ceramic or plaster material, which is allowed to harden before pouring.

The first step in forming the mold is to apply a thin layer of the investment (ceramic or plaster) over the entire surface of the wax pattern. A commercial soap-like liquid material can be applied to the pattern first to reduce surface tension and help eliminate air bubbles that can cause noticeable defects in the finished castings.

After the pattern is coated with the investment and allowed to harden to form a thin shell, additional investment is poured around the pattern for support. After the mold has dried sufficiently, it is placed in an oven; and the wax is melted out of the mold, leaving the finished mold cavity. The gating system, including the sprue, was part of the wax pattern and is formed in the mold cavity (Figure 14–6).

The metal is then heated to temperature and poured into the mold cavity. A centrifugal casting machine is often used with the investment casting process. Centrifugal casting is described in the next section.

Metal	Melting Temperature (°F)	Pouring Temperature (°F)
Nonferrous		
Aluminum	1,220	1,275–1,350
Red Brass	1,700–1,925	2,050
Yellow Brass	1,650–1,725	1,960
Titanium	3,270	3,400–3,600
Lead	620	720
Magnesium	1,200	1,350–1,500
Zinc	1,445	900
Ferrous		
Wrought Iron	2,800	3,000
Steel	2,600	2,800–3,000
Grey Cast Iron	2,360	2,650
White Cast Iron	2,500	2,800
Malleable Iron	2,475	2,800

Table 14–2   Melting and pouring temperatures for ferrous and nonferrous metals

Centrifugal Casting

The centrifugal casting process is used to produce symmetrical castings that are sound, smooth, and clean. Pipe, cylinder liners, piston rings, brake drums, and gun barrels are a few of the products that are cast using this process. This process utilizes centrifugal forces to cause the molten metal to form against the walls of a permanent mold.

The centrifugal force that is produced by rotating the mold at high speeds provides several advantages over gravity-fed processes. The forces produced by rotating the mold on a vertical or horizontal axis increase the density of the metal, improve surface detail and smoothness, and cause impurities to move to the inside of the casting where they can be machined. Wall thickness of the casting is achieved through controlling the amount of metal poured into the rotating mold.

Centrifugal casting is generally more economical than other processes, although it does have certain limitations that other casting techniques do not. Limitations on size and shape are the primary disadvantages of centrifugal casting.

Rotational Casting

Rotational casting is a process used primarily with plastics to form round or cylindrical objects. Plastic in liquid, powder, or granular form is placed in the mold; and the mold is secured inside the rotational molding machine. The temperature inside the rotational molding machine is increased to allow the plastic to flow easily and to coat the inside of the mold uniformly as it is rotated.

The rotation is in more than one direction, and rotational speed is slow compared to centrifugal casting. Gravity rather than centrifugal force causes the uniform coating of the inside of the mold. While this process is generally more expensive than other gravity processes, it is well suited to the production of balls, other hollow plastic toys, and household items.

Slush and Slip Casting

"Slush" and "slip" are words used to describe casting processes that are used to form hollow shaped ceramic, metal, and plastic products. *Slush casting* is a process used in forming metal as well as plastic products that have hollow shaped areas in them. When forming metal products using this process, the molten metal is poured into a cold metal mold; and the metal around the perimeter is allowed to solidify. The metal in the center is poured out while still liquid to form the hollow interior.

Plastic products can be formed in a similar fashion, although there are some differences between the two processes. The metal mold used for slush casting of plastics is first heated for approximately 15 to 30 minutes before pouring the liquid plastic into it. The mold is then removed from the oven, and the liquid plastic is poured into the heated mold. The heat from the mold causes the liquid plastic to begin solidifying around the outside edge.

The wall thickness of the plastic can be controlled by temperature and time to a certain degree. Once the desired thickness is achieved, the remaining liquid plastic is removed and

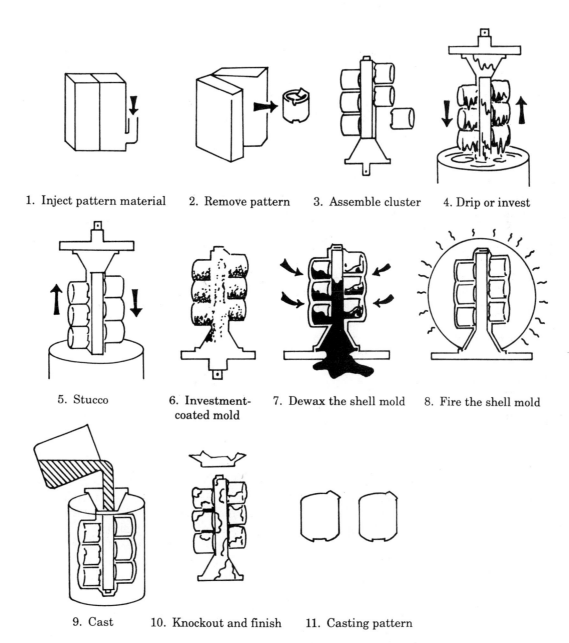

1. Inject pattern material   2. Remove pattern   3. Assemble cluster   4. Drip or invest

5. Stucco   6. Investment-coated mold   7. Dewax the shell mold   8. Fire the shell mold

9. Cast   10. Knockout and finish   11. Casting pattern

Figure 14–6   Preparing a mold for investment casting

the mold placed back in the oven to allow the plastic to cure. The mold is then cooled by running cold water over it, and the product is removed from the mold.

Slush casting for metals and plastics is applicable when the internal detail and exact thicknesses are not critical. Plastic shoes, door stops, and kicking tees are among the products made using this process.

Metal products made using this process include primarily ornamental objects and toys. Low-melting-point alloys are typically used for this process.

Slip casting, a process similar to slush casting, is used to form hollow shaped ceramic products. Ceramic and clay vessels and bathroom fixtures such as sinks and commodes are formed using the slip-casting process.

The solidification process takes place through hydration. The mold that is made of a clay or ceramic material absorbs moisture from the slip (liquid ceramic casting material). As the moisture is absorbed by the mold, the clay begins to form next to the mold surface. When desired thickness is reached, the remaining liquid slip is poured from the mold.

Foam Casting

Casting foams are used to produce a wide variety of plastic products, including flotation devices, duck decoys, and a variety of toys. Polyurethane foams and plastisols are among the most commonly used plastics for foam casting. The process requires the use of a two-piece mold that can be clamped tightly together, a resin of either polyurethane or polyvinyl chloride plastisol, and a suitable catalyst.

Dip Casting

Dip casting is a process used to form small plastic products out of polyvinyl chloride plastisol. This process is also used for coating hand tools such as pliers. Tools that require insulation because of their use in installing and repairing electrical devices are often coated using the dip-casting process.

The process used to form a product using this process is a relatively simple one and can be done with a minimum of equipment. A metal mold, sometimes referred to as a *plug* or *armature,* is first heated to between 350 and 400 degrees Fahrenheit. The mold is then dipped into the liquid plastisol. The plastisol surrounding the mold begins to solidify from the heat. Thickness is controlled by the length of time the mold remains in the plastisol.

Once removed from the plastisol, the mold is held over the container until excess plastic runs off. The plastisol-coated mold is then placed into an oven where it is allowed to cure for 5 to 10 minutes. The mold is then removed from the oven, and the plastic product is stripped from the mold.

The process for coating hand tools is done in a similar fashion. The only differences are that the tool is heated in place of a mold and that the plastic is not stripped after curing.

Full-Mold Casting

The name *full mold* is used to describe this process because the pattern remains in the mold when the casting process takes place. This is a process that has limited use because the pattern is expendable and because of the fumes produced as the pattern vaporizes.

The pattern and gating system are made of polystyrene plastic (styrofoam). Unbonded sand is used to surround the pattern and to support the shape of the pattern as the molten metal begins to vaporize the styrofoam.

The styrofoam pattern with gating system attached is carefully placed in the container, and loose sand is poured around the pattern and gating system. The top of the sprue is exposed to the surface, and the molten metal is poured directly on top of the styrofoam. The hot metal causes the styrofoam to vaporize instantly. In this way, the metal replaces the styrofoam and the product is formed.

Shell Molding

The mold cavity in shell-molding operations is formed through the bonding of a resin-coated sand. Sand with a thermoplastic resin coating is dumped or blown on a heated metal pattern. The heat from the highly polished metal pattern causes the thermoplastic resin to melt and bond the sand particles into a thin shell, thus forming the mold (Figure 14–7).

In general, each half of the mold is separate and then clamped together when pouring takes place. The two halves are supported in a flask-like device, and loose sand is dumped in between the shell mold and the flask to provide support to the thin shell. The gating system, including the sprue, is formed in the shell-molding process.

The principal advantages of shell molding are precision and surface smoothness. However, the process does require rather expensive equipment for making the bonded shell molds.

Vacuum-Molding Process

The vacuum-molding process is commonly referred to as the "V-process" and was developed for practical foundry operations by the Japanese in the early 1970s. The principle behind the process is that sand in an airtight environment will act as a solid and retain its molded shape as long as a vacuum is held in the chamber containing the sand.

This process requires some very special airtight flasks and pattern boards not found in a traditional foundry (Figure 14–8). A double-walled flask is used to form an air chamber for pulling the vacuum. Fine metal screens are placed on the inside wall of the flask to allow the air to be drawn from the sand without pulling sand into the chamber between the flask walls.

The surface of the sand is sealed with a thin plastic sheet to make the complete mold airtight. This prevents any air from being drawn into the mold as forming takes place. The following sequence is used in V-process molding operations:

1. A pattern with riser and sprue is placed on a hollow carrier plate. The pattern has very small vent holes in it to permit air to be drawn through it.
2. A thin plastic film (0.002 to 0.006 inch) is stretched over the pattern and heated until it softens.
3. The film is then allowed to drape over the pattern, and a vacuum (200 to 400 mm Hg) is applied through the hollow pattern carrier to draw the plastic tightly around the pattern.

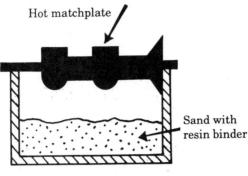

Hot matchplate

Sand with resin binder

A. A heated metal matchplate is placed over a box containing sand mixed with thermoplastic resin.

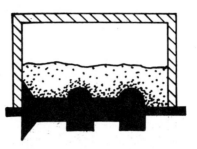

B. Box and matchplate are inverted for short time. Heat melts resin next to matchplate.

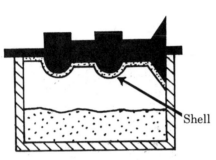

Shell

C. When box and matchplate are righted, a thin shell of resin bonded sand is retained on the matchplate.

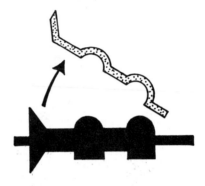

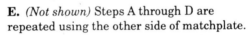

D. Shell is removed from the matchplate.

E. *(Not shown)* Steps A through D are repeated using the other side of matchplate.

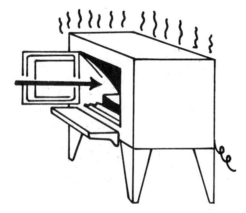

F. The shells are placed in oven and "heat treated" to thoroughly set resin bond.

G. *(Right)* Shells are clamped together and placed in a flask. Metal shot or coarse sand is packed around the shells, and mold is ready to receive molten metal.

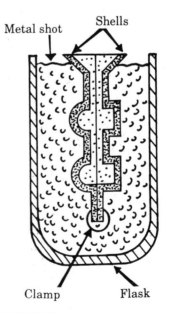

Metal shot

Shells

Clamp

Flask

Figure 14–7 Shell molding process

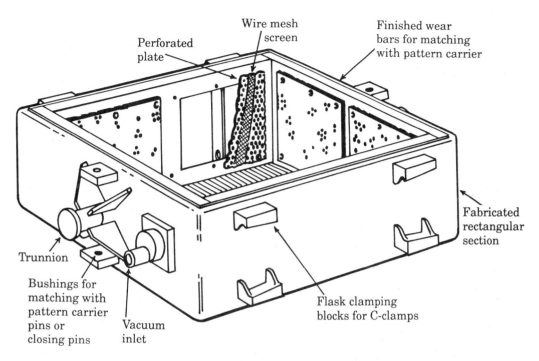

Figure 14–8 V-process flask

4. The special flask is placed over the plastic-coated pattern, and loose sand is dumped into the flask. The sand is vibrated slightly to compact it to maximum density.
5. The pouring basin is formed on top of the sprue, and the cope surface is covered with a thin plastic sheet.
6. Vacuum is applied to the cope part of the flask, and the atmospheric pressure causes the sand to harden around the pattern.
7. The drag is formed in a similar fashion, and the pattern is stripped from the mold. The cope and drag are assembled, and the molten metal is poured into the plastic-lined pouring basin.
8. When pouring is complete, the casting is permitted to solidify; and the vacuum is released. Once the vacuum is released, the loose sand falls freely from the casting; and the process begins again.

The advantages of this process include the following:

1. Improved environmental conditions
2. Smoother finish
3. Little or no shakeout
4. Reduced cleaning time of castings
5. Simplified sand control
6. Better dimensional accuracy

However, the cost associated with purchasing a license and the capital costs associated with equipping the foundry for this process must be considered carefully.

SUMMARY

Casting, one of the oldest methods of forming materials into a shape that has function, is still being used extensively today to produce products of metal, plastic, ceramic, and wood. Regardless of the material, casting requires three main systems: (1) a *pattern* to form the shape, (2) a *mold* to hold the material until it solidifies into the desired shape, and (3) a *gating system* to direct the material to the mold.

Materials for forming the mold are made from numerous materials, including sand, aluminum, steel, and ceramics. The type of material used for the mold depends on the type of material being poured or placed into it. For example, sand with additives is typically used when the material being formed is aluminum or iron. Aluminum molds are typically used for injection molding of plastics, and a ceramic mold is used to form the molds used to make jewelry with intricate detail.

Castings range in size from very delicate jewelry to huge bases for machines that weigh many tons. Although the process has been modernized in some applications, the basic principles of casting have not changed significantly since it was first used.

 REVIEW QUESTIONS _____

1. List ten items commonly found around the house or family garage that are made using a casting process.
2. List the three components of a casting system.
3. What casting process is used to form hollow shaped objects of metal and plastic?
4. What is the name of the process used to form lavatories and other ceramic bathroom fixtures?
5. What is meant by green sand?
6. List the five types of commonly used patterns for metal casting.
7. Why is taper placed on the pattern?
8. What is the basic difference between a natural and synthetic molding sand?
9. List several advantages of synthetic molding sands.
10. Define the following terms:

   Green strength
   Thermal stability
   Permeability

11. What are the principal objectives for a properly designed gating system?
12. Explain why a tapered sprue is preferred over one with straight sides.

13. What is the preferred ratio of sprue choke to gates and runners?
14. List the five types of cores used to develop internal cavities.
15. What advantage does investment casting offer?
16. What is the principal difference between centrifugal casting and rotational casting?
17. Describe the process used in dip casting.
18. What is the major disadvantage of full-mold casting?
19. What binder is used in shell molding?
20. List five advantages of V-process molding.

SUGGESTED ACTIVITIES

1. Bring several objects from home to class that have been manufactured using a casting process described in this chapter. Describe to the class the process used to form each of the objects.
2. Visit a local foundry and try to find out the following:

   a. type of sand used
   b. grain fineness of sand
   c. additives used in the sand
   d. moisture content of sand heap
   e. types of cores used
   f. types of patterns used

CHAPTER 15

Machining Theory and Principles

In today's manufacturing environment, maximum efficiency is critical to being competitive. Efficiency in a highly competitive global market means producing high-quality parts consistently, in the minimum time, and at the lowest cost. To meet these requirements, careful selection of cutting tools, feeds, speeds, cutting fluids, and other elements of machining is essential.

The performance of a cutting tool and the **machinability** (relative ease with which a material may be shaped or machined) are critical to achieving desired results. Equally important is the planned cutting path and setup time for each job. Every move becomes critical to manufacturing high-quality, low-cost parts. What was once an art is now a science.

MACHINING THEORY

Traditional machining is a manufacturing process that uses mechanical force provided by a power-driven device to remove material in chip form from standard and cast forms. The basis for traditional machining is that one material that is harder and stronger than another will cause the softer and weaker material to deform and fail. The conditions under which failure takes place are controlled to produce smooth accurate shapes and surfaces. These controlled conditions include

- Shape of cutting tool
- Relative motion between the tool and the work (cutting speed)
- Relative movement of cutting tool into a new position on the surface being machined (feed)
- Depth of cut
- Cutting tool material

Shape of Cutting Tools

The shape and relative position of the cutting tool greatly influences the surface finish of the part and wear of the tool. If too much of the tool's surface area is in contact with the part, chattering will occur and a rough surface will develop. If too little of the surface area is in contact with the part being machined, burning or breakage of the cutting tool can occur. Therefore, correct angles and shapes must be formed on the cutting tool for effective cutting to take place.

Cutting tools are classified as *single-point* or *multipoint.* Lathe tools for wood and metal are examples of single-point cutting tools (Figure 15–1). Although these tools vary in design, depending on their intended use, the basic principle of material removal with a single-point device is incorporated in each tool's unique shape. Other examples of single-point tools are boring bars and shaper bits.

Figure 15–2A illustrates the various angles incorporated in the design of a lathe bit used for metal cutting. A description of each of these angles is given to explain their purpose in producing the desired results. Figure 15–2B illustrates the common shapes of lathe bits used for shaping and finishing metal parts. Figures 15–3A and 15–3B show the recommended angles for a lathe tool used in shaping wood and commonly used lathe tools for woodworking.

Single-Point Tool Nomenclature

Relief angles (10 degrees). Small angles, both side and front that provide relief for the cutting edge. They prevent the tool surfaces from rubbing on the work.

Rake angles (12 degrees). Rake angles reduce the cutting force and direct the chip away from the material being machined. Rake angles are ground on the face (top surface) of the bit.

Side cutting—edge angle (10 to 20 degrees). The side cutting edge angle directs the flow of the chip away from the cutting area and directs the cutting force back into the stronger portion of the bit. It also controls the thickness of the chip.

End cutting—edge angle (5 to 32 degrees). Limits the tool contact area to promote cutting rather than scraping.

Nose radius. Radius ground at the point of the side and end cutting edge angles. The nose radius is used to control finish quality. Small radii are used for roughing cuts and larger ones for finish cuts.

Lathe tools used for turning wood perform one of three different actions. They are used to (1) shear, (2) scrape, or (3) shape. The diamond point is a lathe tool that can be used to produce these three effects. The angles (see Figure 15–3A) ground on the diamond point produce much of the same effect as those described earlier with regard to metal-cutting lathe bits. The angle provides relief to the cutting edge so that smooth efficient cutting will take place. Figure 15–3B illustrates the shapes and gives the names of other wood-turning tools.

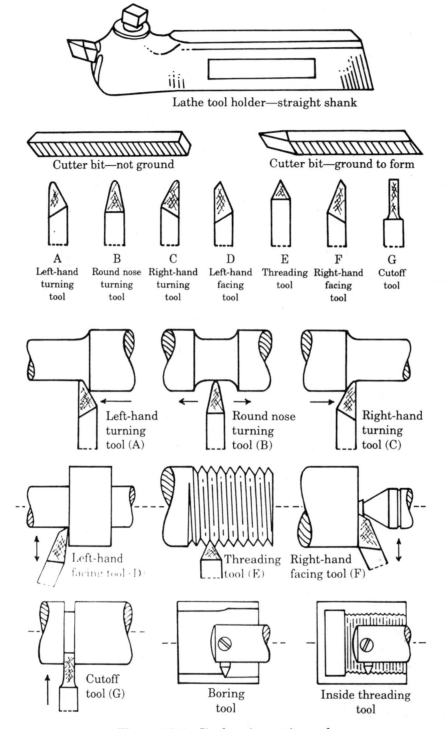

Lathe tool holder—straight shank

Cutter bit—not ground

Cutter bit—ground to form

A
Left-hand
turning
tool

B
Round nose
turning
tool

C
Right-hand
turning
tool

D
Left-hand
facing
tool

E
Threading
tool

F
Right-hand
facing
tool

G
Cutoff
tool

Left-hand
turning
tool (A)

Round nose
turning
tool (B)

Right-hand
turning
tool (C)

Left-hand
facing tool (D)

Threading
tool (E)

Right-hand
facing tool (F)

Cutoff
tool (G)

Boring
tool

Inside threading
tool

Figure 15–1 Single-point cutting tools

290

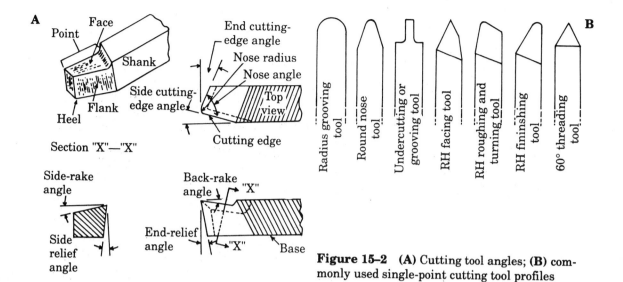

Figure 15–2 **(A)** Cutting tool angles; **(B)** commonly used single-point cutting tool profiles

Multipoint Cutting Tools

Multipoint cutting tools are devices typically used on equipment such as milling machines, saws, wood planers, and jointers. The advantage of multipointed cutters is that faster cutting takes place than with single-point devices. Each individual tooth or point on a multipoint device has essentially the same angles as those of the single-point device. Figure 15–4 shows the various angles ground on a milling cutter for efficient cutting.

The multipoint cutter illustrated in Figure 15–4 has rake angles and relief or clearance angles that provide for controlled failure of the part being shaped. The relief or clearance directly behind the cutting edge prevents drag or friction from occurring.

The materials used for single-point and multipoint cutters must be harder and stronger than the material being shaped. A wide variety of materials have been developed to meet the demands of present-day production requirements. The selection of the cutting tool material is made after careful consideration of many factors related to efficient machining. These factors include but are not limited to

Workpiece composition   Speed and feed rates   Tool life   Initial tool cost

For example, conventional carbides and coated carbides are most economical at speeds under 1,000 surface feet per minute (sfpm); but ceramics have been shown to be more economical at higher speeds (1,000–7,000 sfpm). The initial cost of tooling should be considered when selecting the cutting tool material. In many cases, however, the extended life of a more expensive material will make it more cost-efficient over time.

Many developments have taken place in cutting tool technology recently, and some of the more traditional materials like carbon and high-speed steel are being replaced by carbide and ceramics. The following are brief descriptions of the traditional and newer materials used for single- and multi-toothed cutting tools.

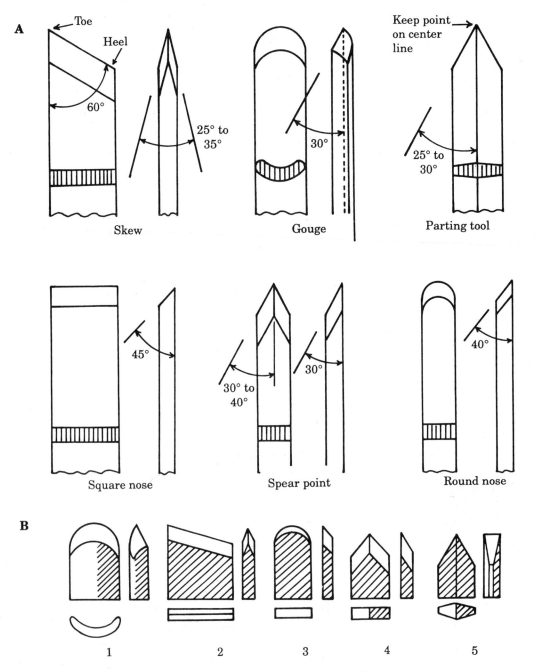

A

Toe

Heel

60°

25° to 35°

Skew

30°

Gouge

Keep point on center line

25° to 30°

Parting tool

45°

Square nose

30° to 40°

30°

Spear point

40°

Round nose

B

1   2   3   4   5

Figure 15–3 **(A)** Cutting angles for woodworking lathe tools; **(B)** common shapes of woodworking lathe tools

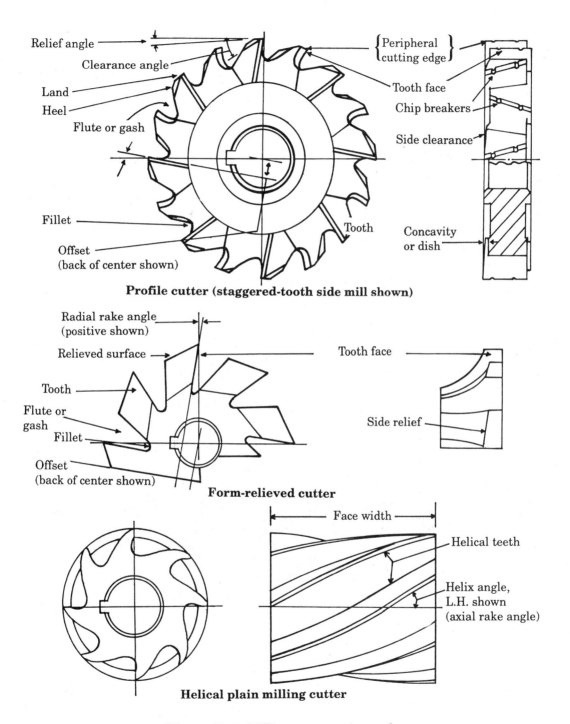

Profile cutter (staggered-tooth side mill shown)

Relief angle

Clearance angle

Land

Heel

Flute or gash

Fillet

Offset
(back of center shown)

Tooth

{Peripheral cutting edge}

Tooth face

Chip breakers

Side clearance

Concavity or dish

Form-relieved cutter

Radial rake angle
(positive shown)

Relieved surface

Tooth

Flute or gash

Fillet

Offset
(back of center shown)

Tooth face

Side relief

Helical plain milling cutter

Face width

Helical teeth

Helix angle,
L.H. shown
(axial rake angle)

Figure 15–4   Milling cutter cutting angles

TOOLING MATERIALS

High-Carbon Steels

The earliest tools were made of steel with high carbon content (0.80–1.20%) and small percentages of manganese and silicon. The depth-hardening ability is poor in this material; and while good hardness can be achieved on the surface, it does not have the wear and heat resistance that the newer materials have. The carbon steels lose hardness at a relatively low temperature (600 degrees Fahrenheit) and are not suitable for high-speed machining. They are still used for cutting woods.

High-Speed Steels

The high-speed steels were made possible through metallurgical advancements in alloying. High-speed steels contain large quantities of carbide-forming alloys such as chromium, tungsten, vanadium, molybdenum, and cobalt. They have excellent hardenability and resist softening at high temperatures (1,200 degrees Fahrenheit).

Carbides

The use of carbide-tipped tools was made possible as a result of developments in powder metallurgy (see Chapter 13). The powders of tungsten carbide and cobalt are pressed together and sintered to form an extremely hard cutting tool that will retain its edge at temperatures exceeding 2,200 degrees Fahrenheit. Titanium and tantalum are added to the tungsten and cobalt base when cutting steel to prevent the steel from fusing itself to the edge of the carbide.

A process known as **physical vapor deposition (PVD)** has helped introduce what are known as coated carbides to the cutting tool industry. Thin coatings of uniformly applied titanium nitride (TIN), titanium carbide (TIC), or aluminum oxide (Al_2O_3) are applied to a tungsten carbide base. The coatings range from 0.0002 to 0.0003 inch in thickness.

The coated carbides are extremely hard (Rockwell Hardness 80 C Scale [R_c80]) and have low coefficients of friction. The coated tools last longer, and chip formation is greatly improved. Lower tool consumption, higher machine feeds and speeds, and better finishes are among the advantages of coated carbides. The disadvantage is that the cutting tool must be sent out for recoating or simply discarded if chipped.

Ceramics

Perhaps the latest development in cutting tool technology has been in the introduction of ceramics to the machining industry. Experts predict that ceramic cutting tools will be a $160-million industry by 2000 and will save the metal-cutting industries in excess of $500 million.

The term *ceramic* has been used in industry to describe alumina-based compositions such as alumina titanium carbide and alumina zirconia. New compositions of silicon nitride under such trade names as Noralide and Quantum 6 are being used on cast iron at speeds as high as 5,000 sfpm while providing exceptionally long tool life.

Diamond

Diamonds have been used in industry for cutting ceramics, graphite, and reinforced plastics for a number of years. They are extremely hard and brittle, which limits their use to a great extent. They must be rigidly supported because of their brittleness and hardness.

CHIP FORMATION

As stated earlier, the force exerted on the workpiece during cutting causes the part to fracture. The result of the fracture or material failure is the removal of material in what is referred to as chips. The shape of the chip is a result of many factors:

Type of material being machined	Shape of the cutting tool
Depth of cut	Feeds and speeds

Chips have been classified as *continuous, discontinuous,* and *continuous with a built-up edge.* The continuous chip is generally the most desirable and is formed on materials that are higher in ductility. The continuous chip is formed in front of the cutting edge in a ribbon-like pattern and slides off the face of the tool without being fractured. The surface finish produced is generally good, but some danger results if the continuous chip is allowed to freely wind around the cutting area. Chip breakers (Figure 15–5) are generally ground into the cutting tool to eliminate this danger.

Discontinuous chips are short segmented chips formed when cutting harder, more brittle material such as cast iron, ceramics, and brass. Rather than forming long, continuous ribbon-like chips, the material ahead of the cutting edge is fractured in short small pieces. This formation is a result of generally low plasticity and high resistance to compressive forces of the harder, more brittle materials.

Continuous chips with a built-up edge generally occur with highly ductile, soft materials such as aluminum. Aluminum is highly plastic and deforms easily. The heat generated

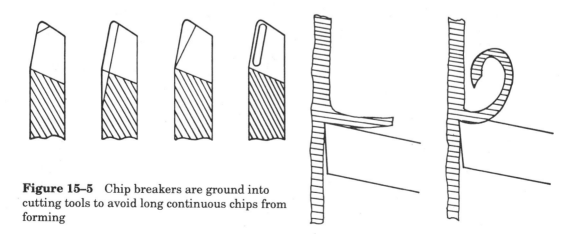

Figure 15–5   Chip breakers are ground into cutting tools to avoid long continuous chips from forming

by the friction between tool and piece can cause pieces of the chip to soften and fuse to the edge and top face of the cutter bit. This buildup of material on the edge will eventually cause a rubbing rather than shearing action to take place. Increased friction and a rough surface will result from these conditions.

The form of the chip can also be a result of the type of cutting action taking place. In woodworking, positioning of the tool in a manner that causes shearing to occur will produce a continuous chip. A scraping action produced by the same tool will cause short segmented chips to be produced.

Machinability

Machinability is a term used to describe the relative ease with which any material may be shaped or machined. Many variables affect the machinability rating of a material, primarily, variables related to the physical properties of the material. Hardness, tensile, and compressive strength are among those physical properties that greatly influence the relative ease of machining.

Machinability ratings for metals are based on AISI-B1112 carbon steel, which has been assigned a value of 100. Tables 15–1 and 15–2 give machinability ratings for some of the more common metals and woods. While machinability of metals is designated using Arabic numbers, the machinability of woods is identified by letters.

While most plastics are formed using other methods, some of the more rigid plastics can be machined using traditional methods (e.g., drilling and turning). Nylon, Acetal, and other rigid forms of plastic can be machined to form with careful control of speed, feed, and depth of cut. Light cuts and reduced speeds are generally used when cutting plastics to avoid overheating due to friction between the cutting tool and part.

Plastics are generally machined without a cutting fluid, and the maximum cutting speed is often determined by testing the individual material. Care must be taken when machining plastics because it is a poor conductor of heat. High speeds can cause the plastic to melt rather than cut. Cutting speeds vary from 250 to 1,500 feet per minute (fpm) for turning

Specification	Rating
Carbon Steel	
C1110	85
C1115	85
C1118	80
C1132	75
C1016	70
B1112	100
Stainless Steel	50–70
Cast Iron	70–110
Brass	150–600
Aluminum	95–600

Table 15–1 Machinability ratings of metal

Specification	Rating*
Ash	G
Birch	A
Cherry	A
Ebony	A
Elm	P
Hickory	G
Mahogany	A
Maple	G
Oak	G
Pine, White	P
Redwood	A
Spruce	A

*Ratings A, G, and P designate (A)verage, (G)ood, and (P)oor.

Table 15-2 Machinability ratings of common woods

and boring operations. The wide range of speeds is based on the type of plastics and the composition of the cutting tool. A range of 250 to 500 fpm is usually recommended for high-speed steel tools and 500 to 1,500 fpm for carbide tools. Acetal, nylon, acrylonitrile butadiene styrene (ABS), and high-density polyethylene are among the plastics that have excellent machinability ratings.

Cutting Speeds

The cutting speed (CS) is the rate at which the work passes the cutting tool. Cutting speed is expressed in surface feet per minute (sfpm) and can be determined using the following formulas:

$$CS = \frac{RPM \times D \times \pi}{12}$$

where

$$RPM = \text{revolutions per minute}$$

$$D = \text{diameter of the workpiece}$$

$$\pi = 3.1416$$

To determine RPM when the desired cutting speed is known, one of the following formulas can be used:

$$RPM = \frac{12 \times CS}{\pi \times D}$$

$$RPM = \frac{3.82 \times CS}{D}$$

$$RPM = \frac{CS \times 4}{D}$$

The cutting speeds for various materials can be found in many textbooks or handbooks. Table 15–3 lists cutting speeds for some of the commonly used materials. Correct cutting speeds will promote efficient cutting and extend the life of the cutting tool. The suggested cutting speeds are based on the material's physical properties, cutting tool material, depth of cut, and especially hardness.

Material	Roughing Cut	Finishing Cut
Low-Carbon Steel	95–120	225–300
Medium-Carbon Steel	70–120	200–275
High-Carbon Steel	50–70	150–225
Free Machining Steel	80–150	250–350
Gray Cast Iron	75–90	120–150
Aluminum	100–150	225–350
Brass	150–225	275–350
Plastics	100–200	300–500

Table 15–3 Suggested cutting speeds for turning operations and boring operations

Note: Values are for high-speed steel cutters.

Feeds

Feed is a term used in machining processes to describe the introduction of new material per cycle of operation to the cutting point. *Feed rate* is distance of feed motion per cycle of operation. The unit used to define the *cycle of operation* is a function of the motion used in the machining operation.

The cycle of operation in turning a spindle on a lathe is a revolution, and the feed rate is expressed in inches per revolution (IPR). If the appropriate feed rate was determined to be 0.010 IPR, then the cutting point would advance $1/100$ inch along the longitudinal axis of the workpiece per revolution of that workpiece (Figure 15–6). Feed can also be expressed in inches per stroke (metal shapers) or inches per minute (band saw).

For machining operations that use multi-toothed cutters, the cycle of operation is defined by the distance each tooth or cutting edge moves into the material. For example, *feed per tooth* is the appropriate cycle of operation for milling cutters. The following formula and example are illustrative of the calculation used to determine correct feed for a multi-toothed milling cutter:

$$IPM = F \times N \times RPM$$

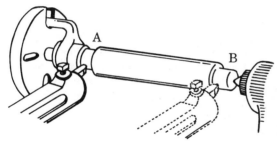

Figure 15–6   Feeding cutting tool along longitudinal axis

where

$$IPM = \text{inches per minute}$$

$$F = \text{feed rate}$$

$$N = \text{number of teeth}$$

$$RPM = \text{revolutions per minute}$$

Calculate the feed rate for a 4-inch plain milling cutter with twenty-four teeth cutting aluminum. The procedure is as follows:

Calculate RPM

$$RPM = \frac{CS \times 4}{D} = \frac{150 \times 4}{4} = 150$$

where

$$CS = \text{cutting speed (see Table 15–3, Aluminum)}$$
$$D = \text{diameter of cutter}$$

Calculate Feed

$$IPM = F \times N \times RPM = 0.007 \times 24 \times 150 = 25.2$$

In this example, the feed rate would be set at 25.2 inches per minute or as close to that value as the machine feed mechanism allows. It is important to remember that the formulas for speed and feed are used to determine acceptable rates. These should be experimented with to determine optimum rates for efficient cutting. The type of material being machined, the cutting tool material, rigidity of machine, and workpiece are a few of the factors that must be considered for optimum feed and speed.

Basic Machine Operation

The basic machining processes described in this chapter can be classified according to the relative motion between the cutting tool and the workpiece. For example, in turning operations, the workpiece turns against a rigid cutting tool; while in drilling, the workpiece is rigid and the drill bit rotates. In some processes, both the cutter tool and the workpiece rotate. Figure 15–7 illustrates relative motions for the basic machining processes.

The selection of the most appropriate machine tool is based on the required shape and surface finish on the finished part. In many cases, the specifications of shape and finish will require more than one machine tool. For example, a flat surface may require a milling operation for efficient economical material removal followed by a surface-grinding operation to meet surface-finish specifications. Table 15–4 compares the relative motion between tool and workpiece for several traditional machine tools.

SUMMARY

Machining, once an art, is now a science requiring careful study and planning. The worldwide competition in today's manufacturing environment requires companies to take every step necessary to ensure maximum efficiency. In machining, careful selection of cutting tools, speeds, and cutting fluids, is essential to being profitable. World-class quality and minimum cycle time means careful planning of cutting paths and reduced setup time with the fewest number of tool changes. The engineer, the technologist, and the machine operator must consider all factors in planning a job, including the following:

- Cutting tool material
- Shape of cutting tool
- Tool wear
- Machinability of material
- Speeds and feeds
- Rigidity of machine
- Rigidity of workpiece

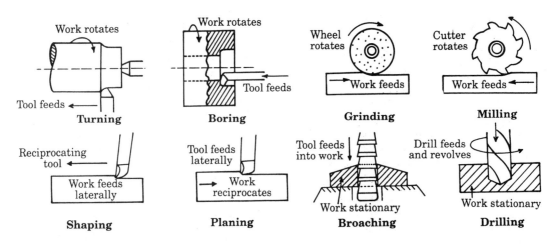

Figure 15–7   Relative motion between work and tool for conventional machining operations

Machine Process	Tool Movement	Feed Movement	Operations
Turning	Stationary	Longitudinal and Traverse	External and Internal Cylindrical
Drilling	Rotates	Vertical	Internal Cylindrical
Milling	Rotates	Traverse and Longitudinal	Flat
Shaper (Metal)	Traverse	Traverse	Flat and Special
(Wood)	Rotates	Traverse	Contoured and Special
Planer (Metal)	Stationary	Traverse	Flat and Special
(Wood)	Rotates	Traverse	Flat
Grinder	Rotates	Rotates or Traverse	Flat, Cylindrical, Special

Table 15–4   Relative motions between tool and workpiece

These factors and others must be considered in determining a starting point for machining operations. Optimum efficiency is achieved from calculating a beginning point and making fine adjustments to maximize tool life, cycle time, and workpiece quality.

 REVIEW QUESTIONS

1. List the conditions that must be controlled to ensure efficient cutting action.
2. What three cutting actions are performed by the different wood-turning lathe tools?
3. Why aren't low-carbon steels suitable for high-speed cutting?
4. List and describe the three types of chips that form when cutting takes place.
5. Calculate the RPM for a finish cut on a piece of aluminum that is 1½ inches in diameter.
6. Describe the relative motion between the tool and workpiece for the following machining operations: drilling, turning, milling, shaping (metal), and shaping (wood).
7. What is machinability?
8. List two examples each of *single-point* and *multipoint* cutting tools.
9. Explain the purpose of the following cutting tool angles and radius:

   relief angle
   rake angle
   side cutting—edge angle
   end cutting—edge angle
   nose radius

10. What three functions do wood-cutting lathe tools perform?

11. List four factors to consider when selecting a cutting tool.
12. What is the disadvantage of using carbon steel–cutting tools?
13. Define *cutting speed.*
14. Define *feed rate.*

SUGGESTED ACTIVITIES

1. Develop a chart of most efficient feeds, speeds, and depths of cut for the lathes and most commonly used materials in your facility.
2. Write a two- to four-page report on a new machining technique that was not discussed in this chapter. Present this information to the rest of the class.
3. Write to manufacturers of cutting tools for metal-cutting and wood-cutting equipment. Ask for samples and literature of new cutting tool materials.

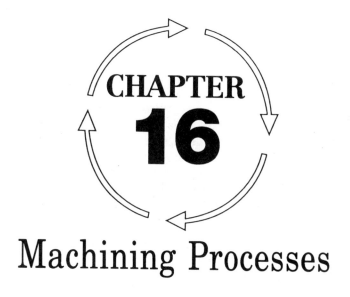

CHAPTER 16

Machining Processes

In Chapter 13, the methods used to transform primary materials to finished products were accomplished without changing the volume. Material was neither added nor removed to form the desired shape. The state of the material was changed during various stages of the process, but the volume remained essentially the same.

In Chapters 16 and 18, the processes described are subtractive or additive; that is, the desired shape of the finished product is achieved by either removing material, as in machining, or adding material, as in welding. The machining processes used in industry to shape various materials are the focus of this chapter.

TURNING

Turning-machine operations are generally selected when the workpiece is cylindrical and when material removal over the circumference of the part is required to produce the desired shape. The workpiece is rotated by the machine while a single-point cutting tool traverses either longitudinally or radially. The equipment used in turning operations must perform the following functions:

1. Rotate the workpiece
2. Secure the workpiece
3. Secure and feed the cutting tool

Lathes

The machines used to perform turning operations are called lathes. They are, perhaps, the oldest of the machine tool family. The concept of rotating a workpiece against a material that will cut it to a desired form has been used for centuries. The modern metal-cutting lathe is the product of several developments that took place during the Industrial Revolution.

Techniques for refining and using steel and iron were developed during the Industrial Revolution; and better more rigid machine tools were needed to shape these harder, stronger materials. Henry Maudslay, a British machinist, is usually given credit for inventing the industrial-type lathe in 1797. In reality, Maudslay refined and improved earlier designs that had been used for instrument making and ornamental work. Jacques Besson, engineer at the court of King Charles IX of France, is credited with inventing the first screw-cutting lathe in 1740.

The historical importance of the metal-cutting lathe cannot be overstated. The lathe is an extremely versatile machine that can be used to perform a wide range of machine operations such as drilling, boring, reaming, milling, threading, and grinding. The early metal-cutting lathe was used to build many of the other machine tools developed during the Industrial Revolution.

Wood-Turning Lathe

The use of a lathe-type device for shaping wood preceded the metal lathe by centuries. This was primarily due to the fact that wood was the primary material of choice for manufacturing for centuries before the Industrial Revolution and the change to metals. Although refinements to the earlier lathes have been made and they have been automated for production purposes, the tools and skill needed to operate them are basically unchanged.

Figure 16–1 shows a basic wood-turning lathe. Lathes are manufactured in various sizes. The size designation is determined by the maximum diameter that can be turned on the inboard side of the headstock and also by the maximum length that can be turned between centers. Typical sizes of lathes used in smaller production shops and school laboratories are listed below.

Swing over Bed	Distance between Centers
10 inches	36 inches
12 inches	38 inches
15 inches	48 inches

Larger production-type equipment with accessories for repetitive manufacturing is used in larger furniture factories. Recently, computer numerical control (CNC) has been added to the wood lathe to automate the production process further and ensure repeatability.

Lathe Tools for the Wood Lathe   There are several lathe tools available for making roughing, finishing, and shaping cuts to the workpiece. A set of turning chisels consists of a gouge, skew, parting tool, round nose, and diamond point. The cutting action used in forming the workpiece is either that of shearing or scraping. The type of cutting action employed depends on the tool being used and the surface finish desired. For example, Figure 16–2 illustrates the two cutting actions using a chisel. Notice that the chisel is rotated slightly when shearing takes place. This reduces the surface area contact between the cutting edge and the part and improves cutting efficiency.

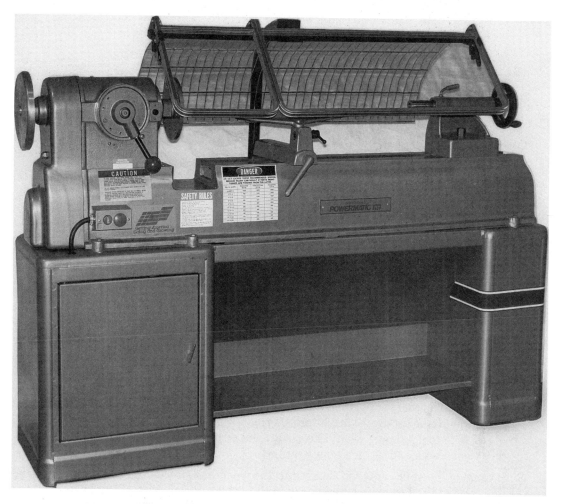

Figure 16-1 Wood lathe *(Courtesy of Powermatic)*

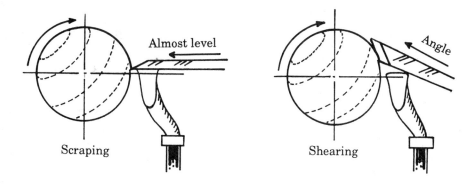

Figure 16-2 A chisel is used to scrape or shear

The five basic wood-turning chisels are used in the following manner:

Gouge: Round nose, hollow turning chisel used for rough cutting. Gouges come in several sizes, and a set of turning chisels generally has more than one.

Parting tool: The parting tool is used to establish required diameters and make recess cuts with straight sides.

Round nose: Using a scraping action, the round nose is used for finishing cuts on flat surfaces and for making contoured cuts such as coves.

Skew: The skew is used as a scraping tool to smooth straight cylindrical surfaces and to form V shapes and beads.

Diamond point: The diamond point can be used to square shoulders, cut bevels, and other shapes that conform to its shape.

Figure 16–3 illustrates the basic shapes and required sharpening angles for the five basic turning chisels previously described.

Accessories   Various accessories can be purchased for the wood lathe for faceplate as well as between-centers turning. The more commonly used accessories are illustrated in Figure 16–4 with explanations of their uses here:

Tool support: The tool support is used to provide a rigid rest for the turning chisel. The tool supports are either straight or at a right angle and are sold in various standard sizes. The tool support should be positioned approximately $1/8$ inch from the rotating workpieces and $1/8$ inch above center.

Spur center: The spur center is secured in the wood and placed in the headstock (driving) end of the lathe. The headstock is the power-driven part of the lathe, and the spur center causes the workpiece to rotate between the headstock and tailstock.

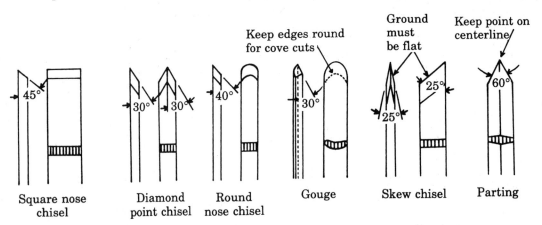

Figure 16–3   Front and side views of six basic woodworking lathe tools

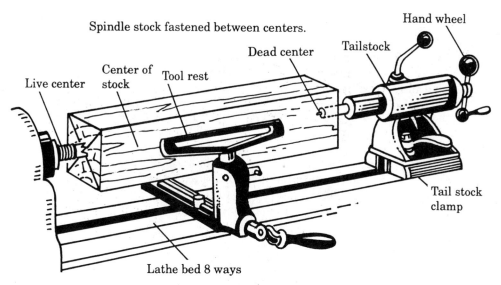

Spindle stock fastened between centers.

Hand wheel

Dead center

Tailstock

Center of stock

Tool rest

Live center

Tail stock clamp

Lathe bed 8 ways

Figure 16–4   Woodworking lathe accessories

Cup or dead center: The cup or dead center supports the workpiece at the tailstock end of the lathe. The cup or dead center does not rotate with the workpiece.

Faceplates: Faceplates come in various diameters and are used for faceplate turning (work supported at headstock only). Methods for attaching work to the faceplate are explained later in this chapter.

Turning between Centers

Longer cylindrical parts such as table legs, baseball bats, and candle sticks are shaped on the lathe while they are supported between the live and dead centers. In this process, square stock is roughed to approximate diameter with a gouge and then turned to shape and size using the other lathe tools described (see Figure 16–3). The use of these chisels by the operator is often restricted to single or small quantities of parts or to the fabrication of patterns. Cylindrical wood parts for mass production are usually made using automatic lathes that are controlled by computers or mechanically using a pattern and follower device.

Faceplate Turning   Faceplate turning on the lathe can be done on the *inboard* or *outboard* side of the headstock. The inboard side is used for smaller diameters such as bowls and coasters. The size of object that can be turned on the inboard side depends on the distance from the center of the headstock spindle to the lathe bed.

Outboard turning is used for turning larger objects such as table tops. The workpiece is mounted on the left-hand side of the headstock where there is no interference or obstructions. A tool rest stand is used in outboard turning operations to support the turning chisel. As with turning between centers, faceplate turning is often automated for mass-production operations.

Metal-Turning Lathe

The metal-cutting lathe, like the wood-turning lathe, has three essential components. It has a headstock and tailstock for holding the workpiece, a power mechanism for rotating the workpiece, and a mechanism for holding and feeding the tool bit. Figure 16–5 shows the basic parts of the lathe and the functions each performs in turning operations.

Metal-cutting lathes are manufactured in a variety of sizes and can be purchased with options that include computer control. The costs range from approximately $40,000 to hundreds of thousands of dollars, depending on size and accessories.

The size of lathe is determined by the maximum diameter that can be machined, referred to as the swing of the lathe, and the length of bed or maximum length of workpiece that can be turned. For example, a 16-inch lathe with a 4-foot bed could turn a diameter slightly less than 16 inches and approximately 40 inches long. The 40-inch length represents the distance between centers.

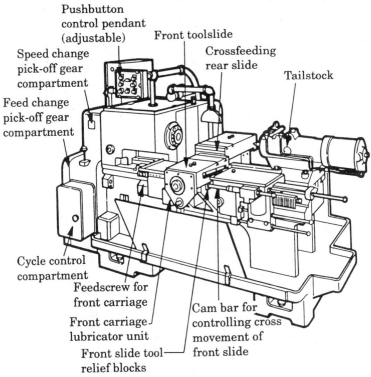

Figure 16–5 Component parts of engine lathe *(From Goetsch, Modern Manufacturing Processes, Copyright 1991 by Delmar Publishers, Inc. Used with permission)*

Lathe Accessories In addition to the lathe tool–cutter bits (see Figure 15–1), there are several accessories for the lathe that are available for various between-center and faceplate-turning operations:

Headstock chucks: Three and four jaw chucks are used to secure work at the headstock end of the lathe. Shorter pieces can be turned without being supported at the tailstock end of the lathe. The universal three-jaw chuck will automatically center the workpiece as the three jaws close at the same time (same principle as a drill chuck). However, the jaws of a four-jaw chuck move independently of one another and require the use of a dial indicator for centering the workpiece. The four-jaw chuck can also be used to turn parts that are not symmetrical.

Faceplates: Faceplates are also attached to the headstock of the lathe. However, the workpiece is attached to the faceplate with bolts or hold-down straps. Flat, cylindrical, and irregular objects can be held on the faceplate and machined this way.

Draw bar and collet set: This consists of a rather long bar with a hand wheel at one end and an internal thread at the other end with a set of collets that will accommodate all standard bar stock diameters. The collets are split, which will allow the stock to be placed in the collet initially. As the hand wheel is turned clockwise, the collet is pulled into a tapered sleeve, causing the collet to squeeze on the workpiece and hold it securely.

Lathe dog: The lathe dog is used in conjunction with the faceplate for between-center turning. The purpose of the lathe dog is to prevent the part from slipping between centers when the cutting point is turned into the rotating work. Lathe dogs are sold in various sizes to accommodate different diameter stock.

Centers: There is a variety of live and dead centers available for the lathe. These are used for turning between centers and are tapered to fit the tailstock and headstock spindles. The headstock spindle rotates as it is driven by the belts or gears in the headstock portion of the lathe. The tailstock spindle remains stationary, but a live center (ball-bearing center) can be used to reduce friction. Bushings are generally required in the headstock spindle to allow the smaller diameter centers to be used.

Tool holders: Various types and sizes of tool holders are available for turning operations. Special tool holders are designed for cutting, parting, boring, knurling, threading, and other operations.

Turning between Centers

A commonly used turning method is between-center machining. The workpiece is supported between the headstock and tailstock using the accessories described earlier. Speeds and feeds are calculated and set for efficient cutting. Parts machined in this fashion are cut from standard sizes of round bar stock. Roughing cuts are made first to bring the part to approximate size and shape. Finished shapes and sizes are accomplished by using different tool geometry and removing less material per pass of the lathe tool. Lathes for cutting metal and the metal machining process are much more precise than those used for wood, since the machine controls are designed for greater precision. Metal lathes are often computer controlled for greater repeatability and efficiency.

Facing

Flat surfaces can be machined using the lathe by attaching the workpiece to a faceplate or placing it in a chuck secured to the headstock of the lathe. This operation is known as facing and uses a facing tool rather than a turning tool. Parts that require facing operations are attached to a holding device (faceplate, chuck, collet) at the headstock end of the lathe. The cutting tool is fed in a radial direction to achieve desired length, shape, and finish. Eccentric parts can be machined on the lathe by offsetting the stock on a faceplate or four-jaw chuck.

Taper Turning

Many parts turned on the lathe require that a taper be machined on the workpiece. Taper turning on a wood lathe is accomplished by removing more material at one end than at the other, and the amount of taper is often controlled by the hands of the lathe operator or tracer attachment. Cutting tapers on the metal lathe often requires the use of a taper attachment or the offsetting of the tailstock from the headstock.

Tapers can be machined on the metal-cutting lathe in one of three ways:

1. Short tapers can be machined using the compound rest and three- or four-jaw chuck.
2. Longer tapers can be machined between centers by offsetting the tailstock. The amount of taper per foot or per inch must be known in order to offset the tailstock the correct distance. The following formulas are used for calculating tailstock offset:

$$\text{Offset} = \frac{\text{TPF} \times \text{L}}{2} \quad \text{Taper per foot and length given}$$

$$\text{Offset} = \frac{\text{TPI} \times \text{L}}{24} \quad \text{Taper per inch and length given}$$

The disadvantage to the tailstock-offset method of taper turning is that internal tapers cannot be cut.

3. Internal as well as external tapers can be machined using the taper attachment. The taper attachment is an accessory to the lathe that can be attached to the carriage, bed, and ways and can be engaged when taper turning is required. The attachment has its own slide or ways and a scale for setting a desired taper in degrees or inches per foot.

The metal-cutting lathe is an extremely versatile machine that can perform numerous other operations, including drilling, boring, knurling, and even milling operations. Some of these will be included in other sections of this chapter. A machine tool technology text outlines each of the procedures used on the lathe in detail.

SHAPING MATERIALS

Metal-Shaping and -Planing Operations

Metal planers and shapers are primarily used for machining flat surfaces. The shaper, however, is used for smaller work, while the planer can accommodate extremely large work surfaces. The cutting tools for both the planer and shaper are identical in shape but vary in size.

The basic difference in principle between the two is related to the relative motion of the cutting bit and workpiece. The shaper uses a reciprocating motion of the cutter bit as the workpiece, attached to a table, traverses across in front of the reciprocating motion of the shaper ram. The shaper, once an important piece of equipment in any machine shop, has all but dissapeared from modern machining facilities.

The planer employs a reciprocating motion of the workpiece against a stationary tool. The cutting tool is fed in a transverse direction as the workpiece reciprocates back and forth in order to introduce new material to the cutting edge.

Wood Planers and Planing Operations

The wood surface planer (Figure 16–6), like the metal planer, is used for machining flat surfaces. However, there are more differences than there are similarities between the two types of machines. The wood surfacer has a series of knife-like cutters attached to a rotating cylinder that shear layers of wood chips from the workpiece as it is fed through the machine.

The planer is manufactured in sizes ranging from 12 to 52 inches wide and is capable of handling stock as thick as 12 inches. The procedure for surfacing stock on a planer is a relatively simple one, but safety precautions must be strictly adhered to as with any machine tool.

Wood Shapers and Wood-Shaper Operations

The relative motion between the cutter and the work on the wood shaper differs substantially from that used on the metal-cutting shaper. The wood shaper has a vertical spindle that rotates a multi-toothed cutter against the workpiece as it is being fed past the cutting edges.

The shaper is used primarily for forming decorative edges on furniture and to prepare edges for joining. The shaper is a relatively simple machine; but because it operates at extremely high RPMs, it can be dangerous.

Irregular shapes can be shaped also with the use of guide pins and collars. Extreme caution should be taken when performing this operation.

Wood Jointer

The jointer (Figure 16–7), like the planer, is used primarily to machine flat surfaces. One difference is that the jointer can be used for the edge, end, and face of the board, while the planer is primarily restricted to machining the face of the board. The relative motion between the cutter and workpiece on the jointer is similar to that of the planer. The difference is that the knife-like cutters on the jointer are fastened to a cutter head that is located be-

Figure 16–6   Wood planer *(Courtesy of Powermatic)*

low the workpiece rather than above and the workpiece is fed across the rotating knives by hand. The depth of cut is controlled by raising or lowering the infeed table.

The jointer is manufactured in a variety of sizes. The jointer size is based on the width of the knife, which is the maximum width of board that can be surfaced. Sizes range from 4 to 36 inches. The jointer is a relatively simple machine with only a few basic parts and controls. Figure 16–8 illustrates the basic parts of wood jointer.

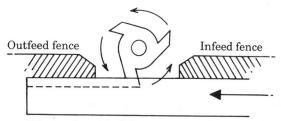

Outfeed fence   Infeed fence

When only part of the edge is shaped, outfeed
fence is aligned with infeed fence.

Outfeed fence   Infeed fence

When all of edge is shaped, outfeed fence
is moved forward to support the work.

Figure 16–7 The outfeed fence of the shaper is
set slightly behind the infeed fence to allow for
the cut

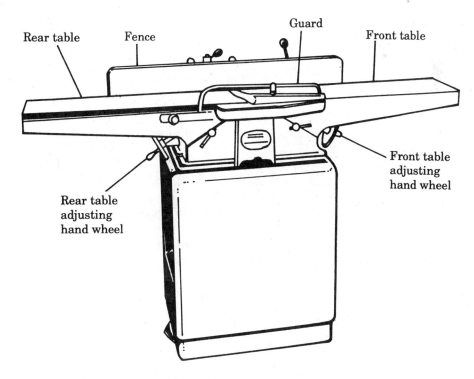

Rear table   Fence   Guard   Front table

Front table
adjusting
hand wheel

Rear table
adjusting
hand wheel

Figure 16–8 Wood jointer parts

DRILLING AND OTHER HOLE-PRODUCING OPERATIONS

Producing precision holes at exact locations is an extremely important process applied to all industrial materials. Mechanical drilling operations using a bit or auger is the conventional process used to produce the initial holes to dimension and location. Follow-up processes such as boring and reaming are employed to improve finish quality and precision. Newer nontraditional processes (e.g., laser and electrical discharge machining) have been used recently in aeronautics, space, and medical technology to produce small, precision holes in exotic as well as conventional materials. These newer processes and others are discussed in detail later.

Drilling Machines

Most machines used for drilling operations are relatively simple devices of varying size and accessories. Their primary purpose is to hold and rotate a drill bit or other hole producing tool. Various attachments such as vises and tapping heads can be added to increase the versatility of the drill press.

The size of machine used for drilling varies from a hand-held portable device to a large, industrial-type radial drilling machine (Figure 16–9).

Machines for tool-room use as well as small and large production operations fall under one of the following categories:

Portable   Sensitive   Upright   Radial   Gang   Multispindle   Turret

The *portable* and *sensitive* drilling machines (see Figure 16–9A) are generally used in tool rooms and maintenance areas but can also be found on assembly lines of larger industries. Portable drills are sized according to the largest diameter drill bit that the drill chuck will hold. Common sizes include $1/4$, $3/8$, and $1/2$ inch. The sensitive drilling machine is perhaps the most common hand feed–type drilling device. It is available in floor and bench models and is sized according to the distance from the spindle to the column. These machines are generally used for smaller, lighter work and are often limited to producing holes $3/4$ inch or less.

Upright, gang, radial, multispindle, and turret drilling machines (Figure 16–9B–E) are used for mass production or larger and heavier applications. The *upright* is similar in appearance to the sensitive drill press discussed earlier. Drilling capacities on the upright drill press range up to 3 inches in diameter, and the spindle is generally gear driven rather than hand fed.

The *gang* drilling machine is a series of drilling heads at a single table. It is extremely useful for mass-production manufacturing. A part requiring a sequence of operations (e.g., drilling, reaming, tapping, and spot facing) can be done at the same station. The fact that the heads are located on one table so close together reduces the transfer time from one operation to the next.

The *multispindle* drilling machine has a number of spindles driven by one motor. The spindles can be positioned to drill multiple holes in one part at the same time. The time to

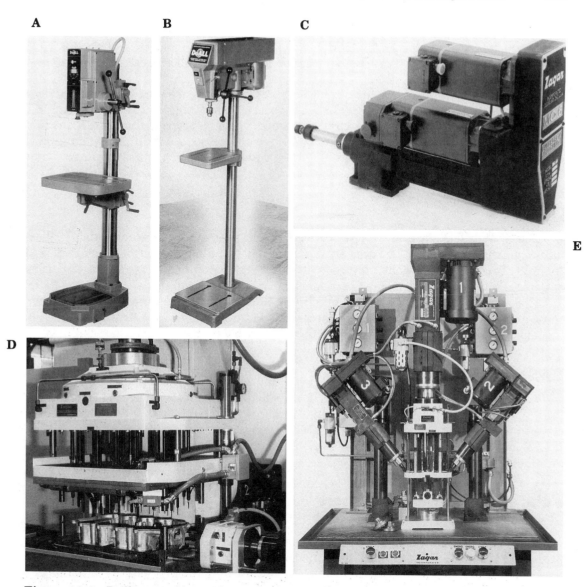

Figure 16–9 Drilling machines: **(A)** sensitive, **(B)** upright, **(C)** radial, **(D)** gang, and **(E)** multi-spindle *(A & B courtesy of DoALL Co.; C, D, & E courtesy of Zagar Inc.)*

set the positions of the spindles is often longer than that required to set up the gang drilling machine. The advantage is that, once positioned, multiple holes can be drilled more quickly in precise locations.

The *radial* drilling machine differs from the upright and sensitive drill in that the head of the radial drill rotates on the column and the table is stationary. It is designed to be used with large, heavy parts that are difficult to move. The radial arm rotates on the column and is positioned over the part in the desired location for drilling. The arm can be raised or lowered using power feed.

The *turret* drilling machine uses a multispindle head that can be indexed to various positions for different operations. Six to ten tools can be mounted on the turret head in a sequence that is suitable for the required processes. The advantage of this machine is that little time is required in going from one operation to the next. A drilling, reaming, and counterboring operation can be done without repositioning the part or using valuable time to change from one tool to another.

Drill Bits

The twist drill (Figure 16–10A) is perhaps the most commonly used hole-producing device. The twist drill is used for a variety of materials, including metal, woods, plastics, and ceramics. Twist drills are available in a wide variety of lengths and diameters and are sized using one of four commonly used systems: fractional inch, metric, wire gauge (numbered), and letter.

The cutting and clearance angles illustrated in Figure 16–10B are standard; however, these angles are modified for different materials. Softer materials such as wood, hard rubber, and some forms of plastic are more efficiently drilled with smaller included angles.

Reamers

Reamers are used after the initial hole is drilled to enlarge the hole and improve the finish. Reamers are classified as either hand or machine and are manufactured from a variety of materials. Reamers can be used on a variety of machine tools, including drill presses, lathes, and vertical mills. Figure 16–11 shows a sample of the numerous reamers available for different applications.

Related Processes and Tools

There are numerous processes that are combined with drilling to shape or enlarge the initial holes. Countersinking, counterboring, and spot facing are a few examples of those processes. Figure 16–12 illustrates some of these commonly used processes.

MILLING OPERATIONS

There is a wide variety of types and sizes of milling machines on the market. They range in size from smaller column-and-knee vertical and horizontal mills used for lighter production and maintenance to huge fixed bed–type machines used in heavy-metals manufacturing and mass-production manufacturing. These machines range in price from less than $20,000 to several hundreds of thousands of dollars. Many are now equipped with computer controls and highly automated tool changers. Larger machine centers are often capable of automatically selecting from among forty or more different tools contained in a carousel that is attached to the machine.

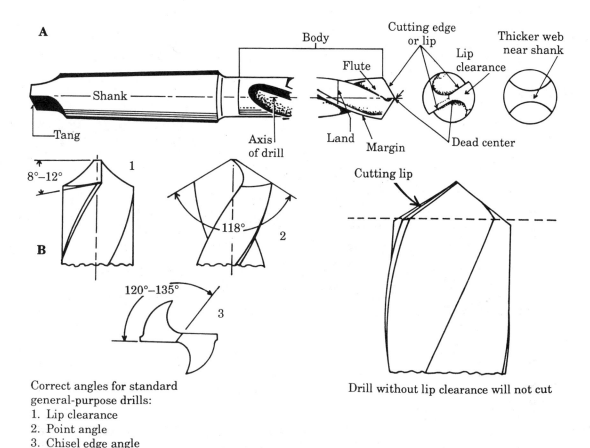

Correct angles for standard general-purpose drills:
1. Lip clearance
2. Point angle
3. Chisel edge angle

Figure 16–10   **(A)** Parts of a twist drill; **(B)** cutting and clearance angles

Equipment and Accessories

Milling machines have been used for well over 100 years to produce precision parts for the assembly of a variety of products, including automobiles, farm implements, and weapons. One of the first milling machines manufactured for sale rather than private use was designed by Frederick Howe in 1848. This machine was designed for the Robbins and Lawrence Company of Windsor, Vermont, who are perhaps best known for producing the Sharpe's rifle. Since that time, the milling machine has been redesigned, refined, and numerous accessories and controls added for more accurate and efficient manufacturing.

Milling machines are classified as either *column-and-knee* or *bed-type* milling machines. The column-and-knee type machines are smaller, more versatile, and less expensive and are used in job shops, maintenance departments, and school laboratories. They are further classified as *vertical* or *horizontal*. The vertical milling machine is designed with a vertical spindle that can be raised and lowered like a drill press spindle.

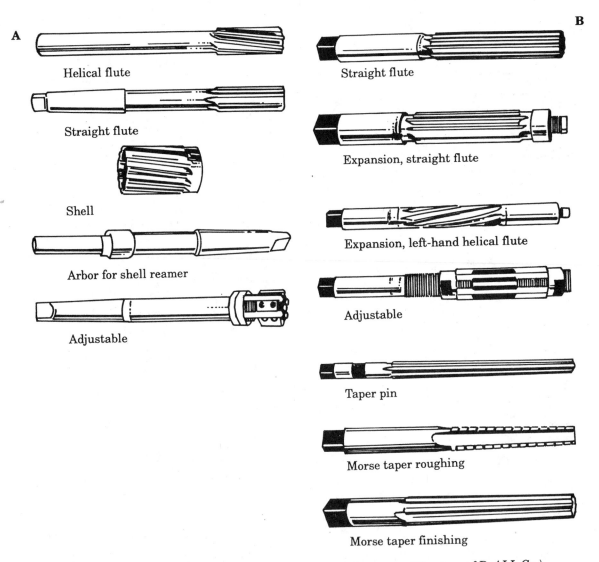

A

Helical flute

Straight flute

Shell

Arbor for shell reamer

Adjustable

B

Straight flute

Expansion, straight flute

Expansion, left-hand helical flute

Adjustable

Taper pin

Morse taper roughing

Morse taper finishing

Figure 16–11 **(A)** Machine reamers and **(B)** hand reamers *(Courtesy of DoALL Co.)*

The vertical mill with accessories is capable of a wide variety of machining operations that include shaping, dovetailing, angular milling, drilling, boring, reaming, circular milling, and others. Computer numerical control has been added to the vertical mill to increase its accuracy and versatility. The vertical mill is sized according to horsepower, table size, and longitudinal and vertical travel of the table.

Horizontal mills, although less versatile than vertical mills, are typically more rigid and capable of heavier and wider cuts. They are excellent for surfacing and are often used for slotting, straddle milling, and form cutting. The horizontal spindle (arbor) is what distin-

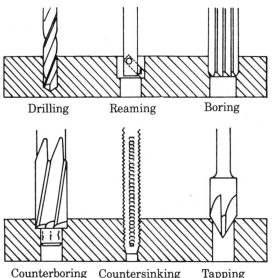

Drilling   Reaming   Boring

Counterboring   Countersinking   Tapping

Figure 16–12 Common operations performed on the drill press

guishes it from the vertical mill, and it uses a circular milling cutter rather than the end mills used on the vertical machine. The horizontal mill is also sized according to horsepower, table size, and travel.

The term *column-and-knee* refers to the general construction of these lighter mills. Figure 16–13 illustrates the major components of the column-and-knee machine. In this illustration, you can see that the rear column provides the support for the head or arm and for the knee-and-saddle arrangement. The column of the mill is typically equipped with dovetail ways to which the knee and table are attached. Lead screws are used to permit vertical, longitudinal, and transverse travel of the table.

The *bed-type* mills are generally heavier and substantially more expensive than the column-and-knee machine. They are less versatile than the column-and-knee milling machine, and the table rests on a stationary bed. This provides the rigidity needed for larger manufacturing. Adjustments are made by lowering or raising the spindles rather than the table. These machines often have more than one spindle mounted either vertically or horizontally.

Milling Machine Accessories

A number of accessories and cutters are available for milling operations. A variety of vises and clamping devices are also available for use on the T-slotted table of the milling machines. Plain, swivel, and universal vises (Figure 16–14) can be purchased from suppliers at costs ranging from approximately $200 to $1,000. The plain-type vise offers the least flexibility. Universal vises are used to secure work that is to be machined at various angles.

Vises are not always the best holding device for larger and uniquely shaped objects. A variety of clamping devices, screw jacks, and fixtures are available for almost any job (Figure 16–15). The holding device is extremely important to the tool operation. The setup must

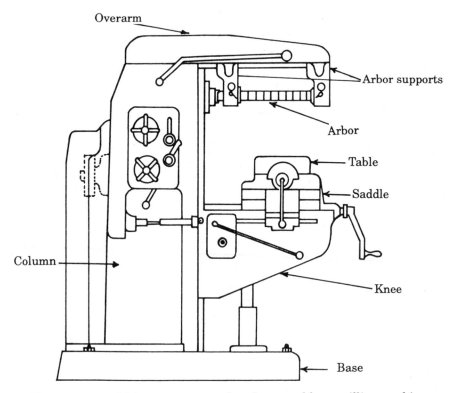

Figure 16–13   Major components of a column-and-knee milling machine

be rigid to produce desirable finishes and to avoid serious injury. Rotary tables and dividing heads are used on milling machines for more complicated shapes and forms. The rotary table (Figure 16–16) is equipped with a T-slotted table for fastening the work to an indexing head for milling arcs or locating holes. This accessory is used on vertical and horizontal milling machines.

The dividing head is used in conjunction with the milling machine to provide rotary motion to the workpiece for the purpose of locating specified angular distances. Applications include gear cutting, locating positions for drilling and slotting, and machining flat surfaces at specified locations.

Milling Cutters

A variety of milling cutters are available for both vertical- and horizontal-type milling machines. The selection of a cutter is based on the requirements of the job to be completed. These requirements are generally related to the finished shape of the specified part. In mass-production manufacturing operations, however, cutting efficiency is always a consideration in selecting the type of cutter and cutter material.

Terminology used to classify milling cutters is taken from the type operation the cutter performs. The most commonly used cutters for milling operations are classified as follows:

A

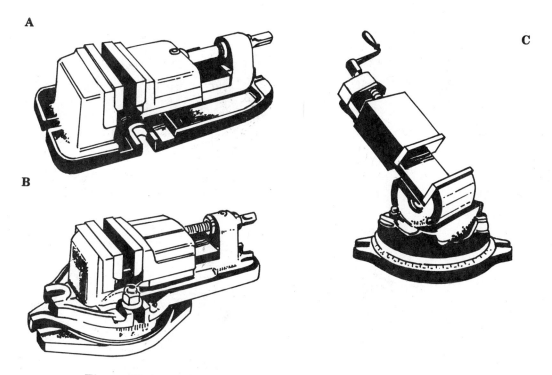

B

C

Figure 16–14 (A) Plain vise, (B) swivel vise, and (C) universal vise

1. *Plain* (Figure 16–17A). These cutters are made in various diameters and widths. They are used to machine flat surfaces. The teeth are only on the circumference of the cutter. These cutters are used on an arbor.
2. *Angular milling cutters* (Figure 16–17B). These cutters are designed to produce angles on the finished part. These include both *single*- and *double*-angled cutters used for forming V-grooves, dovetails, and other angular shapes. Angular cutters are available in 45-degree and 60-degree angles.
3. *Side milling cutters* (Figure 16–17C). The side milling cutter is similar in appearance to the plain milling cutter except that it has teeth on either one or both sides. The side milling cutter is commonly used for straddle milling operations.
4. *Form milling cutters* (Figure 16–17D). The form milling cutters are designed to produce various concave and convex shapes as well as special operations such as gear cutting.
5. *End mills* (Figure 16–17E). End mills are used on vertical mills or horizontal mills with vertical milling attachments. End mills normally have a straight smooth shank that is held in the spindle by a collet or other tool-holding device. Since the spindle is equipped with a morse taper, the collets must be purchased with a matching taper. The split-type collet used with end mills is threaded on the end fitting in the spindle. The end mill is secured in the collet, which is drawn up into the tapered spindle.

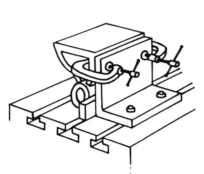

Using angle plates for machining setups.

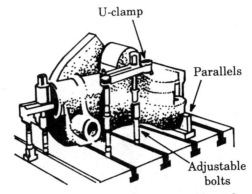

Holding irregularly shaped casting with
U-clamps and adjustable bolts.

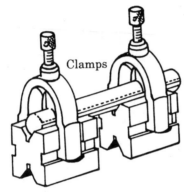

Matched pair of V-blocks.

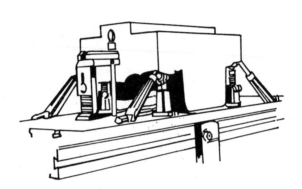

Typical setup tools and accessories.

Figure 16–15 A variety of clamping devices, screw jacks, and fixtures are available for almost any job

Figure 16–16 Rotary table

322

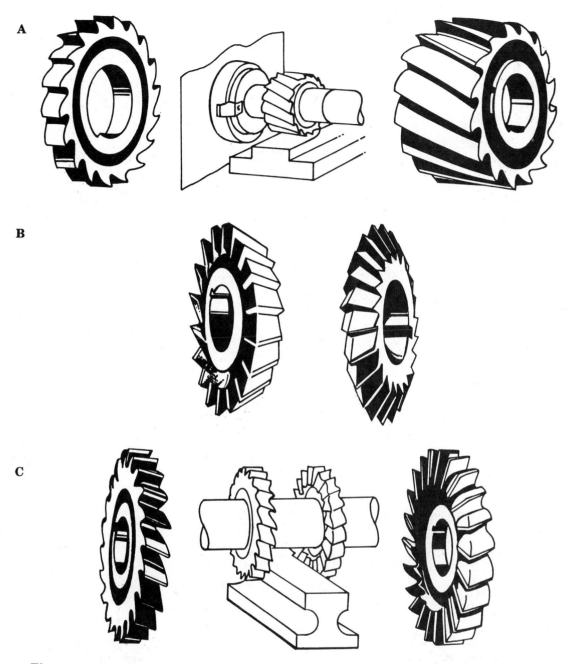

Figure 16–17 *Above* **(A** and **B)** Plain milling cutters, **(C)** single- and double-angular milling cutters, **(C)** side milling cutters, *next page* **(D)** form milling cutters, and **(E)** end milling cutters (1) two-flute single-end, (2) two-flute double-end, (3) three-flute single end, (4) multiple-flute single-end, (5) double-end, (6) two-flute ball-end, (7) carbide-tipped straight flutes, (8) carbide-tipped right-hand helical flutes, (9) multiple-flute with taper shank, (10) carbide-tipped with taper shank and helical flutes, and (11) throwaway insert type *(Courtesy of National Twist Drill Co.)*

D

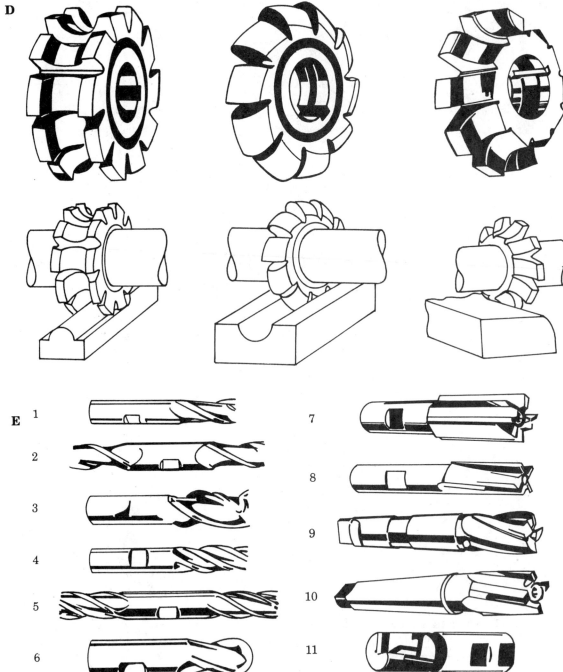

GRINDING PROCESSES

Grinding or abrasive machining processes are principally used to finish surfaces that require a smoother surface quality than milling or turning can produce. One of the first machines designed and built to perform operations similar to our modern surface grinder was completed in 1830. By mid-century, power-driven, heavy-duty grinding machines could be found in machine shops in the United States and Great Britain. Later, machine builders concentrated on achieving greater precision. Brown and Sharpe exhibited the universal grinding machine in Paris in 1876 and demonstrated the precision that others had sought. Today, a variety of grinding machines and grinding processes are used to produce extremely smooth surfaces to precision tolerances of ±0.0002 inch or less.

Equipment

A variety of precision grinding machines are manufactured to perform cylindrical as well as surface grinding operations. The classification system used to describe the grinding machines is derived from the process performed on the type of machine being used. Grinding operations and grinding machines are classified in the following manner:

Surface grinder (Figure 16–18A and B). Primarily used to provide smooth flat surfaces, these machines are available with vertical or horizontal spindles and rotary or reciprocating tables.

Cylindrical grinders (Figure 16–18C). Cylindrical grinding, as the term implies, is used to finish-grind cylindrical parts. The cutting motion is rotary with both the cutting wheel and part rotating against one another. The part being processed is held between centers, much like what was described earlier in the discussion on turning; or the workpiece is rotated between two wheels while being supported underneath in what is referred to as *centerless* grinding.

Internal grinders. Smooth, accurate internal straight and tapered surfaces are machined using the internal grinding operation. During these processes, the wheel is rotated in a fixed position while it is reciprocated back and forth into the work area. The work is generally rotated slowly as grinding takes place.

Universal grinders. Universal grinders have the versatility for grinding operations that include external and internal grinding, surface grinding, and tool sharpening of milling cutters and reamers.

NONTRADITIONAL MACHINING METHODS

Within the last thirty-five years, special processes that are often referred to as "chipless" material removal have been developed and are rapidly becoming accepted practice. The need for lighter better aircraft provided the stimulus for the development of these new processes.

A

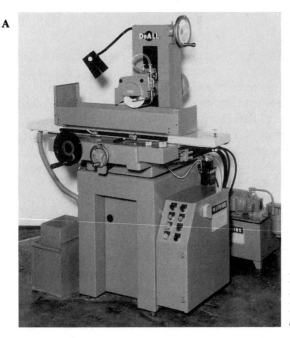

Figure 16–18   (A) Surface grinder, **(B)** cylindrical *(From Goetsch,* Modern Manufacturing Processes, *Copyright 1991 by Delmar Publishers, Inc. Used with permission)*

B

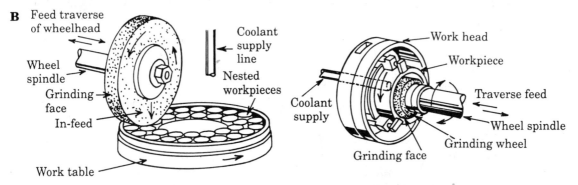

Requirements for more precision in new, harder exotic materials made the traditional processes (i.e., drilling, milling, grinding) impractical.

Nontraditional machining processes use other forms of energy to produce more effectively the results once accomplished through conventional means with mechanical energy. Thermal, electrical, and chemical energy are replacing mechanical energy to machine new harder material now used in the aerospace, electronics, automotive, and aircraft industries. The following is a brief description of a few of the more than thirty nontraditional machining processes in production use today.

Electrical Discharge Machining (EDM)

Electrical discharge machining (EDM) is one of the oldest of the nontraditional machining processes. Originally developed during World War II to cut the tough harder materials, EDM is perhaps the most widely used of the newer processes. EDM machines are available in a variety of sizes, starting with a benchtop model (Figure 16–19). Machining takes place between the workpiece and the electrode that is fed into the work area under servo-control. When the electrode is brought near the workpiece, sparking begins to take place. This rapid-fire sparking action (from 200 to 500,000 sparks per second), fired from a shaped electrode, causes small pieces of metal to be vaporized and removed from the workpiece.

The workpiece is secured inside a tank that is filled with a dielectric fluid (in general, low-viscosity petroleum oils). This low-viscosity oil is directed at the cutting area to remove metal chips in order to provide uniform cutting and improve machined surface finish. Surface finish ranges from 60 to 125 microinches. In general, as the metal removal rate increases, so does the roughness.

Figure 16–19 illustrates the cutting action between a solid-electrode EDM machine and the workpiece. Wire EDM is also available for contouring processes. Wire EDM uses a thin-wire electrode to shape unique contours that generally cannot be formed by conventional machining. The wire EDM uses copper, tungsten, or similar conducting material. Wire EDM is often used to shape electrodes for the ram-type machines. Wire EDM machines are computer controlled and are capable of high accuracy and repeatability. Four-axis systems and special tooling and attachments permit intricate three-dimensional shapes to be produced

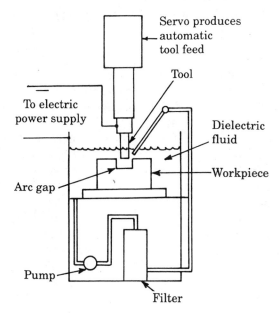

Figure 16–19   EDM process

on these machines. The disadvantage of EDM is that it is a relatively slow process; however, it is capable of producing cavity shapes that are virtually impossible with traditional machine tools.

EDM is restricted to processing materials that conduct electricity. Electrodes are made of high-conductivity materials, including graphite, copper-based materials, tungsten, and zinc alloys. Although a relatively safe process, acetylene gases can form when arcing takes place. The level of the electrolytic fluid should be 1 inch above the workpiece to prevent these gases from forming.

Electrochemical Machining (ECM)

Electrochemical machining (ECM) is used extensively in the aircraft and automotive industries. It is used to machine turbine blades and automobile parts such as connecting rods, pistons, and fuel-injection nozzles. The ECM equipment is comparable in price to the multi-axis CNC milling centers ($200,000 to $1 million).

The ECM process is similar to EDM in that the electrode used to remove the material is shaped like the cavity it is expected to form. The electrode is negatively charged and acts as the cathode. The workpiece (anode) is secured in a tank and submerged in an electrolytic fluid designed to wash away metal particles (Figure 16-20). Cutting proceeds through dissolution in the anode (workpiece). The electrode is fed into the workpiece at a rate equal to the rate of dissolution.

This process in theory is just the opposite of electoplating, in which the electrode is the anode and deposition takes place onto the workpiece (cathode). The rate of metal removal is about 0.1 cubic inch per minute, and surface finishes range from 16 to 60 microinches.

Electrochemical Grinding (ECG)

One of the earliest applications of electrochemical grinding (ECG) was the grinding of cemented carbide tools and dies—very hard conductive materials. Honeycomb metals and turbine tips are also among the types of parts machined using ECG. This process eliminates certain stresses and heat damage that are prevalent in other processes.

ECG is restricted to materials that will conduct electricity. Grinding wheels used in the ECG process must also be made electrically conductive. Conventional abrasives are bonded with conductive adhesives to achieve this. The workpiece and grinding wheel are connected to a low-voltage DC (direct current) power source, and an electrolytic fluid is flushed onto the abrasive wheel to remove metal particles and act as a coolant.

The metal removal takes place through electrochemical decomposition plus abrasive action. An abrasive material, usually diamond abrasive, is used between the wheel and the workpiece to support the cutting action. This abrasive action accounts for about 10% of material removal, with the remainder of the work being accomplished through the electrochemical action. Figure 16–21 illustrates the basic components of the ECG process.

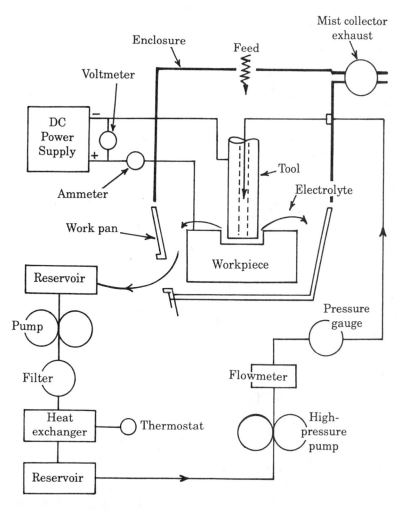

Figure 16–20 Electrochemical machining

Electron Beam Machining (EBM)

Electron beam machining (EBM) is a thermal technique that uses a stream of high-velocity electrons to melt and vaporize the metal (Figure 16–22). This process is used extensively for producing holes that traditionally were drilled or punched. EBM can be used to produce holes of extremely small diameters (0.002 inch) accurately (0.0002 inch).

The EBM process uses an electron gun to fire off electrons that are then accelerated to nearly the speed of light. An electromagnetic coil (lens) focuses the electron beam to converge on the part at the designated location. The power density produced is equivalent to as

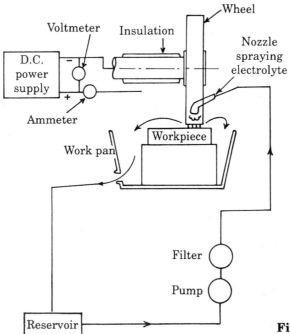

Figure 16–21   Components of ECG system

much as one billion watts per square inch. This vaporizes the metal at rates that are impossible using mechanical processes.

Laser Beam Machining (LBM)

Like EBM, laser beam machining (LBM) is classified as a thermal process. Laser is an acronym for *l*ight *a*mplification by *s*timulated *e*mission of *r*adiation. LBM uses a controlled beam of light to vaporize the material on which it is focused. The advantage of lasers over many other processes is that they are capable of vaporizing any material, not just those that can conduct electricity.

The laser oscillator has three essential parts: (1) an amplifying medium, (2) a source of pump power, and (3) a resonator. A ruby is used quite frequently to amplify a light source (flash lamp or sun). The released energy from the ruby emits photons as the light bounces back and forth. Photons are reflected back and forth between the ends of the ruby long enough to amplify the beam until the intensity is strong enough to pass through the mirrors and onto the workpiece. The beam passing through the mirror is concentrated at the desired cutting point. The high-intensity beam causes the material to evaporate.

LBM has been used since the 1960s. Its use has expanded since its development to include welding, hole production, heat treatment, cutting, and inspection. It is used to pro-

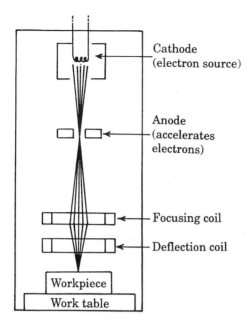

Cathode
(electron source)

Anode
(accelerates
electrons)

Focusing coil

Deflection coil

Workpiece

Work table

Figure 16–22 EBM machining

cess an incredible variety of materials, including the most exotic materials, and to provide a level of precision unknown to conventional machine tools.

SUMMARY

Machining processes used in modern manufacturing operations range from the very traditional to the high-technology state of the art. Traditional machining operatioins are generally identified as those processes that require mechanical force to remove the unwanted material, thus changing the size and shape of the standard stock or blank. These mechanical machining processes are sometimes referred to a chip-producing operations and include turning, milling, drilling, grinding, shaping, and planing.

The nontraditional machining processes include electrical discharge machining (EDM), electrochemical machining (ECM), electrochemical grinding (ECG), electron beam machining (EBM), and laser beam machining (LBM). These processes use electrical, thermal, and chemical energy to remove material rather than the mechanical energy used for traditional machining processes.

Machining processes differ from other manufacturing processes in that material is removed in the course of producing the desired size, shape, and finish. The machines and processes used for conventional machining are not significantly different than those used at the turn of the century. Today, machine tools are often controlled by computers; and the material used for making cutting tools is substantially better. However, the machines are very similar in appearance and function to their forerunners.

REVIEW QUESTIONS _____

1. What is the fundamental difference between machining processes described in this chapter and the molding processes described in Chapter 13?
2. Briefly describe the historical importance of the modern engine lathe.
3. What is the basic difference between metal shaping and metal planing.
4. What is the basic difference between electrochemical milling and electroplating?
5. List the three basic functions performed by components of the lathe.
6. What two dimensions are used for determining the size of a metal lathe?
7. What three functions are performed by cutting tools used with the wood lathe?
8. List three accessories for the metal lathe that can be used to attach a workpiece to the headstock.
9. Calculate the tailstock offset for a part with a specified taper per foot of 0.375 inch. The length is 18 inches.
10. What dimension is used to determine the size of the wood jointer?
11. What type of drilling machine would you select if the part was extremely heavy and large? Explain.
12. List the four systems used to size twist drills.
13. List two uses of reamers.
14. List four processes that are often used in combination with drilling to enlarge or shape holes.
15. What two terms are used for classifying vertical and horizontal milling machines?
16. List four types of grinding machines and their functions.
17. Describe the process used for shaping parts with electrical discharge machining.
18. What advantage does laser beam machining have over other nontraditional machining methods?

SUGGESTED ACTIVITIES _____

1. Write a two- to four-page report on a new machining technique that was not discussed in this chapter. Present the information to the class.
2. Write to manufacturers of machine tools and supplies. Ask for their literature on new cutting-tool materials. Prepare a table comparing tool steel to the latest technology.
3. Acquire samples of machined parts or products from a local manufacturer. Describe the various machining methods used to form the finished geometry of the part (i.e., milling, turning, grinding, etc.).

CHAPTER 17

Machine Control

In today's high-technology computerized environment, any book that includes manufacturing processes is incomplete without a discussion of automation and machine control. Although the basic methods of manufacturing parts by machining have not changed significantly since they were first developed, methods of controlling the cutting tool have. In many instances, computers have replaced the hand of the machinist in guiding the tool along the planned path that produces the desired shape.

The technology used to guide the tool along the desired cutting path is generally referred to as **computer numerical control (CNC)** or computer numerical control (CNC). The invention of the digital computer for computational tasks eventually led to other uses, including machine-tool control.

Initially, NC was available only to the larger manufacturing companies due to the high capital investment required to purchase the technology. Rapid advancements in technology, however, have reduced the costs significantly, improved the product, and made advanced forms of NC and CNC technology available to even the small job shops. The use of computers to control the cutting path of the machine tool has also led to the use of computers for aiding in the design of parts, inspecting in-process and finished goods, maintaining inventory records, scheduling production, and a host of other functions associated with the manufacture and distribution of parts and products.

The purpose of this chapter is to introduce the reader to machine control in modern manufacturing environments. The primary focus is numerical and computer numerical control of machine tools; however, the discussion also includes computer-aided design and manufacturing (CAD/CAM), flexible manufacturing systems (FMS), and computer-integrated manufacturing (CIM).

NUMERICAL CONTROL

The Electronics Industries Association (EIA) defines numerical control as "a system in which actions are controlled by the direct insertion of numerical data at some point." Coded instructions that can include numbers, letters, and symbols are used to establish a cutting path that produces a desired shape.

Numerical control was introduced to the machine tool industry as a result of work done by John Parsons of Parsons Manufacturing Company, Traverse City, Michigan. Parsons realized that the same computer process that was used to define surface contours on complex airframe parts more accurately could be used to control the machining processes used to manufacture them. With the assistance of the Servomechanisms Laboratory at the Massachusetts Institute of Technology (MIT) and support of the Air Material Command of the United States Air Force, the first contouring machine was built in the early 1950s.

The first NC machines were controlled by large mainframe computers and were capable of highly sophisticated multi-axis contouring or continuous-path movements. Complex angles, arcs, and pockets could be formed without the use of cumbersome attachments as needed in conventional machining processes. The first NC machines were extremely expensive and, therefore, restricted to only the largest companies with the capital to purchase them. Later in the 1960s, less complex and less expensive *point-to-point* or *positional* NC machines were made available at prices that smaller shops could afford. Point-to-point machines were restricted to making straight-line movements. Arcs could be formed by these machines but only by making a series of straight cuts and various angles. Today, continuous-path or contouring machines are affordable to most small shops.

The development of the first computer-controlled machine was the result of an identified need for lighter and stronger aircraft. The contouring capabilities of the early NC machines made it possible to manufacture larger individual parts. Machining larger one-piece components rather than producing smaller parts that had to be bolted together to form larger components was found to be less expensive and helped produce lighter aircraft.

Computer Numerical Control

The distinction between numerical control and computer numerical control is not always made. There is, however, a significant difference between the two. NC is the broader of the two terms and includes CNC as a subset. The older machines, many of which are still in operation, were NC rather than CNC machines. NC machines of the earlier type are equipped with tape readers rather than computers. Paper or mylar tapes are punched off-line and fed through the hard-wired tape reader. The tapes are punched to form a pattern of holes that corresponds to a standard code for control of the tool path.

By definition, NC includes any system that uses numbers, letters, and symbols to control actions. The older machines with the tape readers as well as the newer machines with the on-board computers meet the intent of this definition. CNC, however, is normally reserved for the newer systems with more programmable capabilities.

CNC System Components

CNC machines consist of three major components:

1. Machine tool
2. Machine control unit (MCU)
3. Program of instruction

The *machine tool* consists of a milling machine, lathe, punch, or other type of conventional or nontraditional tool. With the possible exception of the control unit, NC machine tools generally do not appear much different than conventional machines (Figure 17–1). Part of the difference between NC and conventional machines is not visible. The lead screws that move machine tables and transmit power are different on contouring NC machines. Referred to as lead (ball) screws, they permit free movement of the table in several planes at the same time to allow the smooth cutting of arcs, circles, and other curvilinear surfaces.

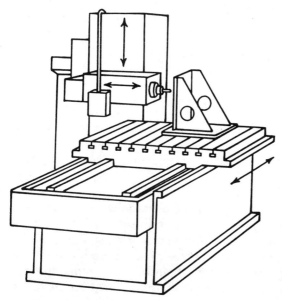

Figure 17–1   A CNC mill looks similar to a conventional mill

A difference between NC and conventional machine tools that is visible is the drive motors. The drive motors, which can be seen in Figure 17–2, move the tables in the programmed direction. The drive motors for computer-aided manufacturing (CAM) equipment are classified as follows:

Stepper motors: Electric motors that move a set amount (a step) every time a pulse is received. Stepper motors require no feedback and are less expensive than others. Power and speed are restricted in stepper motors.

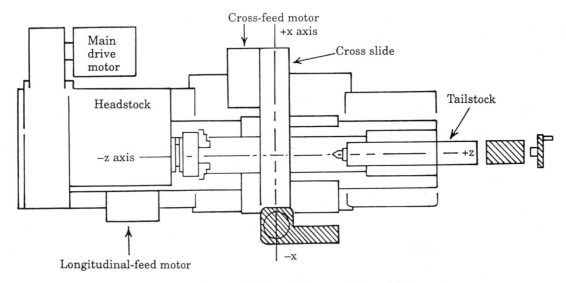

Figure 17–2   Closeup of CNC mill shows table and drive motors

DC servo motors: Extensively used direct-current, variable-speed motors found on small-
and medium-size continuous-path machines. Servo motors are very accurate and
hold their torque when power is removed.

Hydraulic motors: Variable-speed servo motors capable of producing much more power
than electric motors.

The *machine control unit* (MCU) includes the hardware and electronics required to read
and interpret the program of instruction and convert it into mechanical actions. The MCU
is the computer device that typically is attached to the machine tool or resides in a cabinet
beside it. Modern MCUs have cathode-ray tubes (CRTs) (Figure 17–3) that allow program-
mers and operators to verify program accuracy prior to actual machining and monitor cutting actions during the production process. As one can see from Figure 17–3, MCUs also have
a keypad, comparable to a keyboard on a personal computer, to input commands for processing.

The third component, the *program of instruction,* is not visible to the human eye but
resides inside the MCU. The program of instruction directs the movement of the cutting tool
along a predetermined path. It consists of the detailed set of coded instructions that the
programmer inputs into the memory of the controller.

The controller has two types of memory: random-access memory (RAM) and read-only
memory (ROM). The *executive program* (provided by the manufacturer) resides in ROM and
cannot be edited by the CNC programmer. The executive program interprets NC code input
by the programmer and executes canned (preprogrammed) subroutines initiated by a let-
ter-number sequence in the programmer's code.

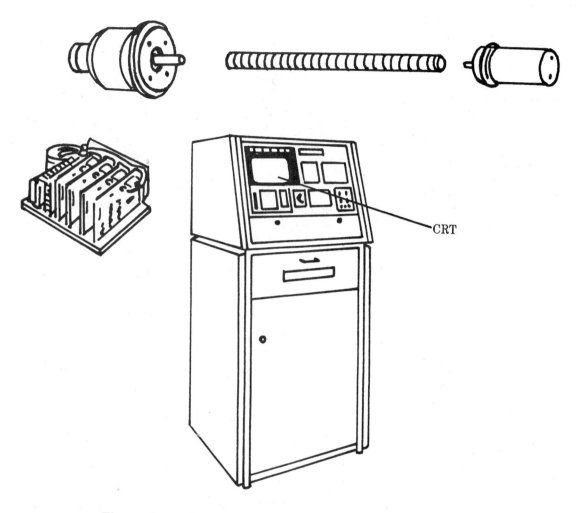

CRT

Figure 17-3   The machine control unit on this CNC mill has a CRT

NC code resides in RAM and can be edited. NC code is prepared by the programmer and loaded into the random-access memory of the MCU. The code consists of letters, numbers, and symbols that, when interpreted by the executive program, direct the cutting tool along the cutting path at the correct feedrate and spindle speed.

NC code can be input into RAM in one of several ways. The programmer can use the keypad on the MCU and directly input the code into RAM. This is referred to as manual data input (MDI) but has at least two disadvantages: (1) It is slow. (2) Long-term storage of programs may not be possible. NC code can also be created on a stand-alone computer, stored on computer disk, and downloaded through a communication port to the RAM in the MCU. Paper tape and magnetic tape are other media used for storing and inputting NC code.

NC Code

As explained earlier, NC code is a series of letters, numbers, and symbols used to program a predetermined tool path. The letters, numbers, and symbols used in NC code have been standardized by the EIA. This does not mean that there is only one NC language, but that certain letters and symbols are universal.

NC commands start with a letter referred to as an *address*. For example, the letter *F* is the address for feedrate. When programming feedrate in NC code, the command line begins with an *F*. A command of F20 would establish the feedrate on a milling machine at 20 inches per minute.

Other commonly used addresses include *N, G, M, S, T, X, Y,* and *Z*. In fact, all but two letters of the alphabet—*H* and *L*—have been assigned as addresses in NC code by the EIA (Table 17–1).

Letter	Description
A	Angular dimension about the x axis
B	Angular dimension about the y axis
C	Angular dimension about the z axis
D	Angular dimension about a special axis, or third feed function, or tool function for selection of tool compensation
E	Angular dimension about a special axis or second feed function
F	Feedrate address
G	Preparatory code
H	Unassigned
I	Interpolation parameter or thread lead parallel to the x axis
J	Interpolation parameter or thread lead parallel to the y axis
K	Interpolation parameter or thread lead parallel to the z axis
L	Unassigned
M	Miscellaneous or auxiliary function
N	Sequence number
O	Sequence number for secondary head only
P	Third rapid-traverse dimension or tertiary-motion dimension parallel to x
Q	Second rapid-traverse dimension or tertiary-motion dimension parallel to y
R	First rapid-traverse dimension or tertiary-motion dimension parallel to z or radius for constant surface-speed calculation
S	Spindle-speed function
T	Tool function
U	Secondary-motion dimension parallel to x
V	Secondary-motion dimension parallel to y
W	Secondary-motion dimension parallel to z
X	Primary X-motion dimension
Y	Primary Y-motion dimension
Z	Primary Z-motion dimension

Table 17–1 Numerical control letter addresses

An *N* word address is used to designate sequence numbers for command lines. Every command line is given a sequence number. For most systems, the executive program automatically generates the sequence number starting with N01 or N001; however, the programmer can elect to start a program at any sequence number.

G word addresses are used for preparatory commands. These are commands that prepare the machine tool to execute a movement. For example, a G2 preceding a dimension prepares the machine for a clockwise circular motion; a G3 would prepare the NC machine for a counterclockwise circular motion. A partial list of other G addresses is defined in Table 17–2.

G Code	Description
G00	Prepares machine for rapid-traverse movement
G01	Linear interpolation
G02	Circular interpolation—clockwise movement
G03	Circular interpolation—counterclockwise movement
G04	Dwell (a time delay)
G41	Cutter compensation left
G42	Cutter compensation right
G80	Drill cycle
G84	Tapping cycle
G90	Absolute input programming
G91	Incremental input programming

Table 17–2   G NC Addresses

The *M* word NC addresses are reserved for miscellaneous functions. Sometimes referred to as auxiliary functions, M addresses are used for start–stop actions. For example, M07 starts coolant; M09 stops coolant. Table 17–3 provides additional examples of M addresses standardized for NC code.

M Code	Description
M00	Automatically stops the machine
M02	End of program stop
M03	Spindle start in clockwise direction
M04	Spindle start in counterclockwise direction
M05	Stop the spindle in normal manner
M06	Stop machine for tool change
M07	Start coolant
M09	Stop coolant
M10	Automatic clamping of workpiece
M11	Unclamping of workpiece

Table 17–3   M NC Addresses

The *S* address is for spindle function, and *T* is for selecting tools on machines with turrets or automatic tool changers. The *X, Y,* and *Z* addresses are used for movement of the cutting tool in the horizontal and vertical axes.

Cartesian Coordinate System

The path or position the cutting tool follows on a CNC machine is based on the geometric points and lines that define the part. The system used to establish the geometry that defines the part or tool path is known as the **Cartesian coordinate system**. The simplest system is a two-axis system that is commonly used in basic mathematic courses (Figure 17–4). In a two-axis system, the x and y axes form four quadrants. The origin is the intersection of the x and y axes, and positioning is initiated from that origin. For example, positioning for a drilling operation at location one of the part described in Figure 17–5 using the Cartesian coordinate system would require a movement along the x axis +3 inches and along the y axis +2 inches. Since the positioning of the drill center at this location is the desired result, the moves translate to a longitudinal move of the table to the left 3 inches and a transverse (cross-feed) 2 inches forward. Note that the x and y axes refer to movements in the horizontal plane.

If a third dimension is added (Figure 17–6), it represents movement in the vertical plane (up and down) and is defined by the z axis. The z axis on a mill, for example, would most likely be the movement of the spindle up or down. In some cases, the knee is automated as well; and the table can be programmed to move up and down. Any vertical movement above the origin of the z axis is positive, and any movement below the origin is negative.

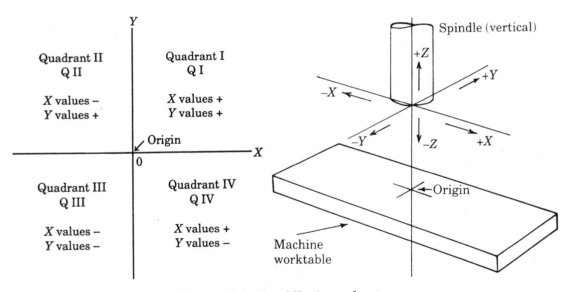

Figure 17–4   X and Y axis quadrants

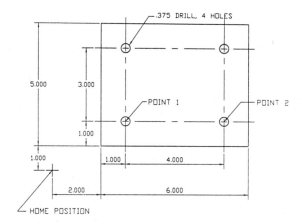

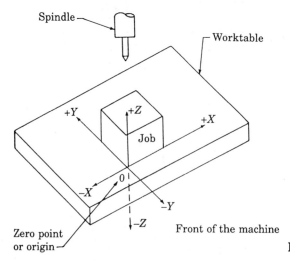

Figure 17–5 Holes located for NC drilling

Figure 17–6 Z-axis coordinate

Absolute and Incremental Positioning

The positioning of the drill bit to the locations defined by the blueprint can be accomplished in one of two ways. Either *incremental* or *absolute* positioning can be used to locate all points for processing. Absolute coordinates are given in terms of the distance from origin. In Figure 17–5, the move from location 1 to location 2 would be defined using the distance from the origin if absolute positioning were used. Therefore, point 2 is defined by a movement of a +5 inches along the x axis and +1.5 along the y axis.

Incremental coordinates are defined as a dimension or a movement with respect to the preceding point. Using Figure 17–5 again as an example, the incremental positioning for point 2 would use point 1 as a point of origin. Therefore, point 2 is defined by its distance on the x and y axis from point 1. In programming the position of the drill for point 2, point 2 is defined as +2 inches along the x axis and –0.5 inch on the y axis.

An understanding of the Cartesian coordinate system and a background in trigonometry is necessary to become a competent CNC programmer. Although there are several programming languages, each with its own codes, an understanding of trigonometric functions and coordinate systems will make it easier to adapt to one or the other.

NC Programming

NC programming is based on the binary number system (base one). Programming formats allow NC code to be written using letters, base ten numbers, and symbols and then converted to binary form when read from the programming media (tape, disk). Table 17–4 shows the conversion from base ten to binary numbers. Every binary digit represents a "1" or a "0," which corresponds to a negative or positive charge. The negative and positive charges open and close circuits, which control machine tool movement and auxiliary NC functions.

Many NC languages have been developed since the early and mid-1950s when numerical control was first successfully demonstrated. The number most likely exceeds 100 by now. The leading languages, however, are among the first developed. They include APT, COMPACT II, and ADAPT (adaptation of APT).

Binary	Decimal
0000	0
0001	1
0010	2
0011	3
0100	4
0101	5
0110	6
0111	7
1000	8
1001	9
1010	10
1011	11
1100	12
1101	13
1110	14
1111	15
10000	16
10001	17
10010	18
10011	19
10100	20

Table 17–4   Binary code

Manual Part Programming

Manual part programming is generally considered the most basic of the programming methods and can become quite time consuming for all but the most simple parts. The following is a typical manual part program for Figure 17–7:

Sequence No.	Command	Comment
001	F20.0	Set feedrate at 20 IPM
002	M03	Spindle on clockwise
003	M08	Coolant on
004	/Y1.0	Rapid move to start of slot 1
005	−Z.500	Set depth of end mill
006	X4.750	Mill slot 1
007	/Z.500	Retract end mill, rapid move
008	/−X4.750/Y3.00	Rapid move to start of slot 2
009	/Z−.500	Set depth of cut
010	X4.750	Cut slot 2
011	/Z.500	Retract end mill, rapid move
012	M98	Return to start
013	M05	Spindle off
014	M09	Coolant off
015	M20	Remove end mill
016	M23	Install .625 drill
017	M03	Spindle on clockwise
018	M08	Coolant on
019	/X1.0/Y2.0	Rapid move to center of hole 1
020	−Z.875	Drill hole, allow for clearance
021	/Z.875	Retract drill, rapid move
022	/X1.00	Rapid move to center of hole 2
023	Z−.875	Drill hole, allow for clearance
024	/Z.875	Retract drill, rapid move
025	/X1.00	Rapid move to location of hole 3
026	Z−.875	Drill hole, allow for clearance
027	/Z.875	Retract drill, rapid move
028	G98	Return to start
029	M05	Spindle off
030	M09	Coolant off
031	M02	End of program, rewind

The manual part program written for Figure 17–7 is a simple point-to-point program. This type of program is well suited for point-to-point and less complex contoured parts; however, it may not be the first choice for complicated multisurface parts that require programming in two or more axes.

APT

APT (automatically programmed tools) is a product of the MIT's developmental work on NC programming in the mid-1950s. It was first used in production about 1959 and is probably the most extensively used NC language today.

APT is an English-like NC language used to command or guide the cutting tool through a predetermined sequence of machining operations. APT statements are used for one of the following purposes:

1. To define the geometric elements of a part
2. To describe the cutting path
3. To describe auxiliary functions (e.g., tool offset definitions)
4. To define machinery parameters (i.e., operation of spindle, feedrate)

The following short APT program with explanation is designed to introduce the reader to command statements. The part description is shown in Figure 17–7.

APT Sample Part Program

		Column	
1 6 8		**10**	**Comments**
PARTNO		SAMPLE PART	
		MACHIN/VERTMILL 4	
		CUTTER/.500	
		DRILL/.625	
P0	=	POINT/0,−1.0,0	ESTABLISHES STARTING POINT
P1	=	POINT/0,3,−.500	STATEMENTS 6–10 USED TO
P2	=	POINT/0,3.0,−.500	DEFINE PART GEOMETRY
P3	=	POINT/1.0,2.0,0	
P4	=	POINT/2.0,2.0,0	
P5	=	POINT/3.0,2.0,0	
		SPINDL/600,CLW	SET SPINDLE SPEED TO 600 RPM CLOCKWISE
		FEDRAT/20.0	SET FEEDRATE TO 20 INCHES PER MINUTE
		COOLNT/ON	TURN COOLANT ON
		FROM/P0	FROM STARTING POINT
		GOTO/P1	
		GORGT/4.750	FROM P1 MOVE RIGHT TO CUT FIRST SLOT
		GOTO/P2	START OF SLOT 2
		GORGT/4.750	CUT SECOND SLOT
		COOLNT/OFF	
		STOP	STOP MACHINE TO REMOVE END MILL AND INSTALL DRILL

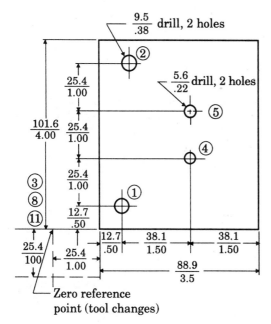

$\frac{9.5}{.38}$ drill, 2 holes

$\frac{5.6}{.22}$ drill, 2 holes

Zero reference
point (tool changes)

Figure 17–7   Sample part for NC program

SPINDL/552,CLW	SET SPINDLE SPEED TO 552 RPM
	CLOCKWISE DIRECTION
COOLNT/ON	TURN COOLANT ON
GOTO/P3	LOCATION OF FIRST HOLE
GODLTA/0,0,−.875	DRILL THROUGH .500 PART, ALLOW FOR CLEARANCE
	TOP AND BOTTOM
GODLTA/0,0,+.875	RETRACT DRILL
GOTO/P4	MOVE TO LOCATION OF SECOND DRILLED HOLE
GODLTA/0,0,−.875	DRILL HOLE 2
GODLTA/0,0,+.875	RETRACT DRILL
GOTO/P5	LOCATION OF THIRD HOLE
GODLTA/0,0,−.875	DRILL HOLE 3
GODLTA/0,0,+.875	RETRACT DRILL
GOTO/PO	GO TO START POSITION
COOLNT/OFF	
SPINDL/OFF	
FINI	FINISH

This brief introduction to APT and manual part programming and the sample programs is not intended to prepare one to be an NC programmer. It merely touches the surface of the knowledge needed to be a competent programmer. The purpose of the brief introduction to manual part programming and APT is to reinforce the discussion of Cartesian coordinate systems and demonstrate the simple English-like statement used in APT. It is important to remember that all NC languages are based on the binary number system and incorporate EIA standard codes.

Advantages of NC Control

Although CNC equipment is more expensive than conventional machine-tool equipment, the benefits generally pay larger returns. The following is a list of some of the advantages of CNC equipment over traditional machine tools:

1. Reduces tooling requirements
2. Improves accuracy and repeatability
3. Minimizes potential human error
4. Reduces need to train machinists
5. Substantially reduces setup time

CAD/CAM

CAD/CAM (computer-aided design/computer-aided manufacturing) is a term used to describe the direct link between product design and manufacturing. Initially, the activities associated with the design aspects of the product were separated from manufacturing. The goal of CAD/CAM is to link and automate the transition from design to manufacture directly.

The CAD/CAM process begins with the use of the computer-aided design station (Figure 17–8). The part or product is created using CAD software. The geometry from the final design is then used to create a tool-path program for an NC machine.

The translation from the CAD software to the tool-path program is accomplished through the use of another software program called a postprocessor. The job of the postprocessor is to accept the input from the CAD package and reformat it to meet the requirements of the NC machine. This involves defining the tool path according to the geometry used to describe the part and adding the necessary parameters to control the NC machine, coolant, spindle speed, and so on. Table 17–5 is an example of the parameters established through the use of the postprocessor for a vertical mill. Notice that coolant, radius compensation, spindle speed, feedrate, and other parameters needed to define the cutting process are set through the use of the postprocessor.

FLEXIBLE MANUFACTURING SYSTEMS

Flexible manufacturing systems (FMS) often utilize the technology of CAD/CAM; however, they extend the automation to other areas of the production process (e.g., material handling). A flexible manufacturing system typically consists of a group of NC machines and automated material-handling devices operating as an integrated work cell (Figure 17–9).

The distinct difference and advantage of the FMS is that it is not dedicated to one part. Flexible manufacturing systems have the ability to process many different parts at various workstations using computer control. FMS is of particular value in medium-size companies with several products and medium volume runs. The high-production, high-volume lines of large companies are usually rigid and inflexible. Production lines and their automation are dedicated to a single product or part. The small companies or job shops often do not produce the volume required to justify the capital investment in FMS equipment.

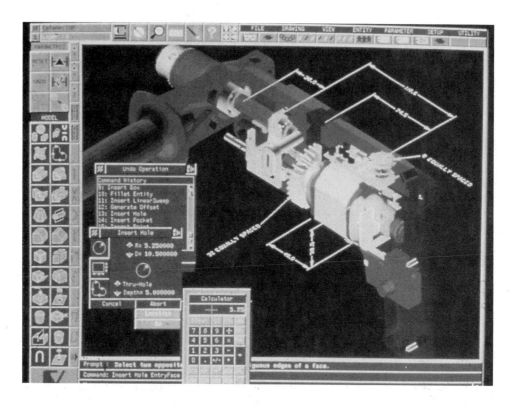

Figure 17–8   CAD station screen showing part geometry *(Courtesy of Computervision Inc.)*

COMPUTER-INTEGRATED MANUFACTURING

The concept referred to as computer-integrated manufacturing (CIM) extends automation to operations management and other functional areas of producing parts and products. CIM incorporates inventory control, production scheduling, quality assurance, financial reporting, distribution, as well as the design and manufacture of a product into the automation arena.

The ideal CIM system would initiate all functions of production from a customer order. Raw materials would be ordered and scheduled for delivery, production schedules generated, material handling planned and operationalized, programs loaded into production equipment, inspection scheduled, distribution arranged, and reports generated.

Although various levels of CIM are being used around the world, few have accomplished full implementation. The potential for achieving maximum effectiveness is, however, an achievable goal.

Pocket Parameters

Cutting Method = Spiral inside out
Pocket Depth = −0.625
Roughing Angle = 0.000
Roughing Cut Spacing = 0.1000
Number of Finish Passes = 1
Finishing Pass Spacing = .1000
Machine Finish Passes At: All Depths
Roll Cutter Around All Sharp Corners
Tool Library: 1   Material: 1140
Tool Number = 3   Diameter Offset = 0   Length Offset = 0
Cutter Diameter = 0.375
Amount of Stock to Leave = 0.015
Feedrate = 20.00   Plunge Rate = 10.00   Spindle Speed = 1750
Coolant = off
Rapid Depth = 0.10
Starting Sequence Number = 100   Increment = 10   Program N. = 5
No Rotary Axis
Linear Array: Nx Ny=1 Dx, Dy = 0.0000 0.0000
Depth Cuts: Rough: 3 Finish: 1
Home Position = X0.0000 Y0.0000 Z0.0000
Misc. Real [1] = 0.0000   Misc. Integer [1] = 0
Mill in the XY Plane
Display: No Tool Display   Toolpath

Table 17–5   Postprocessor Parameters for CAD/CAM

SUMMARY

The need for lighter, better aircraft and space vehicles has provided new machining methods and control systems that provide intricate shapes with great accuracy and repeatability. The use of computer control in manufacturing has now been extended beyond designing and machining of parts. The design and machining operations have not only been automated but integrated to link one to another directly. More recently, systems (i.e., manufacturing and operations) have been integrated and computerized to monitor and control the efficiency of production more effectively.

Factories of the future may, in fact, be large "black boxes" (raw materials in one end and finished goods out of the other) with greatly reduced human effort and increased high-level automation. If so, history will surely give credit to John Parsons and others for developing the first NC machines. The ideas of Parsons and the development work of researchers at MIT marked the first use of programmable computers to control machine-tool movement. This work and further advancements in digital electronics and computer technology prepared the way for automation of many production functions (i.e., material handling, inspection, design, scheduling).

Figure 17–9   FMS work cell *(Courtesy of Cincinnati Milacron Co.)*

The potential for computer control in the manufacturing arena has not yet been reached. The need to compete in a worldwide market will surely increase the use of automation at all levels.

 REVIEW QUESTIONS_____

1. Who has been given the credit for the development of NC control?
2. What academic department and institution developed the first contouring NC machine?
3. What specific need provided the impetus for the development of NC technology?
4. Explain the difference between NC and CNC.
5. List the three major components of a CNC system.
6. List three types of drive motors described in this chapter and describe their features.

7. What are two disadvantages of manual data input?
8. What organization has standardized NC code?
9. What two letters of the alphabet have not been assigned an address in NC code?
10. What is the name of the system used in NC programming to establish the geometry that defines a tool path?
11. Describe the difference between incremental and absolute positioning.
12. What number system is used as the basis for NC programming?
13. APT statements are used for one of four purposes. List the four purposes.
14. List five advantages of CNC equipment over conventional machine tools.
15. Describe the purpose of the postprocessor in CAD/CAM applications.
16. What advantage does FMS have over conventional systems for medium-size manufacturing plants?
17. In what way is CIM different from CAD/CAM?

SUGGESTED ACTIVITIES _____

1. Select an object from your room, house, or garage that has basic geometric form (i.e., straight lines, full or half circles). Sketch the views of the object required to define it. Using the Cartesian coordinate system, locate x, y, and z coordinates for all major features of the object (e.g., center points of holes, arcs, and circles). Select a starting point that is 1 inch from the southwest corner of the front view in both the x and y axes. Use incremental positioning to define the locations of the features of the object.
2. Visit a local manufacturing plant with CNC equipment. Find out what programming language they are using and whether incremental or absolute programming is being used. If possible, obtain a sample of a part that was machined using CNC programming. Describe how the part could be machined using conventional processes. List the advantages of using CNC to machine the part.

CHAPTER 18

Fabrication Processes

Fabrication processes are used to twist, bend, stretch, combine, or compress materials into a desired shape. Varying amounts of pressure and heat are used in conjunction with special fixtures and machines to accomplish the task of fabricating, depending on the type and size of material.

Unlike the casting or machining processes described earlier in this text, fabrication processes do not add or remove material but simply change shape or form. This chapter describes the various processes used to shape wood, metal, plastic, and ceramic materials. First, it is necessary to understand some basic engineering concepts of what happens to material when we twist, bend, pull, combine, or compress it into shape.

BASIC ENGINEERING CONCEPTS

When material is fabricated into shape, a common relationship between tension and compression is achieved whether the material is heated or worked cold. In general, materials that are put into tension will become longer and thinner, while materials that are put into compression will become shorter and thicker. Figure 18–1 shows how material is always in tension or compression when it is fabricated into a new shape.

The relationship between tension and compression is called **stress** and can be predicted on a stress diagram to show the yield point (the point when the material begins to change shape). Figure 18–2 shows the relationship of stress and strain for mild carbon steel.

Some materials will reach their yield point at normal room temperature, while others will require the use of heat or chemicals. More detailed information on the characteristics of materials is presented in Section 2 of this text and, therefore, not discussed in this chapter. However, it is important for the reader to understand that the fabricating of materials

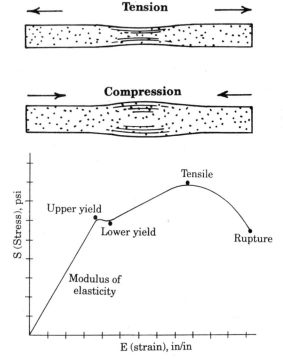

Tension

Compression

Figure 18–1   All materials are always in tension or compression

Figure 18–2   The relationship between stress and compression can be plotted on a stress–strain diagram

does cause strain and stress and that the yield point of materials must be reached before a permanent change in shape can occur.

Because each material requires slightly different processes for fabrication, this chapter describes the hot and cold processes according to the material being worked.

METAL FABRICATION

Hot Working

Some metals are inheritably soft, while others are hard and tough. Soft metals such as low-carbon steel, aluminum, copper, and brass can usually be fabricated in their normal (cold) state. Others such as cast iron, high-carbon steel, and a variety of alloyed metals usually require the use of heat to help reach a yield point for fabrication. Heating softens most metals but not all. For example, tungsten and cobalt are stiff even when white hot. Even though most metals can be fabricated cold or hot, these special metals require a white-hot state before a new shape can be achieved.

Body-centered cubic structures in metals, such as in most carbon and alloy steels, permit hot or cold deformation (change in shape). Face-centered cubic structures, such as austenitic (nonmagnetic) steels, deform easily but harden rapidly when cold worked. Annealing (heating) between forming operations is, therefore, frequently required. Hexagonal close-packed metal such as magnesium requires heating in order to deform and prevent rupturing of the material. Figure 18–3 shows the families of metals.

The major purpose of fabricating metal is to change its shape. The decision on whether to fabricate or use some other method to shape the material is based on its purpose, and cost of procedure. For instance, it would not be necessary or cost-effective to cast metal to make a tin can. The material is difficult to cast into such a thin structure, while it is simple to fabricate with a punch-and-die process. Sometimes, the secondary benefits of fabricating metal are desired because they add certain benefits that cannot be achieved through other processes. For example, the forging of engine parts can increase strength over casting of the same parts. Figure 18–4 shows how the shaping of a part by heat and pressure (forging) can cause the grain structure to change, which actually increases strength.

Cold Working

When metal is formed into different shapes below the recrystallization temperature, it is referred to as a cold-working process. During this process, both stretching and compression occur. When the amount of pressure exerted exceeds the amount of resistance of the material being formed, the yield point is reached and a change of shape occurs.

Sheet-metal forming makes up a large area of the metal-fabrication industry that uses cold-working processes. Most of these processes make use of machines designed to bend special joint designs. Bar folders and box and pan brakes are usually used to bend flats, angles, and special seams.

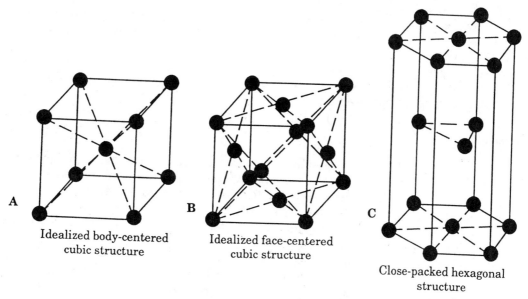

A   Idealized body-centered
cubic structure

B   Idealized face-centered
cubic structure

C   Close-packed hexagonal
structure

Figure 18–3   Basic structures of metals

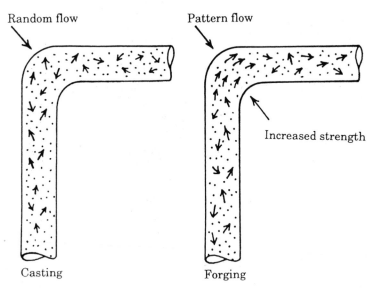

Random flow

Pattern flow

Increased strength

Casting

Forging

Figure 18–4   Forging can actually increase strength by changing grain structure

Metal that is less than 25 gauge thick does not need a special bend allowance because there is enough stretch in the metal to allow accurate work. Metals heavier than 24 gauge do require bending allowance.

A press brake is used for bending heavier sheet and steel plate. A 1,000-ton press brake can bend metal up to 1 inch thick. It is used for heavy fabrication applications as shown in Figures 18–5 and 18–6.

Complex shapes can also be pressed by a special process known as sheet metal stamping. By using a male and female die, the press can shape one piece items such as automobile parts, household goods, tools, toys, and the like.

Roll Forming

To form a radius with metal, a set of rolls is used. Sheet metal and heavy gauge metal can be shaped by roll forming using the same basic concept. As shown in Figure 18–7, the curvature is controlled by the distance roll #3 is from roll #1. Rolls #2 and #3 are adjustable. Roll #2 controls the space allowed for the thickness of the metal. The piece is formed by rolling the metal back and forth until the correct diameter is obtained. Each time the metal passes through the rollers, the follow roll #3 is adjusted tighter until the operation is complete.

For special applications on sheet metal, a rotary machine can be used to form or crimp the edge of the metal. Special combinations of rollers and crimpers allow the rotary machines to conduct more than one operation at a time. Figure 18–8 shows a pipe receiving a bead and crimp at the same time.

Figure 18–5 To bend heavy sheet and plate metal, a large press brake is used *(Courtesy of Niagara Machine & Tool Works)*

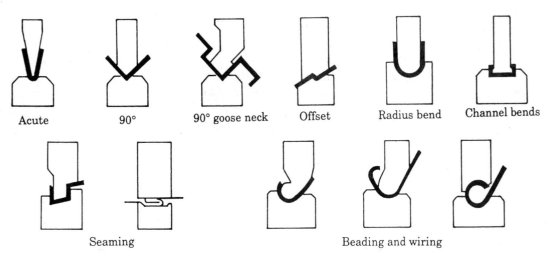

| Acute | 90° | 90° goose neck | Offset | Radius bend | Channel bends |

| Seaming | | | Beading and wiring | | |

Figure 18–6 Press brakes with a few standard dies can produce many bends and shapes

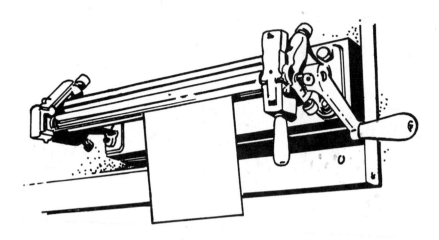

Figure 18–7   A form roll is used to bend circles

Larger cold-roll-forming machines can form metal into complex shapes such as garage door rails, roof gutters, spouts, and other special shapes.

Die Forming

Flexible die forming uses rubber, oil, or water as the energy-transmission source to form metal. The upper part of the mold is made of rubber and acts as the female to stretch metal over the male mold. As the mold closes, the rubber forces into the trapped space and pushes the metal around the mold. To help increase the life of the mold (usually up to 20,000 parts), sharp corners are removed and a lubricant such as powdered talc or graphite is used.

Products such as license plates, furniture, and aircraft parts are made by the flexible die forming process. Hubcaps, pots, electrical fixtures, and other types of products which require stretching the material outward can by produced by bulging, a reverse process of flexible die forming. Figure 18–9 shows the major difference between the two processes.

Stretch Forming

Difficult large shapes such as airplane wings, automobile parts, trailer parts, etc. can be formed by stretch draw forming. The material is pulled into tension and stretched over a mold. Because the tension is close to the yield point of the metal, very little spring back or thinning of metal occurs. In fact, the stretch forming actually conditions and increases the tensile strength of the metal by as much as 10 percent. Figure 18–10 shows a diagram of how the metal is molded to the proper shape by stretch forming.

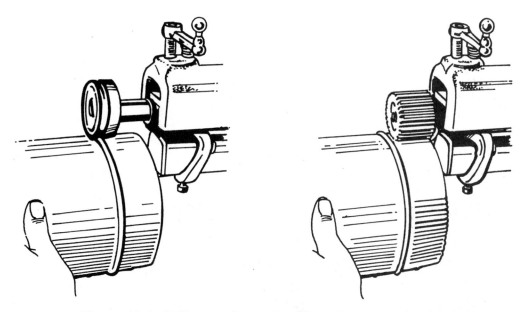

Figure 18–8   Rollers are also used to crimp edges of sheet metal

Metal Spinning

Like stretch forming, the process of metal spinning wraps the metal around a mold. However, in the case of metal spinning, the metal is clamped to a chuck on a lathe and rotated while the forming tool is pressed against the disc, forcing it to conform to the shape of the chuck (mold).

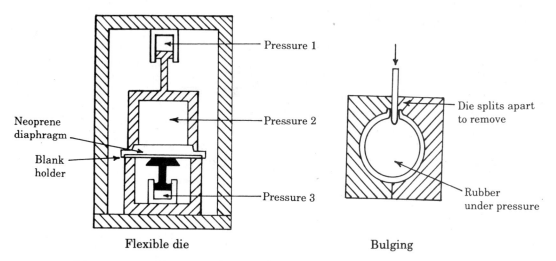

Figure 18–9   How metal is formed using flexible rubber-molding processes

Spinning operations are used on odd-shaped parts such as bowls, pans, trays, and other items which require a cylindrical shape. Aluminum, copper, brass, pewter, and stainless steel are the most common materials used for metal spinning. Larger shapes are executed on large, power spinning machines and may reach diameters of 12 feet with thicknesses of up to ¼ inch. Figure 18–11 shows methods for metal spinning.

Rigid-Die Forming

Many production parts are formed by pressure and special dies. Such products as automobile bodies, structural shapes, stoves, refrigerators, and cooking utensils are made from the rigid-die process.

The blank (metal) is placed between a male and female die and forced by power into the shape desired. Special deep-drawing dies are used to get high-quality finishes on deep draft shapes. Metal can be stretched into shape, causing some work hardening to occur; and on

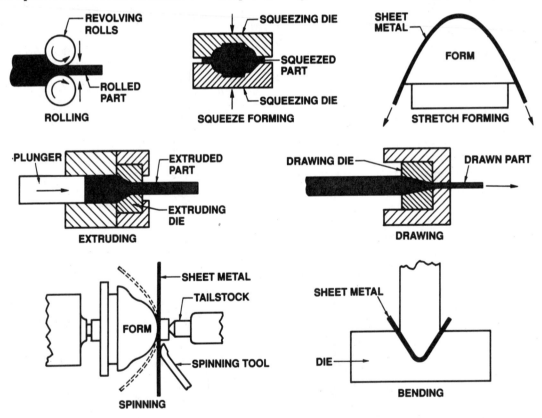

Figure 18–10 How metal is stretched over forms *(From Komacek,* Production Technology, *Copyright 1993 by Delmar Publishers, Inc. Used with permission)*

thicker items, it may be necessary to anneal the metal and conduct a second stage of drawing until the final shape is produced.

Cold Forging

Forging metal cold usually can be classified as an upsetting, swaying, and conditioning process. The shaping of a rivet head is an example of a cold forging process. High pressure and some annealing procedures can allow the cold forging of a number of parts. Many screw fasteners, bolts, and ball bearings are made in this manner. The process is fast and cost-effective for such items.

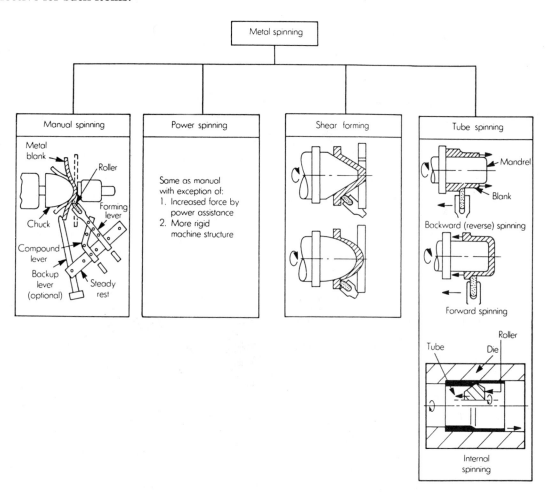

Figure 18–11   Types of metal-spinning processes *(From Goetsch,* Modern Manufacturing Processes, *Copyright 1991 by Delmar Publishers Inc. Used with permission)*

Explosive Forming

Using dies covered with an epoxy facing material, difficult materials such as titanium and tungsten can be formed by releasing a charge into the mold and using the resulting shock waves to press the material around the mold (Figure 18–12).

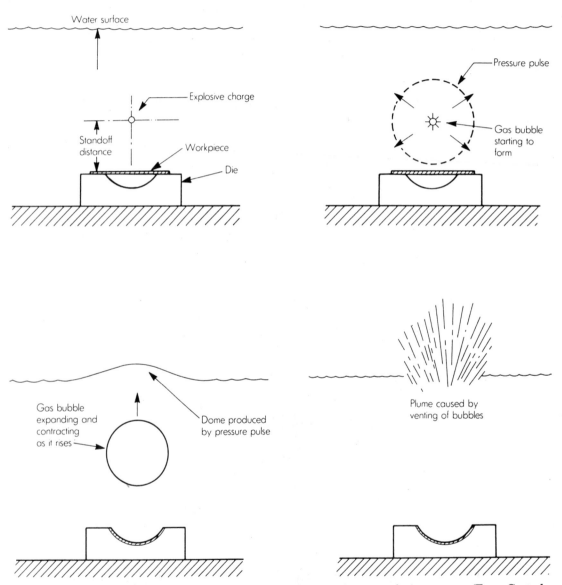

Figure 18–12   Difficult metals to shape such as titanium require an explosion process *(From Goetsch, Modern Manufacturing Processes, Copyright 1991 by Delmar Publishers Inc. Used with permission)*

Combining

Another way to fabricate metal into desired shapes is to cut, fit, and combine using various welding, mechanical fastener, pressure fitting, and adhesion techniques. These fastening processes are discussed in further detail in Chapter 19. However, such items as truck bodies, water tanks, wrought-iron railings, airplanes, bridges, and automobiles use a variety of forming and fastening techniques in combination to fabricate metal into usable products.

WOOD FABRICATION

Wood is one of the oldest materials used by humankind to make utensils, weapons, boats, furniture, and shelters. Many of these items are now fabricated from other materials such as metals, plastics, and ceramics. However, wood still remains popular for use in furniture and home construction.

Wood is an organic material made up of long, tubular woody cells which are bonded together by the glue-like substance lignin. Chapter 6 provides a more detailed look at wood as a polymer. Because of the unique cell structure of wood, the material has an affinity for water in both vapor and liquid forms. By controlling the moisture content of wood, we are able to fabricate the material into various shapes by bending, twisting, and laminating.

Hot Working

When we raise the temperature of wood, the material becomes weaker; and when we lower the temperature back to normal, the material regains its original strength. The effect on strength is immediate, and its magnitude depends on the moisture content of the wood and how long the material is exposed to high temperatures. Research shows that air-dried wood can be exposed to temperatures up to 150 degrees Fahrenheit for one year without significant loss of strength when it is returned to normal temperature.

For each increase of 1 degree Fahrenheit above 70 degrees Fahrenheit, wood strength will decrease by 0.33–0.50%. If moisture content is also raised, the change in strength will decrease even more significantly; conversely, if lowered, the strength will be less affected.

Because of this principle, wood materials can be bent and twisted into shape by applying heat and moisture. The use of a mold as shown in Figure 18–13 allows the material to be pressed into shape using steam heat and pressure. After several hours in the mold, the material will retain the shape and return to most of its original strength. This same heat–moisture–pressure concept can also be applied to veneering of various shapes. It is actually possible to increase strength and beauty by selecting veneers (thin strips of wood) that are placed in the molds with glue to cool and dry. Such products as water skis, chair seats, rocking chair parts, and structural laminated beams are made using heat and moisture to fabricate the wood into a desired shape.

Cold Working

Most wood materials are fabricated without using a heat source. Some are twisted and bent, but most are multiple parts joined together by some type of fastener.

Because wood has a cellular structure, it is possible to soak the material in a chemical and make the material elastic. This procedure causes plasticization, a temporary condition caused by the loosening of the lignin-cellulose structure. Once the material is shaped, it is possible to change the strength characteristics of wood by using different wood modification processes. One of the most important chemical altering processes is known as wood plastic composition (WPC). The workpiece is placed in a vacuum chamber, and a plastic monomer such as methacrylate is introduced into the chamber and enters the wood cells. The wood is then subjected to radioactive isotopes, causing polymerization and making the material plastic-like with improved dimensional stability.

Another popular method for improving the dimensional stability of wood is known as the polyethylene glycol process (PEG). This treatment requires the material to be soaked in the PEG chemical for several weeks. Once treated and shaped, the material resists shrinking during the drying process, thereby improving its dimensional stability.

The majority of wood fabrication occurs in the wood shop by joining machined parts together or on the construction site where structural components are joined together. In both cases, the

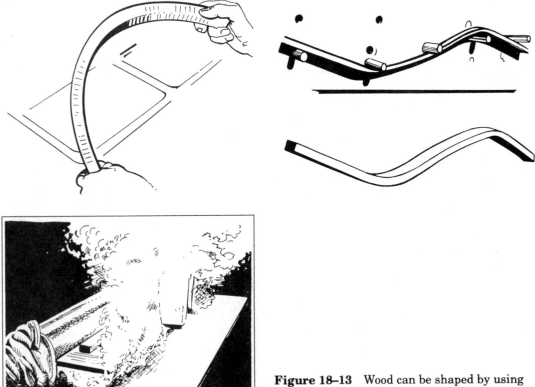

Figure 18–13 Wood can be shaped by using moisture and pressure

material is not actually shaped but combined to created desired shapes. Joining processes for wood material are discussed in detail in Chapter 19. As can be seen in Figure 18–14, wood products that were once solid wood may be a combination of new materials and technologies.

Figure 18–14   Many beautiful products of wood are shaped and fabricated for consumer use

PLASTIC FABRICATION

As discussed earlier in this text, there are two major classifications of plastics: thermoforms and thermosets. Thermoforms (sometimes called thermoplastics) refer to those plastics which can be shaped and formed many times due to their linear molecular pattern. Thermosets, on the other hand, refer to those plastics which can be shaped only one time because of their cross-linked molecular pattern. The cross-bonding, because of its complex molecular chains, prohibits plasticity and plastic flow in cured thermosets. Following is a list of the most common thermoform and thermoset plastics:

Thermoform	**Thermoset**
Acrylics	Silicones
Acetals	Phenolics
Cellulosics	Urethanes
Polyamides	Epoxies
Styrenes	Aminos
Vinyls	
Polyolefins	
Fluorocarbons	

Most plastics are formed into their final shape by casting, thermoforming, reinforcing, foaming, and molding. Since these processes are covered earlier (see Chapters 13 and 14), they will not be discussed here. Instead, this chapter section concentrates on how to fabricate plastic by twisting, bending, and shaping as well as combining parts together.

Hot Working

The shaping of plastic when heated is very easy and can be done with a simple heat source following some basic hand-forming procedures. The acrylics, cellulose acetates and butyrates, high-impact styrenes, and rigid vinyl plastics are most commonly used for fabrication. Cellulose nitrates have a high flammability and are not recommended for hot working processes.

The most popular type of plastic to be fabricated is the acrylic (see Figure 18–15) because of its clear beauty. It has good plastic memory and will easily return to its original shape if a forming mistake is made by simply reheating it to the original forming temperature. Calendered sheet is more economical than cast sheet and, if the clarity is not important, will work very well and have unique beauty.

The most common sources used to heat plastic are a kitchen oven; electric hot pot; heat gun; soldering gun; strip heater; infrared radiant heating; and, in some cases, an open flame. Table 18–1 shows the various temperature ranges used to heat plastic. Some plastic materials must be heat-soaked for upwards to one hour before they can be easily shaped. Thin sheets also require special heating beyond their pliable state because they cool very rapidly.

Shaping, twisting, and bending plastic materials in their heated pliable state can be done by hand with gloves or by using special fixtures designed to shape and hold the material in place until it cools and retains its new shape. If jigs and fixtures or special molds are to be used, they should be made of smooth and soft material such as wood, ceramics, or plastic. If the mold is not smooth, the surface of the plastic will be marked because of its pliability.

The use of localized heat (heat gun or strip heater) can also help create special shapes and can provide reheating for larger projects. The heat gun provides constant heat which is moved along as the plastic is formedalized; heat may also be used for 90-degree bends in plastic using the heat gun and heat shields or a strip heater.

Cold Working

The processes related to the fabrication of plastic in its natural cold state are much more limited than in hot working. Those plastics which can be shaped cold must be held in place because of their tendency to return to their original shape. Cylinders, various angle bends, and twisting can be accomplished also by bending the material beyond the deformation yield point so that the spring back results in the desired shape. However, tight bends such as 90 degrees can result in weakness and, therefore, are not desirable. Cold fabricated plastic parts can also be processed by cutting and joining various shapes together. Both mechanical fasteners and welding procedures work well with plastic. Plastic also responds exceptionally well to a number of solvents and glues.

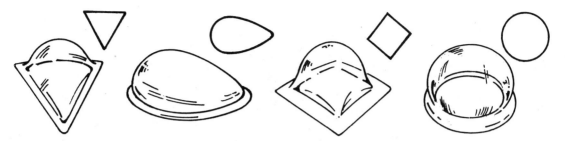

Figure 18–15 Acrylic plastic shapes very well

Plastic	Heat range
Acrylic	260 to 360°F
Cellulose acetate butyrate	265 to 320°F
Acrylonitrile butadiene styrene	300 to 350°F
Polyethylene	325 to 425°F
Cellulose acetate	270 to 325°F
Polyvinyl chloride	255 to 355°F
High-impact polystyrene	365 to 385°F
Polycarbonate	440 to 475°F
Acetal	365 to 390°F
Styrene acrylonitrile	430 to 450°F
Ethyl cellulose	225 to 290°F

Table 18–1 Heat ranges for sheet plastics

A popular method for fabricating plastic materials for products without heat is to laminate various clothes and mats using thermosetting resins as a binder and special molds to create the shape. The mold is usually a slightly smaller version of the actual object to be made. It may be made from most materials, but those which are porous will require a film or film-forming type of mold release to prevent the fiberglass coating from adhering to the mold itself. Once the necessary preparation is made for the mold and the parting agent has been applied (see Figures 18-16A and B), the process of applying material and resin begins. Alternate applications of resin and cloth or resin and mat (chopped cloth) are applied until the appropriate thickness is achieved.

The smooth finish produced on a fiberglass product is called the gel coat. It is a pigmented (for color) thixotropic resin, which leaves a very smooth finish. In general, this coat is applied first to the mold (remember that molds work in reverse) and should not exceed 0.010 to 0.020 inch in thickness. Excessive thickness will result in cracking and crazing. If you look inside a boat hull, you will easily see the difference that the gel coat makes on the finished surface of the hull.

Fiberglass products (as they are usually referred to) make up a large part of the plastic-fabricating industry. The most popular products are boats (see Figure 18–16C), automobiles, building structural parts, chairs, desks, and fences. The ability to create unique shapes with

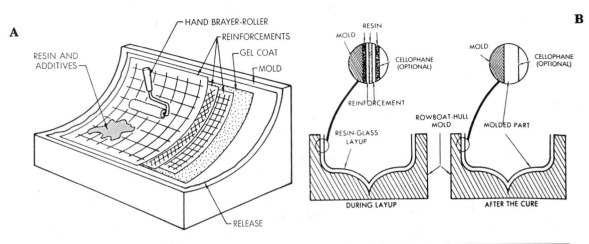

Figure 18–16 **(A)** and **(B)**, parts are fabricated using a laminated process with a mold *(From Richardson,* Industrial Plastics: Theory and Application, *2nd ed., Copyright 1989 by Delmar Publishers, Inc. Used with permission);* **(C)** most boat hulls are made of fiberglass *(Courtesy of Photri, Inc.)*

impressive strength and quality finishes makes fabricated reinforced-plastic materials a very important material for both the manufacturing and construction industries.

CERAMIC FABRICATION

The processing of ceramic materials into usable products is easily accomplished using similar methods employed to process metal or plastic. However, the fabrication of ceramic materials is simplified because of their natural plasticity and, with the exception of glass, usually performed in a green (unfired) state.

Hot Working

Clay and concrete materials are fabricated while cold (in a green state); and the use of heat is designed to harden and strengthen the material. Glass, on the other hand, is fabricated while hot and cooled to gain its hardness and strength.

Because glass does not crystallize when it is cooled, it has been called a hard liquid. This is because the atomic pattern in hardened glass resembles the atomic pattern in liquid glass. It remains in this noncrystalline state regardless of how hot or cool the material is. Therefore, while some limited processes can be applied to working glass while it is cold, most processes are applied while it is hot.

Glass can be cut and shaped easily while it is hot. Both the shear and flame cutting methods can be used to separate glass.

Most glass is shaped by molding processes discussed earlier in this text. Specialized fabrication of glass is done by laboratory technicians and craftspeople. We have all been amazed by the glassblower who heats glass material to a plastic state and shapes it at will (see Figure 18–17). Bottles, art forms, glasses, mugs, and a number of other shapes can be accomplished by the fabrication of hot glass. While not as glamorous as the glassblower at the county fair, the laboratory technician is also an artist/technician making scientific glassware to be used for research.

Cold Working

Clays and concretes make up the ceramics that are primarily fabricated while cold. Clay is a very plastic and shapable material while in its green state. The material can be twisted, bent, stretched, and compressed using simple tools. It can also be shaped while it is turned on a potter's wheel (Figure 18–18) in a similar fashion to cutting wood on a wood lathe or spinning metal on a metal-spinning lathe. The major difference, of course, is that the potter can use his or her hands and liquids to control the shape.

Clay can also be shaped by cutting tools, such as wire cutters, which shape the various types of bricks. Ceramic pipe, drains, roof tiles, bathroom tiles, and other types of construction products are fabricated like bricks in a green state and then fired in a kiln for strength.

Concrete products rely on a different principle to achieve shape and strength. Basic materials (aggregate and cement) are combined with water to chemically cause curing (hard-

Figure 18–17   The professional glassblower can make any shape from glass *(Courtesy of Photri, Inc.)*

ness and strength). While their basic shapes are usually achieved through some type of molding process, it is possible to shape concrete while it is in a semihard state.

Fabrication of clay and concrete products extends beyond the actual shaping of the basic product into the use of such products to create other products. Bricks, concrete blocks, prestressed concrete beams, flooring and roofing tiles, drains, and a variety of other ceramic products can be used to fabricate homes, offices, bridges, highways, dams, drainage systems, and a host of other products.

Ceramic materials are an important part of our everyday life. From the cups and plates that we use daily to the heat shield that protects the space shuttle (see Figure 18–19) during reentry, ceramic materials are everywhere.

SUMMARY

Fabricating materials into different shapes requires a variety of hot and cold working processes. An understanding of how materials are made and how they behave under different conditions will help one predict what will happen when force is applied.

Figure 18–18 The professional potter shapes many objects on a potter's wheel *(Courtesy of Photri, Inc.)*

While metals can be worked hot or cold, the trade-off for using heat is in strength and uniformity. Wood also gives up strength when worked with heat or chemicals, yet special injected agents in wood can make it stiff and strong during the conditioning process.

Some plastic materials, called thermosets, can be formed only once when heated. Others, called thermoforms, can be heated and worked into different shapes several times. In a sense, that is also true of ceramics which, while in a green state, can be worked and reworked until desired shapes are formed. Unlike wood, metal, and thermoform plastic, once the ce-

Figure 18–19   High-tech ceramics are used as a heat shield for the space shuttle *(Courtesy of NASA)*

ramic is fired, heat effects are not able to create a state of "plastic" for further fabrication. Instead, the ceramic must be worked cold just like plastic thermosets.

Fabrication is the process of molding, bending, pulling, twisting, and forcing materials into a desired shape. This chapter introduces the readers to the basics of fabrication used for manufacturing.

 REVIEW QUESTIONS _____

1. What is the primary difference between fabricating processes and casting, machining, or molding processes?

2. List and describe two basic engineering principles that are basic to all fabricating processes.
3. When you bend or twist a material, it will tend to return to its original shape unless the _____ point is reached. Why must this point be reached before a permanent position or shape is achieved?
4. Give a brief one-line description of the following terms:
   a. body-centered cubic structures
   b. face-centered cubic structures
   c. hexagonal close-packed structures
5. How much allowance should be provided for making a 90-degree bend using $1/8$-inch hot-rolled steel?
6. What types of products are made using the flexible dies forming process?
7. Describe the wood plastic composition process for fabricating wood materials.
8. List and describe the two major categories of plastic materials. How do they differ in the way in which they can be fabricated?
9. List and describe three heating sources used to heat plastic materials for fabrication.
10. Why is thixotropic resin used in fiberglass-molding processes?
11. What type of ceramic materials are fabricated while hot? Why aren't more ceramics wrought in the hot state?
12. Why is glass called the hard liquid?

SUGGESTED ACTIVITIES

1. Design and build a project or product that has all four materials fabricated into one.
2. Develop a tank and chemically treat wood to demonstrate how it may become plastic-like and shape it into a product.
3. Using a heat gun, bend and twist plexiglass into a napkin holder.
4. Invite a glassblower to give a demonstration to the class.

CHAPTER 19

Material Fastening Processes

The concept of joining two materials (similar or dissimilar) together is called fastening. Many products are made up of subparts which (once processed) need to be assembled together using some type of fastening technique.

For products that must be shipped and assembled on site, mechanical fasteners such as bolts and screws are very popular. **Mechanical fasteners** are usually special devices or joint designs that allow almost all materials to be joined together quickly in permanent, semi-permanent, or temporary arrangements. However, some products require a more permanent or stronger bonding procedure than can be afforded by mechanical techniques and need to take advantage of cohesive or adhesive bonding processes.

Cohesive bonding techniques make use of fusion (melting of materials to flow together) processes such as welding to bond the materials together. *Adhesive* bonding techniques do not require the melting of the base materials but instead make use of a bonding material such as glue or silver solder to diffuse and adhere the materials.

This chapter concentrates on fastening techniques for wood, metal, ceramic, plastic, and combinations of these materials. The selection of an appropriate procedure can make a significant difference in the reliability and quality of the product.

MECHANICAL FASTENING

Mechanical fasteners made of metal or plastic are used extensively to fasten wood, metal, plastic, ceramics, and combinations of these materials. The most common metal fastener is the nail. It is used extensively in the construction industry and in home maintenance. Most everyone knows what a nail is, and most of us have had to use a nail for some project. Nails come in a variety of sizes and styles (Figures 19–1 and 19–2).

372

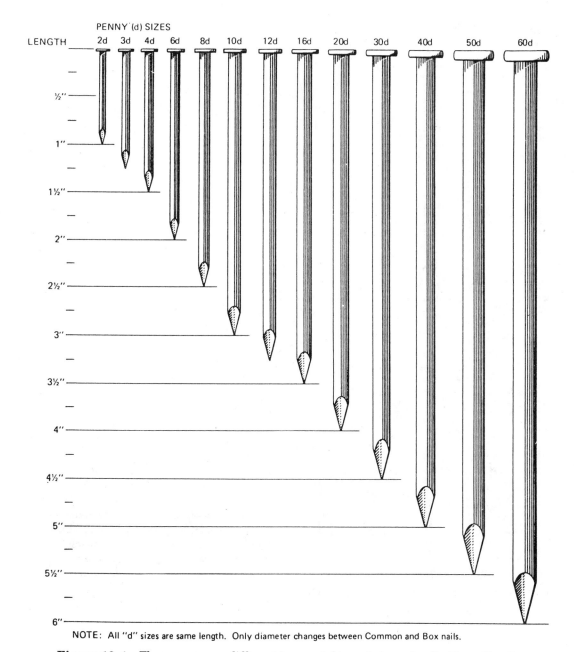

NOTE: All "d" sizes are same length. Only diameter changes between Common and Box nails.

Figure 19–1 There are many different types, styles and sizes of nails *(From Van Orman, Estimating for Residential Construction, Copyright 1991 by Delmar Publishers, Inc. Used with permission)*

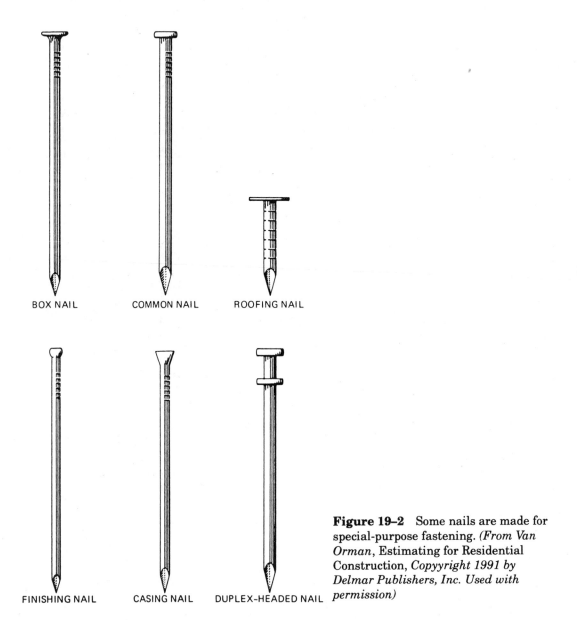

BOX NAIL   COMMON NAIL   ROOFING NAIL

FINISHING NAIL   CASING NAIL   DUPLEX-HEADED NAIL

Figure 19–2   Some nails are made for special-purpose fastening. *(From Van Orman*, Estimating for Residential Construction, *Copyyright 1991 by Delmar Publishers, Inc. Used with permission)*

Nails

Common nails are made with heavy flat heads. They are used extensively for fastening the structural parts of a house. Because of their driving strength, they have excellent holding power. Box nails look like common nails except that they have thinner heads and smaller shank diameters. Box nails are used for siding and trim applications. Because of their smaller dimensions, they are less likely to split the wood; because of the flat head, they have good holding power.

Finishing nails have small heads and are usually driven below the surface with a nail set and covered with a wood filler. Brads are small finishing nails (¼ to 1¼ inches in length) used for very thin materials. A casing nail is made with a cone-shaped head (see Figure 19–2) and has better holding power than the finish nail, making it ideal for cabinet work.

Nails are sized according to their penny symbol represented by a lowercase *d*. A 2d (2 penny) nail is 1 inch in length. For each additional penny, add ¼ inch in length upward to 3 inches. The size of the shank and head is automatically increased according to the d classification. Figure 19–1 can be used to estimate the length of nail needed. A simple rule of thumb is this: The length of a nail should be three times the thickness of the board being nailed. Special nails may also be purchased to fasten wood to other materials or fasten other materials to wood. For example, short nails with very large heads are used to fasten asphalt roof shingles to wood. They also are excellent for attaching plastic materials to wood. Nails designed to fasten wood to concrete are available. They are made of special metal alloys to increase the strength needed to penetrate the ceramic material. Figure 19–2 shows the family of special nail fasteners that are available on the market today.

Screw Fasteners

The second-most common fastener in use today is the screw. Screws can be used to fasten materials together tightly with excellent strength and have the added feature of easy disassembly. Screws can be used with all materials but are classified as wood screws, metal screws, and sheet rock screws.

Wood screws have an aggressive incline and come in flat, round, and oval head shapes. Most wood screws are slot heads (see Figure 19–3) or Phillips heads and require pilot holes for the thread and shank. The size of a wood screw is designated by the diameter (in gauge size) of the shank and the length of the screw in inches. For example, a 1½ inch x 10 FH screw is larger than a 1 inch x 8 FH screw. The following designations apply to head styles:

FH = Flat Head
OH = Oval Head
RH = Round Head

Metal screws are sometimes called sheet metal screws. They are usually made from a tougher material than wood screws and have the feature of a self-starting thread. Even so, most metal screws are used with a pilot hole equal to the shank of the screw for ease of use and assistance to ensure a tight joint between materials.

New to the mechanical fastener market is the sheet rock screw. Available in plain and galvanized form, it has become popular as a general fastener because of the developments in screw-drill technology. It usually has a Phillips flat head configuration, making it ideal for sheet rock applications with automatic screw guns.

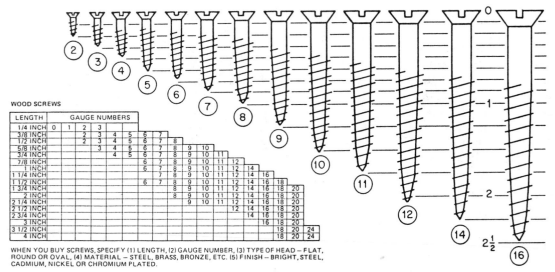

WOOD SCREWS

LENGTH	GAUGE NUMBERS																
1/4 INCH	0	1	2	3													
3/8 INCH			2	3	4	5	6	7									
1/2 INCH			2	3	4	5	6	7	8								
5/8 INCH				3	4	5	6	7	8	9	10						
3/4 INCH					4	5	6	7	8	9	10	11					
7/8 INCH							6	7	8	9	10	11	12				
1 INCH							6	7	8	9	10	11	12	14			
1 1/4 INCH								7	8	9	10	11	12	14	16		
1 1/2 INCH							6	7	8	9	10	11	12	14	16	18	
1 3/4 INCH									8	9	10	11	12	14	16	18	20
2 INCH									8	9	10	11	12	14	16	18	20
2 1/4 INCH										9	10	11	12	14	16	18	20
2 1/2 INCH													12	14	16	18	20
2 3/4 INCH														14	16	18	20
3 INCH															16	18	20
3 1/2 INCH																18	20	24
4 INCH																18	20	24

WHEN YOU BUY SCREWS, SPECIFY (1) LENGTH, (2) GAUGE NUMBER, (3) TYPE OF HEAD — FLAT, ROUND OR OVAL, (4) MATERIAL — STEEL, BRASS, BRONZE, ETC. (5) FINISH — BRIGHT, STEEL, CADMIUM, NICKEL OR CHROMIUM PLATED.

Figure 19–3   Wood screws have several head styles and different gauge sizes *(From Lewis, Carpentry, Copyright 1984 by Delmar Publishers, Inc. Used with permission)*

Machine Screws and Bolts

Machine screws and bolts are essentially the same type of fasteners. The major difference is their size, with machine screws being smaller (less than ¼ inch in diameter) and bolts being larger. They have the same type of head styles as wood screws but cannot be used without a proper receiver such as a threaded hole or threaded nut.

Machine screws are designated by the diameter of the shank, number of threads per inch, and overall length. Most machine screws have fine threads which are good for products that may have movement and vibration. Bolts, on the other hand, have both fine and course threads. Sometimes bolts have slotted heads like machine screws, but most bolts have square or hexagonal heads. As shown in Figure 19–4, machine screws and bolts have a variety of head styles. They also make use of several types of washers and nuts depending on the requirements of the fastener.

One interesting cross between a bolt and a wood screw is called the lag screw or bolt. It is used like a heavy-duty screw but has the head of a bolt and is tightened like a bolt with a wrench. It is used for heavy-duty fastening requirements of structural wood applications. Figure 19–5 shows various types of lag bolts, machine bolts, and carriage bolts.

Rivets

Rivets are used primarily to fasten metal and plastic materials. They are a permanent type of fastener and should not be used if assembly or disassembly is anticipated in the future. Many of the former riveting applications are now done by welding. However, rivets are still popular when excessive heat (caused by welding) would cause problems.

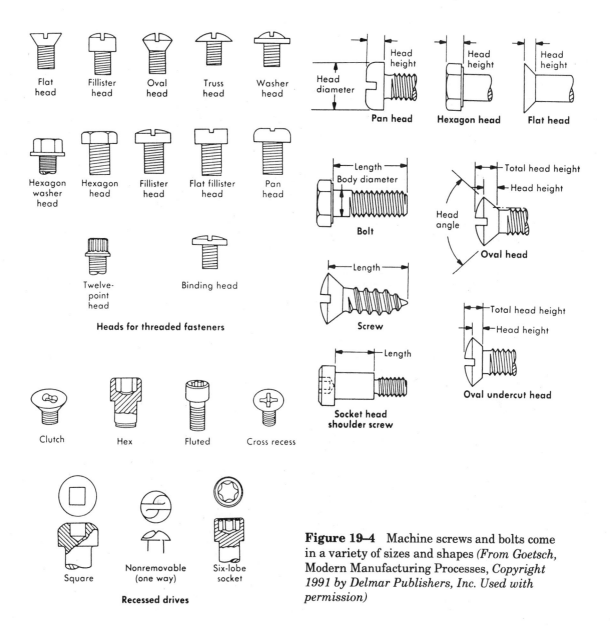

Figure 19–4   Machine screws and bolts come in a variety of sizes and shapes *(From Goetsch,* Modern Manufacturing Processes, *Copyright 1991 by Delmar Publishers, Inc. Used with permission)*

Rivets are metal pins that look like bolts without threads, and heads without slots. They are usually made of soft iron or steel, aluminum, copper, and brass. Standard rivets are solid, while special rivets may be hollow and contain a pull shank or small explosive.

Standard rivets have a variety of head styles (as shown in Figure 19–6) and sizes. The size of a rivet is measured by the diameter and length of the body. The head is not included

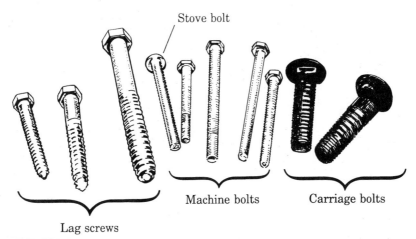

Figure 19–5   Lag screws and machine and carriage bolts are used for special applications for structural purposes

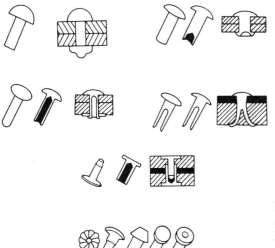

Figure 19–6   Rivets are little metal pins with different style heads *(From Goetsch,* Modern Manufacturing Processes, *Copyright 1991 by Delmar Publishers, Inc. Used with permission)*

in the length except on those designed to be countersunk. Rivet sizes range from $\frac{1}{8}$ to $\frac{3}{8}$ inch in diameter for normal applications with lengths of $\frac{1}{4}$ inch to 3 inches.

The length of a rivet should be the thickness of the metal or plastic, plus $1\frac{1}{2}$ times the diameter of the shank. The shape of the backside head can be round (using a rivet set) or flat (using a peening hammer), depending on preference. As a general rule, rivets should not be spaced closer together than three times the diameter of the rivet.

It is sometimes necessary to use riveting in situations where there is access to only one side of an assembly. As mentioned earlier, special-applications rivets which provide alternative methods for forming the backside head are available. Figure 19–7 shows two types for blind riveting.

Explosive rivets are activated by the application of a heated tool to the rivet head, causing the charge to explode and expand the end of the shank into a head. The pull type (pop rivets) mechanically expands the shank, after which the pull pin is broken off flush with the head.

DOMED HEAD   COUNTERSUNK HEAD
                         (120°)§ §

Figure 19–7   Special-purpose rivets are also used to fasten materials together *(From Zinnigrabe and Schumacher,* Sheet Metal Hand Processes, *Copyright 1974 by Delmar Publishers, Inc. Used with permission)*

Staking

Staking is a common method for permanent fastening when a pin or shaft extends from one piece of metal into another. By punching (expanding) the end of the protruding piece, the metal deforms sufficiently to squeeze tightly against the second piece so that they are mechanically locked together.

Special Joints

So far, we have discussed the use of mechanical fasteners which are applied to various joint designs. However, in woodworking and metal working, it is popular to design special joints which fasten pieces together. They can be used with or without a special adhesive such as glue or solder. When they are complete, neat and strong joints can fasten pieces together very effectively.

Perhaps the most popular material to use special-designed joints is wood. The selection of the proper joint design is based on the stress load, amount and size of the material, and overall appearance. The eight basic joint designs are butt, dado, rabbet, lap, tongue-and-groove, miter, dovetail, and mortise-and-tenon. Each joint design has special attributes and applications. In each case, the selection of the proper design should be based on its design strength rather than on the strength of glue or special mechanical fastener.

The *butt* joint is the simplest and easiest to use. It offsets a pushing force very well and is usually reinforced by dowels, screws, nails, braces, and splines (see Figure 19–8). It is widely used in kitchen cabinet and upholstered furniture work.

Dado joints are used widely to support shelves in cabinets and fasten ends of furniture that has a shear load. Dado joint design may also include a rabbet-and-tongue design which helps to strengthen the joint when it is used on a corner (see Figure 19–9).

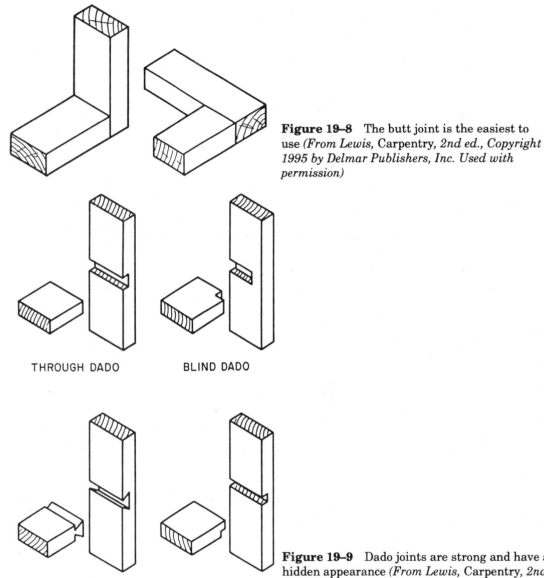

Figure 19–8   The butt joint is the easiest to use *(From Lewis,* Carpentry, *2nd ed., Copyright 1995 by Delmar Publishers, Inc. Used with permission)*

THROUGH DADO   BLIND DADO

DOVETAIL DADO   RABBET AND DADO

Figure 19–9   Dado joints are strong and have a hidden appearance *(From Lewis,* Carpentry, *2nd ed., Copyright 1995 by Delmar Publishers, Inc. Used with permission)*

Rabbet joints are used to close end pieces and have good frontal and side thrust resistance. They can also be used in pairs or in combination with dado and dado tongue designs. In addition, splines can be used to strengthen the joint for pulling forces (see Figure 19–10).

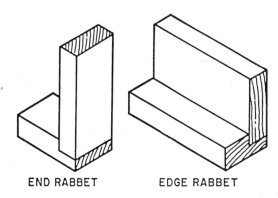

END RABBET   EDGE RABBET

Figure 19–10   Rabbet joints are modified dado joints used for end pieces *(From Lewis,* Carpentry, *2nd ed., Copyright 1995 by Delmar Publishers, Inc. Used with permission)*

The *lap* joint is sometimes referred to as the structural strength joint. They are used to support legs, cabinet frames, and shelving. The basic design of a lap joint is simple and good looking. Its major drawback is end grain appearance when used for corner construction (see Figure 19–11).

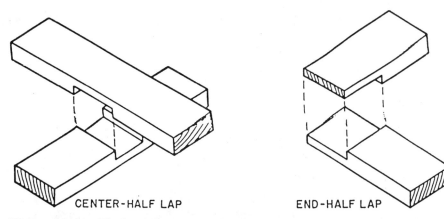

CENTER-HALF LAP   END-HALF LAP

Figure 19–11   The lap joint is exceptionally strong *(From Lewis,* Carpentry, *2nd ed., Copyright 1995 by Delmar Publishers, Inc. Used with permission)*

Tongue-and-groove joints are used to align boards as they are joined together to form larger panels of wood. The joint resists upward and downward thrusts very well and helps prevent large sections from warping. When combined with the alternation of grain patterns, the tongue-and-groove panel becomes exceptionally stable. Tongue and grooves are also used extensively for wanescoatings, hardwood flooring, and roof sheathing for cathedral designs (see Figure 19–12).

The *miter* joint is very popular for picture frames, paneled doors, and fine furniture because it allows the wood grain to flow together in the corner. While not an exceptionally strong joint, it can be reinforced using splines, dowels, and a number of decorative mechanical fasteners. Miter joints also allow the designer to create corners without the unsightly end grain showing (see Figure 19–13).

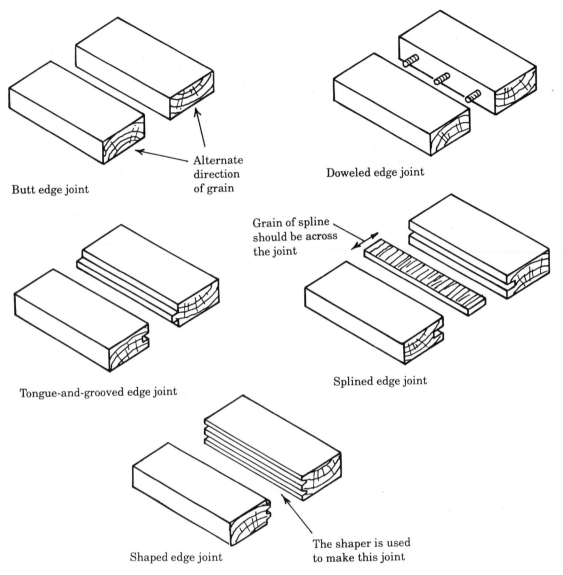

Butt edge joint

Alternate
direction
of grain

Doweled edge joint

Tongue-and-grooved edge joint

Grain of spline
should be across
the joint

Splined edge joint

Shaped edge joint

The shaper is used
to make this joint

Figure 19–12   Tongue-and-groove joints are used extensively for hardwood flooring

Dovetail joints are used for corners which require both pull and thrust strength. The most common use of dovetails is in drawer construction of fine furniture. The design is similar to the concept of locking your fingers together. The wedge-type fingers interlock creating the exceptionally strong joint (see Figure 19–14).

The *mortise-and-tenon* joint was once very popular for house and barn construction. It consists of a projecting tenon and a receiving mortise which, when joined, creates a strong

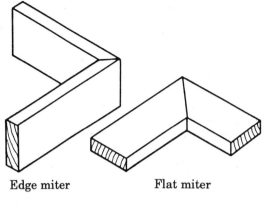

Edge miter   Flat miter

Figure 19–13 Miter joints allow ends to be joined without showing the end grain of wood *(From Lewis,* Carpentry, *2nd ed., Copyright 1995 by Delmar Publishers, Inc. Used with permission)*

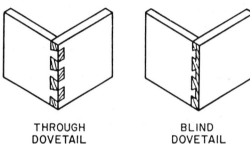

THROUGH
DOVETAIL

BLIND
DOVETAIL

Figure 19–14 Dovetail joints are used for strength *(From Lewis,* Carpentry, *2nd ed., Copyright 1995 by Delmar Publishers Inc. Used with permission)*

joint. It is similar to the tongue-and-groove joint but usually requires the use of a pin or wedge to secure it mechanically (see Figure 19–15).

In addition to the joints mentioned, mechanical devices such as wood blocks, metal braces, corner braces, flat corner irons, flat T-plates (see Figure 19–16), splines, and dowels are used. Such devices are usually used when appearance is not a factor or they can be hidden behind the joint.

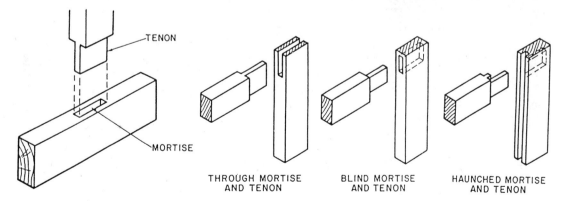

TENON

MORTISE

THROUGH MORTISE
AND TENON

BLIND MORTISE
AND TENON

HAUNCHED MORTISE
AND TENON

Figure 19–15 The mortise and tenon joint is very strong and used for structural applications *(From Lewis,* Carpentry, *Copyright 1995 by Delmar Publishers, Inc. Used with permission)*

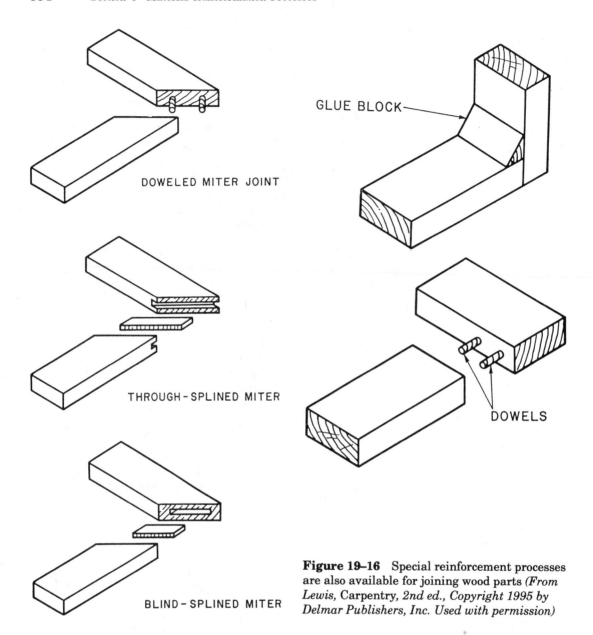

DOWELED MITER JOINT

GLUE BLOCK

THROUGH-SPLINED MITER

DOWELS

BLIND-SPLINED MITER

Figure 19–16 Special reinforcement processes are also available for joining wood parts *(From Lewis,* Carpentry, *2nd ed., Copyright 1995 by Delmar Publishers, Inc. Used with permission)*

Metal working also makes use of some special joint designs to fasten sheet metal together. The most popular sheet metal joints used to join sheet metals mechanically are the standing, folded, grooved, and double-seam joints.

Standing joints are used to join two pieces of sheet metal together when the metal on one side must be flush. The mechanical grip is provided by pressure to the outside fold which engulfs the single turned up piece (see Figure 19–17).

The *folded* joint provides pulling strength for two pieces of sheet metal. Its simple single fold design allows the joint to be strengthened by pressure. Its drawback resides in its appearance and disruption of smoothness (see Figure 19–18).

Grooved joints are the strongest and most difficult to make. It requires two folds and an offset which serves to lock the joint. It has exceptional pull and pushing strength and the advantage of allowing one surface to remain flat (see Figure 19–19).

Double-seam joints can be used to secure a corner in sheet metal. The seam is strong and allows the inside corner to remain free from obstruction (see Figure 19–20).

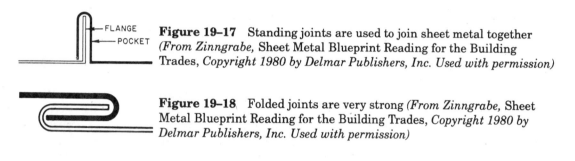

Figure 19–17   Standing joints are used to join sheet metal together *(From Zinngrabe,* Sheet Metal Blueprint Reading for the Building Trades, *Copyright 1980 by Delmar Publishers, Inc. Used with permission)*

Figure 19–18   Folded joints are very strong *(From Zinngrabe,* Sheet Metal Blueprint Reading for the Building Trades, *Copyright 1980 by Delmar Publishers, Inc. Used with permission)*

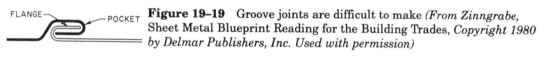

Figure 19–19   Groove joints are difficult to make *(From Zinngrabe,* Sheet Metal Blueprint Reading for the Building Trades, *Copyright 1980 by Delmar Publishers, Inc. Used with permission)*

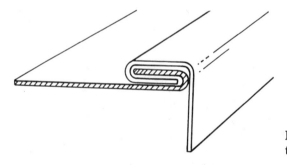

Figure 19–20   Double seam joints are exceptionally strong

Unlike wood joints, which usually require the use of an adhesive-like glue, the grooved and double-seam joints in sheet metal may be used as they are. However, the use of a standing or folded joint requires the addition of solder to gain the necessary strength for use.

It might also be noted that the aforementioned joints may also be accomplished using plastic sheets and a heat source. The design and concepts are the same, but the handling of materials differs because of the general softness of plastic when heated. In terms of an adhesive, the plastic material can be bonded using a variety of glues in a similar fashion to wood; or cohesive reducers may be used to fuse (weld) the joint together. These concepts are further discussed in the next section of this chapter.

Pressure Fitting

The use of pressure to force the pieces together is widely used in the manufacturing machine tool area. They are generally called force fits (FN) and include several classes of fits which involve interference between mating parts. The basic idea is that the shaft is slightly larger than the hole and the pressurized joining of the parts creates a tightness and fastening process.

Force fits are used where parts must be held together very tightly. With these fits, parts can be fastened together almost as tightly as though they were made from one piece. *Shrink fits* are similar to force fits except that heat is used to cause the hole to expand slightly before the shaft is pressed into place. When the piece cools, the metal shrinks around the shaft, causing a tight bond. Figure 19–21 shows how force fits are used in the machine tool industry.

ADHESION

Adhesive fastening (bonding) has recently received a great deal of attention by the aerospace industries because of its ability to bond lightweight honeycomb structures into high-strength configurations (Figure 19–22). However, adhesion techniques have been used successfully for many decades in the past. This type of bonding is a mechanical process of providing filler material between two similar or dissimilar materials (parts), creating strong attraction (bonding) and holding power without melting or diffusing the base materials. The most popular adhesion processes used today are the various glues, brazing, and soldering.

Glues

Various glue adhesives are available on the market to bond almost any material. The most commonly used adhesives are as follows:

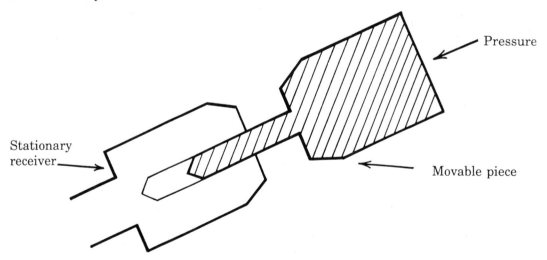

Figure 19–21   Force fits are also very strong but do not require heating

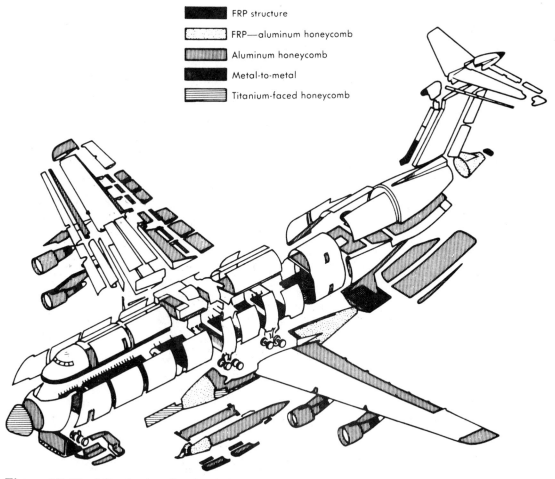

■ FRP structure
▢ FRP—aluminum honeycomb
▥ Aluminum honeycomb
■ Metal-to-metal
▤ Titanium-faced honeycomb

Figure 19–22   Adhesive bonding is used extensively in the aerospace industry *(From Goetsch, Modern Manufacturing Processes, Copyright 1991 by Delmar Publishers, Inc. Used with permission)*

Thermoplastic	**Thermosetting**
Polymides	Epoxies
Vinyls	Isocyanate
Nonvulcanized neopreme rubbers	Phenolic-rubber rubbers
	Thidkol-epoxy

The *thermoplastic* glues are the easiest to use because they rely on evaporation to cure. They are very popular for bonding wood and plastic materials. Their major drawback is in low resistance and failure of holding power in elevated temperature applications.

Thermosetting glues are usually activated by a chemical hardener, heat, or pressure. Once the resin and hardener (catalysts) have been mixed, the adhesive begins to cure. If catalysts and accelerators are to be used together, care must be exercised to make sure that the first additive is mixed thoroughly with the resin before adding the second chemical. Otherwise, the material could explode and cause injury. Thermosetting glues are used extensively to bond glass, plastic, and metal materials. They have good resistance to heat (500 degrees Fahrenheit), provide a corrosion barrier between dissimilar metals, can reach the strength of low-carbon steels in a rolled condition, and can exceed the strength of steel in a honeycomb configuration (see Table 19–1).

Plastic Adhesives	Available Forms
Thermosetting	
Casein	Po, F
Epoxy	Pa, D, F
Melamine formaldehyde	Po, F
Phenol formaldehyde	Po, F
Polyerster	Po, F
Polyurethane	D, L, Po, F
Resorcinol formaldehyde	D, L, Po, F
Silicone	L, Po, Pa
Urea formaldehyde	D, Po, F
Thermoplastic	
Cellulose acetqte	L, H, Po, F
Cellulose butyrate	L, Po, F
Cellulose carboxymethyl	Po, L
Cellulose ethyl	H, L
Cellulose hydroxyethyl	Po, L
Cellulose methyl	Po, L
Cellulose nitrate	Po, L
Polyamide	H, F
Polyethylene	H
Polymethyl methacrylate	L
Polystyrene	Po, H
Polyvinyl acetate	Pt, D, L
Polyvinyl alcohol	Po, D, L
Polyvinyl chloride	Pa, Po, L

Note: Po = powder; F = film; D = dispersion; L = liquid;
Pa = paste; H = hot melt; Pt = permanently tacky.   **Table 19–1**   Plastic adhesive forms

As with all adhesives, the type of joint to be glued will make a significant difference in the strength of the material. Consequently, joints should be designed to utilize the superior lap design whenever possible, as shown in Figure 19–23.

Adhesive-glue bonding has many advantages. The curing temperatures seldom exceed 350 degrees Fahrenheit, allowing the bonding of heat-sensitive materials to be joined without distortion. Thin materials can be bonded to thick materials, and dissimilar materials can be bonded together with an insulation effect. The importance of clean surfaces and proper joint design cannot be overemphasized when working with adhesive glues.

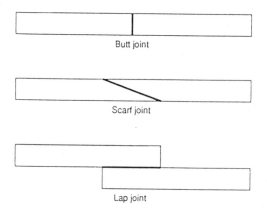

Butt joint

Scarf joint

Lap joint

Figure 19–23   Joints should be designed to use the lap advantage whenever possible *(From Richardson,* Industrial Plastics: Theory and Application, *Copyright 1989 by Delmar Publishers, Inc. Used with permission)*

The major drawback to adhesive glues is their limited ability to withstand situations in elevated temperatures above 180 degrees Fahrenheit to 500 degrees Fahrenheit.

Solders

Soldering is an adhesion process where metal surfaces are bonded by a filler metal which flows between the surfaces using temperatures less than 800 degrees Fahrenheit. Usually, the filler metal is made up of a mixture of lead and tin and is an alloy ratio:

Tin : Lead
  40 : 60
  50 : 50
  60 : 50

When additional strength is needed, silver is added to the tin and lead to form the alloy silver solder. This alloy combination increases both strength and heating requirements. Normally, solder melts at about 350 to 450 degrees Fahrenheit. Silver solder melts at about 600 degrees Fahrenheit and is usually applied using an open-flame torch rather than the traditional soldering copper or electric gun. Even if a torch is used, the base metal (piece to be soldered) is never heated to a coalescence or melting state, thereby preventing the flow (fusion) together.

As with all adhesives, the preparation of the joints to be soldered is very important. The surfaces must be clean, and the joint design should make use of the lap procedure mentioned earlier in this chapter to increase strength. In addition, the surfaces may need to be precoated (called tinning) with solder before the joint is assembled for soldering to assure maximum adhesion.

Brazing

Brazing is an adhesion process similar to soldering which requires a heat range above 800 degrees Fahrenheit and below the melting point of the base metal. The filler metal is distributed between the base metal surfaces by capillary attraction. The brazing metals and alloys commonly used are listed below:

Material	Heat or Range in Fahrenheit
Copper	1,892 degrees
Copper alloys	
Brass	1,700 degrees
Bronze	1,675 degrees
Silver alloys	1,165 to 1,550 degrees
Aluminum alloys	1,060 to 1,185 degrees

Brazing produces good-looking smooth joints with excellent strength. The key to success is to make sure that the brazed joint has correct clearance for the material (about 0.02 to 0.03 inch) to flow between the surfaces. Joints that are too tight or have a loose fit are not recommended for brazing.

To apply the braze material, surfaces require the use of fluxes (to clean surfaces and reduce oxides) and a heat source, which may be an oxygen–acetylene torch or furnace, depending on the type of joint to be brazed. The flux is usually some type of borax-based material which can be applied directly to the joint by brush (wet mixture) or attached to the end of the brazing rod by heating the rod and dipping it into the flux can.

The oxygen–acetylene torch should be set with a slight carburizing flame (see following section on welding for more details) and the metal heated until coalescence (dull red, with loss of magnetism for iron-based materials) is reached. The rod is then melted to form a molten puddle and continuously dipped into the puddle as forward movement is made with the torch along the joint or seam. If long seams are to be brazed, it is helpful to tack (make several ½-inch braze spots) the material in advance of brazing the seam. This will help prevent distortion or movement caused by the heat.

Brazing and braze welding (an adhesion process) are used extensively in industry for thin, lightweight tubular materials because of their ease of application, and with heavy iron-casting repairs because of the ability of a brazed joint to withstand shock. Indeed, brazing is one of the most common repair processes used to maintain production equipment in the industrial setting.

COHESION

Cohesion fastening (bonding) represents a type of bonding that allows the joint materials to liquefy and flow together. Most cohesive-fastening processes are referred to as welding processes, and they make up a significant portion of industrial fastening in modern industry.

Plastic Welding

Only thermoplastics can be welded with the fusion process. Very thick as well as very thin pieces may be welded together using one of four methods: hot-gas welding, hot-plate welding, spin welding, and ultrasonic welding. Table 19–2 shows the various methods for fastening plastic materials.

Plastics	Solvent	Boiling Point C°	F°
ABS	Methyl ethyl ketone	40	[104]
	Methyl isobutyl ketone		
	Methylene chloride	40	[104]
Acrylic	Ethylene dichloride	84	[183]
	Methylene chloride	40	[104]
	Vinyl trichloride	87	[189]
Cellulose plastics			
Acetate	Chloroform	61	[142]
Butyrate,			
propionate	Ethylene dichloride	84	[183]
Ethyl cellulose	Acetone	57	[135]
Polyamide	Aqueous phenol		
	Calcium chloride		
	in alcohol		
Polycarbonate	Ethylene dichloride	41	[106]
	Methylene chloride	40	[104]
Polyphenylene	Chloroform	61	[142]
oxide	Ethylene dichloride	84	[183]
	Methylene chloride	40	[104]
	Toluene	110	[232]
Polysulfone	Methylene chloride	40	[104]
Polystyrene	Ethylene dichloride	84	[183]
	Methyl ethyl ketone	80	[176]
	Methylene chloride	40	[104]
	Toluene	110	[232]
Polyvinyl	Acetone	57	[135]
chloride and	Cyclohexane		
copolymers	Methyl ethyl ketone	80	[176]
	Tetrahydrofuran	65	[149]

Table 19–2   Common solvent cements for thermoplastics *(From Richardson,* Industrial Plastics: Theory and Application, *Copyright 1989 by Delmar Publishers, Inc. Used with permission)*

Hot-gas welding makes use of a stream of hot air or water-pumped nitrogen that is heated as it passes through the weld gun (Figure 19–24). In a similar process to gas welding, the welding rod is melted along with the base metal and flows into a fusion bond. The welding rod is made from the exact same materials as the base plastic in order to create smooth flow. As with some metals, dissimilar types of plastic material cannot be welded together.

Large tanks, pipes, and sheets can be welded in this manner economically. This process works exceptionally well with the polyvinyl chloride (PVC) family of plastics.

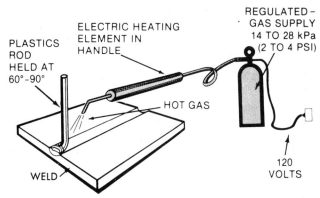

Figure 19–24   Hot-gas welding is used to fasten plastic materials *(From Richardson,* Industrial Plastics: Theory and Application, *Copyright 1989 by Delmar Publishers, Inc. Used with permission)*

Spin welding can be done with thermoplastic materials. Sometimes called friction welding, the cylindrical pieces may be rotated and brought together with pressure until the heated edges melt and fuse together. Usually, only one piece is rotated while the other is held stationary. When the fusion occurs, the stationary piece breaks away from its temporary stop and rotates with the other piece. This method prevents distortion to the piece being welded.

Many plastic drinking containers are welded together in this manner to allow the injection molding process to be simplified. Figure 19–25 shows a simple drink container being spin welded.

Hot-plate welding is a method of heating the edges of a sheet or piece of plastic pipe using a hot plate or electric frying pan. Once the edges have been heated to the point where they begin to melt (see Figure 19–26), they are quickly forced together causing a fusion bond.

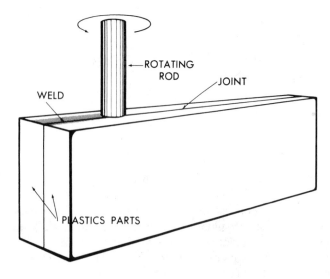

Figure 19–25   Spin welding plastics uses friction just like metallics *(From Richardson,* Industrial Plastics: Theory and Application, *Copyright 1989 by Delmar Publishers, Inc. Used with permission)*

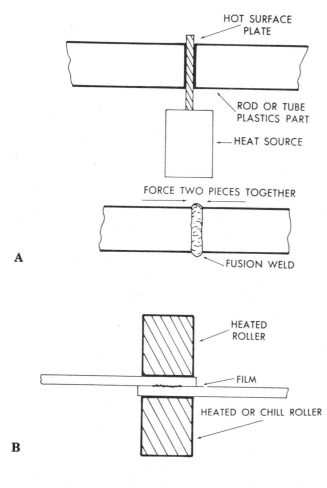

HOT SURFACE PLATE

ROD OR TUBE PLASTICS PART

← HEAT SOURCE

FORCE TWO PIECES TOGETHER

FUSION WELD

A

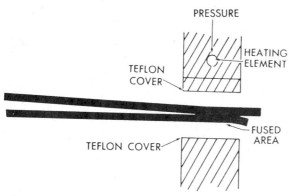

HEATED ROLLER

FILM

HEATED OR CHILL ROLLER

B

PRESSURE

HEATING ELEMENT

TEFLON COVER

FUSED AREA

TEFLON COVER

C

Figure 19–26 Hot-plate welding is used for custom plastic working. **(A)** Heated tool bonding of plastics parts by fusion welding; **(B)** heated tool bonding of plastics film using rollers; **(C)** heated tool bonding of plastics film using press *(From Richardson, Industrial Plastics: Theory and Application, Copyright 1989 by Delmar Publishers, Inc. Used with permission)*

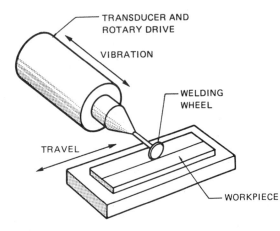

TRANSDUCER AND
ROTARY DRIVE

VIBRATION

WELDING
WHEEL

TRAVEL

WORKPIECE

Figure 19–27 Ultrasonic welding of plastic makes use of vibrations to cause fusion *(From Jeffus,* Welding Principles and Applications, *3rd ed., Copyright 1993 by Delmar Publishers, Inc. Used with permission)*

Ultrasonic welding makes use of vibrations to cause the material to fuse together. The two parts to be welded are vibrated at 20,000 cycles per second by the ultrasonic horn. Because this causes the edges to melt, a slight bit of pressure unites the two pieces into a cohesive fusion bond. Figure 19–27 shows how an ultrasonic plastic weld is made.

Ultrasonic welding is exceptionally fast and economical. As shown in Table 19–2, not all thermoplastics can be welded in this manner. *Wood welding* is not a cohesive-bonding process. Instead, it is an adhesive gluing process that makes use of a heat source called a wood welder.

Ceramic Welding

The ceramics which are easily welded are clays and glass. Both respond to cohesive (fusion) bonding quite well and make extensive use of welding processes. However, they differ in that all clay welding is done in a green (cold) state, and all glass welding is done in a hot state.

Clay welding is a fusion process where the parts to be welded are in a green state (not fired). The surface of the base piece must be moistened with water to make it into a semislurry state. The piece to be welded is also moistened at the joint, and the two pieces are pressed together.

Using a molding tool (or finger), the clay joint is pressed and molded into a permanent bond. Additional clay material may also be added to strengthen the welded joint. When the piece is fired in a kiln, the welded joint becomes as strong as the base clay.

Glass welding is a fusion process that requires heat and occurs after the glass material has been fired and cooled. Using a gas flame, the edges of the two pieces to be joined are heated until the edges are in a molten state. The edges are pressed together, and the molten glass edges are cohesively bonded into one piece. The welded joint is as strong as the base glass material. Glass welding is used widely in scientific laboratories when special glassware must be fabricated to conduct research experiments.

Metal Welding

Metal welding is the joining of metal in a fusion process and is the most popular welding (cohesive) procedure in use today. Five major metal-welding processes (see Figure 19–28) are used in industry today to weld a variety of ferrous and nonferrous metals.

Gas welding (OAW) is the process of joining various metals together by melting and fusing adjoining edges together using a gas-air or gas-oxygen heat source. It consists of applying a concentrated flame on to the area to be welded until a molten puddle, cohesive fusion of materials occurs and the joint is welded.

Gas welding torches which make use of atmospheric air are primarily used for soldering or brazing because of their lower heat range. Torches which make use of oxygen have a much higher heat range and are used to weld all ferrous and some nonferrous metals. The most popular fuel for gas welding is acetylene because of its clean and hot burning characteristics. Other fuels such as hydrogen, MAPP (artificial), and liquid petroleum are also used for special applications.

The choice of welding flames is important to the welder and is dictated by the mixture (ratio) of gas to oxygen. Figure 19–29 shows the three types of flames used for the oxyacetylene welding process.

The neutral flame is used for most applications of gas welding. It is a 1:1 mixture of acetylene (or other gas) and oxygen. If too much oxygen is provided, an oxidizing flame will be achieved and is harmful to the weld because of the oxides it provides. On the other hand, too little oxygen will create a carburizing flame which is dirty and sooty. Sometimes, a slightly carburizing flame is used for brazing cast iron or welding aluminum.

A neutral oxyfuel welding flame will produce a temperature of 5,600 to 6,300 degrees Fahrenheit. Figure 19–30 shows the typical oxygen–acetylene welding equipment needed for gas welding. The skill needed to gas weld is in the control of the puddle, speed, and consistent dipping of the filler metal (welding rod) into the puddle. It requires the use of both hands doing different operations at the same time.

Arc welding (SMAW) is also a fusion (cohesive) process for joining both ferrous and nonferrous metals. The heat source for arc welding is created by passing electricity through a gaseous gap from one electric conductor to another. The temperature of the spark (arc) jumping between the two conductors is approximately 6,500 to 7,000 degrees Fahrenheit.

As the arc is heating the metal to be welded, it is also heating the welding rod, which provides a gaseous shield and filler material. Figure 19–31 shows how the molten puddle is protected by the gaseous shield and receives additional metal fillers from the welding rod. The amount of heat provided by the arc welder (measured in amperes) determines the amount of penetration (depth of the puddle) and the speed of the weld. The coating on the welding rod provides fluxes which help stabilize the arc and release protective gases which provide a shield around the weld area to keep unwanted oxides out.

Two types of arc welding are alternating current (AC) and direct current (DC). The AC welders are designed as general purpose and, because of the advancements made in welding electrode technology, are used extensively for general welding applications. DC welders are more heavy duty and can deliver very high amperage for special applications. The ma-

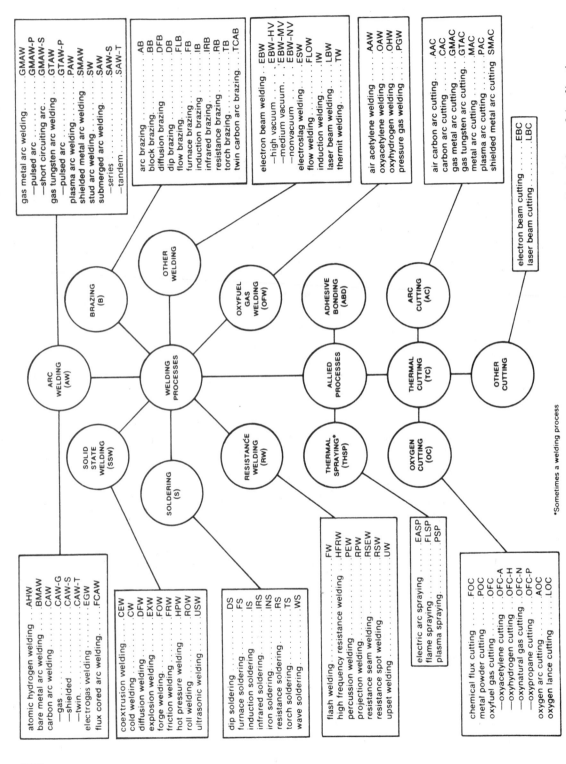

gas metal arc welding	GMAW
—pulsed arc	GMAW-P
—short circuiting arc	GMAW-S
gas tungsten arc welding	GTAW
—pulsed arc	GTAW-P
plasma arc welding	PAW
shielded metal arc welding	SMAW
stud arc welding	SW
submerged arc welding	SAW
—series	SAW-S
—tandem	SAW-T

arc brazing	AB
block brazing	BB
diffusion brazing	DFB
dip brazing	DB
flow brazing	FLB
furnace brazing	FB
induction brazing	IB
infrared brazing	IRB
resistance brazing	RB
torch brazing	TB
twin carbon arc brazing	TCAB

electron beam welding	EBW
—high vacuum	EBW-HV
—medium vacuum	EBW-MV
—nonvacuum	EBW-NV
electroslag welding	ESW
flow welding	FLOW
induction welding	IW
laser beam welding	LBW
thermit welding	TW

air acetylene welding	AAW
oxyacetylene welding	OAW
oxyhydrogen welding	OHW
pressure gas welding	PGW

air carbon arc cutting	AAC
carbon arc cutting	CAC
gas metal arc cutting	GMAC
gas tungsten arc cutting	GTAC
metal arc cutting	MAC
plasma arc cutting	PAC
shielded metal arc cutting	SMAC

| electron beam cutting | EBC |
| laser beam cutting | LBC |

*Sometimes a fusion process

atomic hydrogen welding	AHW
bare metal arc welding	BMAW
carbon arc welding	CAW
—gas	CAW-G
—shielded	CAW-S
—twin	CAW-T
electrogas welding	EGW
flux cored arc welding	FCAW

coextrusion welding	CEW
cold welding	CW
diffusion welding	DFW
explosion welding	EXW
forge welding	FOW
friction welding	FRW
hot pressure welding	HPW
roll welding	ROW
ultrasonic welding	USW

dip soldering	DS
furnace soldering	FS
induction soldering	IS
infrared soldering	IRS
iron soldering	INS
resistance soldering	RS
torch soldering	TS
wave soldering	WS

flash welding	FW
high frequency resistance welding	HFRW
percussion welding	PEW
projection welding	RPW
resistance seam welding	RSEW
resistance spot welding	RSW
upset welding	UW

electric arc spraying	EASP
flame spraying	FLSP
plasma spraying	PSP

chemical flux cutting	FOC
metal powder cutting	POC
oxyfuel gas cutting	OFC
—oxyacetylene cutting	OFC-A
—oxyhydrogen cutting	OFC-H
—oxynatural gas cutting	OFC-N
—oxypropane cutting	OFC-P
oxygen arc cutting	AOC
oxygen lance cutting	LOC

*Sometimes a welding process

Figure 19–28 Metal also uses the fusion process to join parts together *(From Jeffus, Welding Principles and Applications, 3rd ed., Copyright 1993 by Delmar Publishers, Inc. Used with permission)*

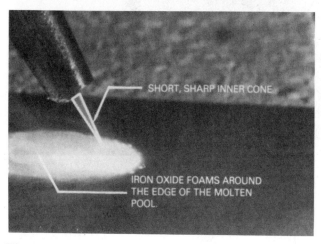

Figure 19–29 Three flames are used for oxyacetylene welding: **(A)** carburizing, **(B)** neutral, and **(C)** oxidizing *(From Jeffus,* Welding Principles and Applications, *3rd ed., Copyright 1993 by Delmar Publishers, Inc. Used with permission)*

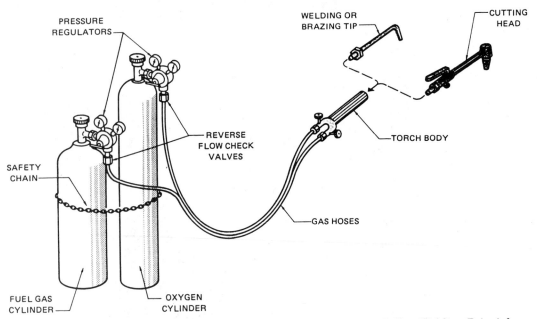

Figure 19–30   Equipment needed for oxyacetylene welding. *(From Jeffus,* Welding Principles and Applications, *3rd ed., Copyright 1993 by Delmar Publishers, Inc. Used with permission)*

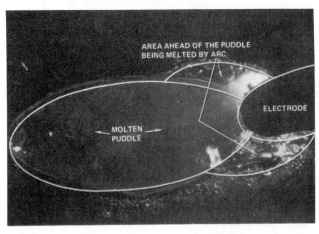

Figure 19–31   The welding puddle determines the size, depth, and appearance of the weld *(From Jeffus,* Welding Principles and Applications, *3rd ed., Copyright 1993 by Delmar Publishers, Inc. Used with permission)*

jor advantage of a DC welder is in its ability to switch polarity, that is, change the direction of electrons from the electrode to the workpiece (straight polarity) and the workpiece to the electrode (reverse polarity). Changing polarity is helpful for welding very heavy pieces of metal and for helping the welder apply welding beads in difficult positions. Figure 19–32 illustrates the concept of changing polarity for DC welders.

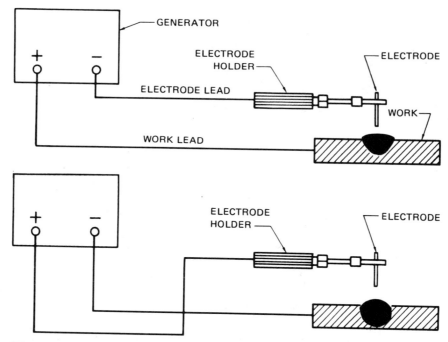

Figure 19–32 DC welding uses both straight and reverse polarity to maximize current flow during the welding operation *(From Jeffus,* Welding Principles and Applications, *3rd ed., Copyright 1993 by Delmar Publishers, Inc. Used with permission)*

    The selection of a welding electrode is as important to arc welding as the selection of a machine. Some welding rods are designed only for AC or designed only for DC welding. They also have fast-freeze (for position welding) or fast-fill (for flat production welding) features and stabilizers which help the arc maintain consistent heat and penetration. Figure 19–33 provides application information for the most popular welding rods. For special applications, the reader should refer to a detailed welding guide.

    The basic equipment for arc welding is shown in Figure 19–34. The welder can adjust the equipment for each special application and select special welding rods to achieve desired characteristics. The skill needed to arc weld is in the control of the arc gap (pattern), speed of the weld, and ability to maintain arc length while the rod continues to get shorter.

    Gas arc welding (GMAW) is like a combination of gas welding and arc welding except that the gas is used as a protective shield rather than the fuel. Welds have much better chemical and physical properties if oxygen can be kept away from the weld area. In gas arc welding, the electric arc is protected by a special gas shield as illustrated in Figure 19–35.

AWS Classification	Type of Covering	Capable of Producing Satisfactory Welds in Positions Shown[a]	Type of Current[b]
E60 Series Electrodes			
E6010	High cellulose sodium	F,V,OH,H	dc, reverse polarity
E6011	High cellulose potassium	F,V,OH,H	ac or dc, reverse polarity
E6012	High titania sodium	F,V,OH,H	ac or dc, straight polarity
E6013	High titania potassium	F,V,OH,H	ac or dc, either polarity
E6020	High iron oxide	H-fillets	ac or dc, straight polarity
E6022[c]		F	ac or dc, either polarity
E6027	High iron oxide, iron powder	H-fillets, F	ac or dc, straight polarity
E70 Series Electrodes			
E7014	Iron powder, titania	F,V,OH,H	ac or dc, either polarity
E7015	Low hydrogen sodium	F,V,OH,H	dc, reverse polarity
E7016	Low hydrogen potassium	F,V,OH,H	ac or dc, reverse polarity
E7018	Low hydrogen potassium, iron powder	F,V,OH,H	ac or dc, reverse polarity
E7024	Iron powder, titania	H-fillets, F	ac or dc, either polarity
E7027	High iron oxide, iron powder	H-fillets, F	ac or dc, straight polarity
E7028	Low hydrogen potassium, iron powder	H-fillets, F	ac or dc, reverse polarity
E7048	Low hydrogen potassium, iron powder	F,OH,H,V-down	ac or dc, reverse polarity

[a]The abbreviations F,V,V-down,OH,H, and H-fillets indicate the welding positions as follows:

F = Flat
H = Horizontal
H-fillets = Horizontal fillets
V-down = Vertical down
V = Vertical
OH = Overhead

[b]Reverse polarity means the electrode is positive; straight polarity means the electrode is negative.

[c]Electrodes of the E6022 classification are for single-pass welds.

For electrodes 3/16 in (4.8 mm) and under, except
5/32 in (4.0 mm) and under for classifications E7014, E7015, E7016, and E7018.

Figure 19–33 Welding rods are determined by numbers and color codes *(From Jeffus,* Welding Principles and Applications, *3rd ed., Copyright 1993 by Delmar Publishers, Inc. Used with permission)*

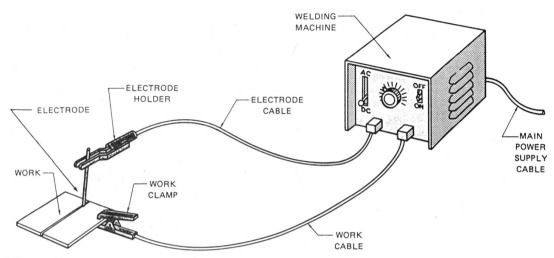

Figure 19-34   The type of equipment needed for arc welding *(From Jeffus,* Welding Principles and Applications, *3rd ed., Copyright 1993 by Delmar Publishers, Inc. Used with permission)*

The two most popular types of gas arc welding are

1. Gas tungsten arc welding (GTAW)
2. Gas metal arc welding (GMAW)

Both processes make use of shielding gases such as carbon dioxide, helium, or argon.

The GTAW (usually called tungsten inert gas [TIG]) process is very similar to gas welding because a puddle is developed and a filler rod is dipped into the puddle to add metal (see Figure 19-35A). The electrode is made of tungsten and is not consumed in the weld. This type of gas arc welding is used mostly for repair work.

GMAW (usually called metallic inert gas [MIG]) process is similar to TIG except that the tungsten is replaced by a consumable rod (see Figure 19-35B) that melts as it is fed into the puddle.

The amount of current and rate of wire feed determines the speed and size of the weld. Because it is much faster than the TIG process, it is used in production as a semiautomatic weld process.

The skills needed for gas arc welding are approximately the same as for gas or arc welding. Control of the electrode and filler rod require a certain amount of dexterity and control. Gas arc welding is not only popular for welding standard metals but also has the advantage of being able to weld special alloy metals that would otherwise be very difficult to weld.

Resistance welding (RSW) consists of passing electrical current through two pieces of metal, causing them to heat and melt together and forming a cohesive fusion bond. The control of the welded area is affected by the size of the contacts (electrodes) and location of pressure. It is a fast and accurate production method for welding sheet metals with minimum warpage.

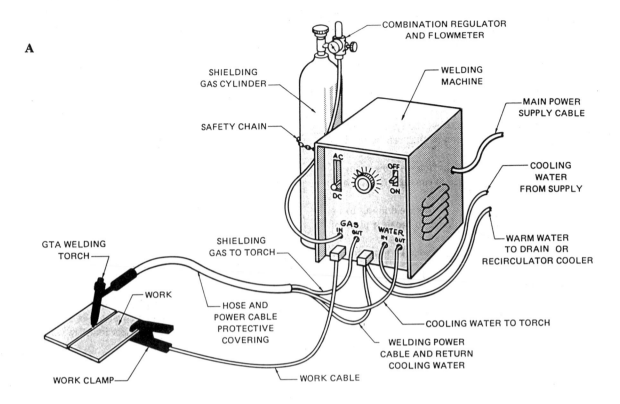

A

COMBINATION REGULATOR AND FLOWMETER

SHIELDING GAS CYLINDER

WELDING MACHINE

SAFETY CHAIN

MAIN POWER SUPPLY CABLE

COOLING WATER FROM SUPPLY

AC

DC

OFF

ON

GTA WELDING TORCH

SHIELDING GAS TO TORCH

GAS IN OUT

WATER IN OUT

WARM WATER TO DRAIN OR RECIRCULATOR COOLER

WORK

HOSE AND POWER CABLE PROTECTIVE COVERING

COOLING WATER TO TORCH

WORK CLAMP

WELDING POWER CABLE AND RETURN COOLING WATER

WORK CABLE

B

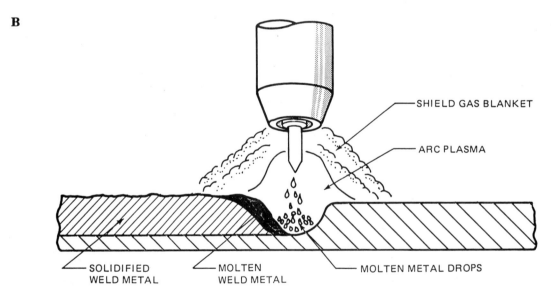

SHIELD GAS BLANKET

ARC PLASMA

MOLTEN METAL DROPS

SOLIDIFIED WELD METAL

MOLTEN WELD METAL

Figure 19–35 The gas (inert) shield protects the weld from oxygen. **(A)** GTAW equipment; **(B)** GMAW equipment *(From Jeffus,* Welding Principles and Applications, *3rd ed., Copyright 1993 by Delmar Publishers, Inc. Used with permission)*

The most popular types of resistance welding are

Spot welding	Seam welding
Gun welding	Flash welding
Shot welding	

Spot welding is the most popular term used to describe resistance welding. Its principle is similar to the other types of resistance welding, with the pieces to be welded being held with pressure between two electrodes. A predetermined time and amount of current passes through the metal; and it becomes molten and flows together, causing a fusion weld.

Gun welding is a portable type of spot welding. The flexibility of the gun welder makes it ideal for sheet metal production work and automobile assembly. Today, many gun welders are controlled by industrial robots, as shown in Figure 19–36.

Shot welding is used to weld aluminum sheet metal. It is similar to spot welding except it is much faster because a high amount of current is applied in a shorter period of time. This will minimize the oxidation of the material and require less heat-treating requirements.

Flash welding is used to resistance weld round pieces together. The two rods are held together end to end, and a current is passed through the material. The resistance of the joint

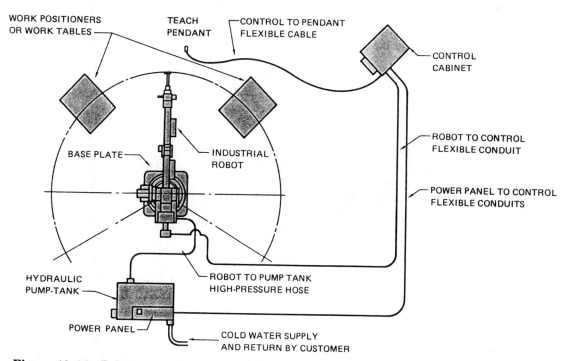

Figure 19–36 Today, many types of welding are performed by robots *(From Jeffus,* Welding Principles and Applications, *3rd ed., Copyright 1993 by Delmar Publishers, Inc. Used with permission)*

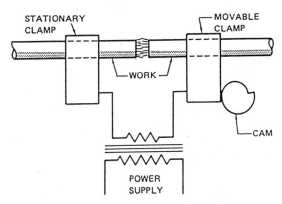

Figure 19–37 Flash welding is a resistance process of heat and pressure *(From Jeffus, Welding Principles and Applications, 3rd ed., Copyright 1993 by Delmar Publishers, Inc. Used with permission)*

causes the ends to heat up; and when they become plastic, they are forced together, causing fusion. Figure 19–37 illustrates the flash welding process.

In *seam welding,* the copper electrodes are replaced by wheels; and the material is rolled along the seam as the welding takes place. This process can weld a solid leakproof seam or just spot weld as it moves along (see Figure 19–38).

Special-application welding processes are those that are designed for a particular welding application that requires special preparation or equipment. The list of special-applications welding is quite large today, and only a few of the most popular methods are discussed in this text. For further information, the reader is referred to the American Welding Society.

The most popular special-applications welding processes are as follows:

Submerged arc welding   Forge welding
Electroslag and electrogas   Ultrasonic welding
Metal arc underwater welding   Friction welding
Plasma arc welding   Laser welding

Submerged arc welding (SAW) is used for welding heavy steel applications where strength, appearance, and speed are important. The method (hand-held or automatic) uses a flux covering which prevents the oxides from contaminating the weld and blocks the bright arc light from the operator. The flux forms an airtight slag covering that can be used over again by a special vacuum device which returns the material to the hopper. Figure 19–39 illustrates the submerged arc welding process.

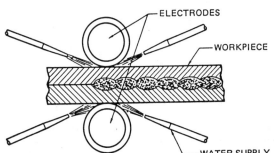

Figure 19–38 The weld of a seam can be spotted or solid, based on need *(From Jeffus, Welding Principles and Applications, 3rd ed., Copyright 1993 by Delmar Publishers, Inc. Used with permission)*

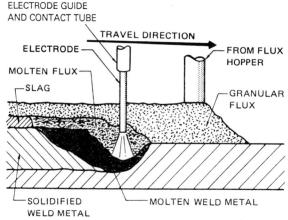

Figure 19–39 Submerged arc welding uses a very heavy flux material to keep oxides out of the weld *(From Jeffus,* Welding Principles and Applications, *3rd ed., Copyright 1993 by Delmar Publishers, Inc. Used with permission)*

Because the operator cannot see the weld process, special gauges are used (voltmeter and ammeter) to determine the correct length of the arc. Excellent penetration and strong clean welds make this heavy welding process popular in industry.

Electroslag and electrogas welding (EW) is used to weld very thick seams without having to make multiple passes. The weld joint is put in a vertical position and special copper supports (called shoes) are used to retain the molten liquid in the weld joint. Because the process is more of a resistance method of welding, there is no arc or flame present. Figure 19–40 illustrates how the flux and filler wire are added to the weld area as the copper shoes are moved along the seam.

The addition of a shielding gas is called electrogas, and it is similar to the electroslag process except that a flux core wire feed is used and an electric arc is continuously main-

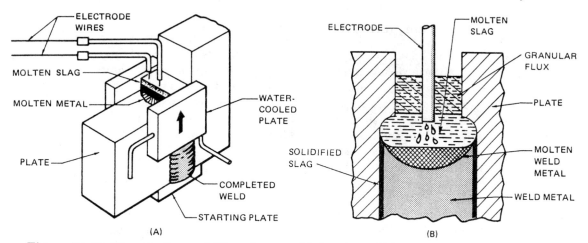

Figure 19–40 How the flux and filler wire are added to the weld area as the copper shoes are moved along the seam *(From Jeffus,* Welding Principles and Applications, *3rd ed., Copyright 1993 by Delmar Publishers, Inc. Used with permission)*

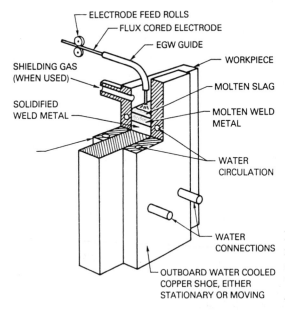

ELECTRODE FEED ROLLS
FLUX CORED ELECTRODE
EGW GUIDE
WORKPIECE
SHIELDING GAS (WHEN USED)
MOLTEN SLAG
SOLIDIFIED WELD METAL
MOLTEN WELD METAL
WATER CIRCULATION
WATER CONNECTIONS
OUTBOARD WATER COOLED COPPER SHOE, EITHER STATIONARY OR MOVING

Figure 19–41 The addition of a shielding gas improves the welding process for EW operations *(From Jeffus,* Welding Principles and Applications, *3rd ed., Copyright 1993 by Delmar Publishers, Inc. Used with permission)*

tained between the electrode and the weld puddle. Figure 19–41 illustrates a typical setup for electrogas welding.

Metal arc underwater welding (MUW) requires special equipment and safety precautions. A special welding rod which has a coating that will not be affected by the water yet allow the weld area to generate enough heat for a puddle is used in underwater applications. The welding rod holder is insulated to prevent electrical shock. Unlike conventional welding processes, the underwater welder must also be a skilled diver in order to maintain the steadiness needed for good welding.

Plasma arc welding (PAW) makes use of ionized particles to create the weld puddle. A filler metal is added at the leading edge (called keyhole for plasma arc welding) and creates an oxide-free and uniform weld bead. To create the plasma arc, a tungsten electrode (nonconsumable) and copper nozzle (water-cooled) are used to create an electric arc. Gases such as helium or hydrogen are forced through the nozzle, become heated and ionized, and create a stream which is used to melt the metal. Because the ion impingement has the ability to remove oxide films from the base metal surface, plasma arc welding of aluminum and other space-age materials is very practical.

Plasma arc welding (and cutting) makes a screeching sound which requires the operator to protect his or her ears. Most plasma arc welding operations are done with automatic equipment.

Forge welding (FOW) is the oldest fusion welding process. The metal to be welded is heated in a blacksmith's forge (a tabletop furnace with forced air) until the material is white hot. Flux is applied to the joint just prior to the hammer blow that forces the metal to fuse together. Figure 19–42 shows a typical forge. Because of the skill needed by the operator and

Figure 19-42 Forge welding is one of the oldest forms of welding used *(Courtesy of Photri, Inc.)*

the advancements in other welding processes, forge welding has almost entirely been replaced as a metal-fastening process.

Ultrasonic welding of metal is similar to that of plastic mentioned earlier in this chapter. By combining vibration and pressure, heated joints can be fused together with very little distortion to the product.

Friction welding of metals is also similar to friction welding of plastics. One piece is held stationary while the other is revolved. When sufficient speed is reached, the ends are forced together. The friction between the pieces creates heat, melting the edges together. A special release is designed into the friction welder to allow the piece to stop turning once the weld is complete.

Laser welding (LBW) is very popular for welding very small or thin materials. Because the light beam can be focused and enhanced, powerful and deep penetrating welds can be accomplished in narrow areas where conventional welding can not be done. Figure 19-43 illustrates the principles of laser welding.

Fastening is an important concept in the processing of material goods. This chapter has discussed many of the more popular types of fastening but by no means all of them. The product designer and engineer must be aware of the options for fastening materials together and select the appropriate method that will do the job best.

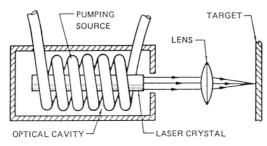

PUMPING SOURCE

TARGET

LENS

OPTICAL CAVITY

LASER CRYSTAL

Figure 19–43 Lasers are used for special-purpose welding (From Jeffus, Welding Principles and Applications, *3rd ed., Copyright 1993 by Delmar Publishers, Inc. Used with permission)*

SUMMARY

Fastening of materials is always a challenge for manufacturers. The decision of whether to use permanent, semipermanent, or removable devices or processes is dependent upon function.

As the reader has learned in this chapter, mechanical devices are used for some applications and materials, while adhesive and cohesive techniques are used for others. Adhesion requires the materials to bond by use of special agents and chemicals that cause the materials to become attracted to each other by use of glues or special solders. Cohesion requires materials to fuse together by melting the material. Both adhesive and cohesive bonding processes are considered permanent even though a special material-removal process can be used to separate them.

The manufacturer must be aware of the various choices available for material fastening. The technology ranges from simply driving a nail into a board to complex, laser-driven fusion processees.

REVIEW QUESTIONS

1. What is cohesive bonding?
2. What is adhesive bonding?
3. Identify six popular mechanical fasteners used to join wood materials and explain their advantages and disadvantages.
4. Compare and contrast the welding (fusion) processes used to weld plastic and metal.
5. Of the mechanical fastener family, describe and explain the differences between permanent and semipermanent devices.
6. Describe the advantages and disadvantages of the eight basic joints discussed in this chapter.
7. What are the five major classifications of metal welding?
8. Describe two ceramic welding processes and compare them to metal welding processes.
9. Are *soldering* and *brazing* welding processes?
10. Describe one special-applications welding process and indicate why it is used as a fastening device.

SUGGESTED ACTIVITIES _____

1. Have students design a test machine to determine the holding power of wood screws versus nails in hardwoods and softwoods.
2. Have students prepare several wood joints and destructively test them for strength.
3. Prepare a sample that will demonstrate at least five types of welding and conduct an inspection and analysis in class.
4. Design and develop five glued joints using at least three different materials to demonstrate adhesive holding power.
5. Design a chart that will contrast the difference between cohesive and adhesive bonding.

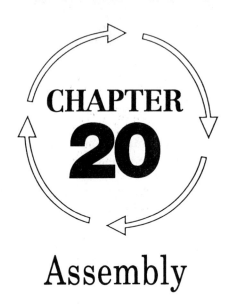

CHAPTER 20

Assembly

In manufacturing, the term **assembly** refers to the joining of various parts and subassemblies together. Although there are some single-component products that are complete when the basic materials-processing procedures have been performed, most products are made up of multiple components.

Assembly has traditionally been labeled as high-cost nonvalue-added activity that reduces profit significantly. In general, very labor-intensive assembly can account for up to 50% of the cost of a product. In recent years, manufacturers have been increasingly more serious about automation for the assembly of products, in part, because of the high labor costs and, in part, because of better quality control.

The assembly process is a complicated part of the overall manufacturing operation. It is based on the concept of interchangeable parts and a system to organize an efficient method to assemble parts and materials. It includes intentional assembly-friendly parts design, special fixtures and assembly systems, parts-feeding technology, fastening techniques, inspection, packaging, and warehousing. In a typical modern manufacturing plant, the technology will range from manual assembly by workers on a line to high-speed, automated materials-handling systems with programmed robots.

MANUFACTURING

Interchangeable Parts

In 1798 when Eli Whitney won a contract to produce 10,000 muskets for the U.S. Army, a new era in manufacturing began. The idea of using jigs and fixtures so that each part was made to a conforming standard provided the basis for interchangeable parts. Whitney's muskets could be assembled by nonskilled workers because each part that was produced could fit into any gun assembly without further adjustment work. The basic Whitney concept is still used today by design engineers throughout the world.

A second major development occured at the turn of the twentieth century when Henry Ford created his assembly line for the manufacture of automobiles. Building upon the knowledge of process jigs and fixtures, Ford developed assembly jigs and fixtures and organized the assembly process into a continuous line (see Figure 20–1). The idea of moving parts to an assembly line and designing special fixtures and transport systems that would allow the product to "grow" as it moved along was indeed a sign of the times ahead and a giant step in assembly technology.

Design for Automation

Essential to the idea of the interchangeable parts is the control of manufacturing using precise measurement and acceptable *standards*. An exact fit between two subparts is not very practical for assembly procedures. Therefore, a system of *tolerances* has been developed that allows the parts to be assembled easily yet maintain quality standards for good product performance.

Figure 20–1   Henry Ford organized the assembly process into a line *(Courtesy of Photri, Inc.)*

In addition to a system of tolerances, the design engineer is concerned about design for assembly (DFA). This concept occurs during the early design phase and is focused on how the various parts can be produced so that they are assembled more easily. Perhaps the most common design solutions for assembly are based on simplification during manufacture. In Figure 20–2, one can see how a simple bracket can be redesigned to reduce the part to a simple stamping and still meet product specifications.

IBM has provided a great deal of leadership in DFA by reducing the number of parts in its typewriters and computer printers during recent years. They have also used design to make it easier for robotics assembly by considering the limitations of the robot in the work cell. DFA also has advantages for product reliability and serviceability.

Size and shape of parts can be a very important aspect of DFA. Manufactured parts (particularly smaller parts) that will be subjected to automated assembly will most likely end up in some type of feeder device. Therefore, the concept of symmetrical parts such as spheres, cylinders, pins, and rods becomes important so that they will position themselves with proper orientation when being fed from a parts feeder. Weight is also an important factor with special designs that make one end heavier than the other so that it is always fed to the assembly line in the same position (heavy end first). By designing with automated assembly in mind, major savings can be realized during assembly.

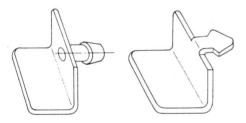

Figure 20–2   DFA can be used to simplify a part *(From Goetsch,* Modern Manufacturing Processes, *Copyright 1991 by Delmar Publishers, Inc. Used with permission)*

Assembly

The assembly of components (parts) to create quality products is one of the most competitive areas in manufacturing today. It is also one of the most challenging areas because of the new technologies that have emerged during the past two decades. Three methods are used by industry to assemble products:

1. Custom (handcrafted)
2. Batch (or lot)
3. Continuous

Custom assembly is generally used for one-of-a-kind-type products which have been handcrafted and require special assembly techniques. Some foreign automobile manufacturers still use this method to produce luxury automobiles.

Batch assembly is used for small lots of products when the cost of custom or continuous assembly would be too high. For example, if 5,000 kitchen garbage disposals were to be assembled, it would be too expensive for each unit to be put together by hand yet not enough work to invest in a continuous line assembly. Instead, a semisystem could be set up using cost-effective assembly techniques, jigs, and fixtures to assemble the 5,000 products efficiently. The key to the batch technique is flexibility. Many companies who manufacture more than one product or have a product with many options will set up a flexible assembly line or cell that can easily be changed to meet the "batch needs" of product demand.

Perhaps the newest approach to batch manufacturing and assembly is the *automated cell*. By setting up several automatic machines in a circle or block, using a robot to load, unload, and palletize, several products can be manufactured or assembled by simply changing the tooling and computer programs. This allows the manufacturer to produce a "batch" of widgets for ten days and then switch to gidgets for the next ten days. Such technology also helps the manufacturer plan more accurately for raw materials and low inventory, a concept known as **zero inventory**, to allow the product to be moved into the marketplace more quickly. This reduces the cost of storing the product, making it more competitive on the marketplace.

Continuous assembly is used best when large volumes of the same product are needed. This technique for assembly can produce large amounts of the product, but the tooling costs for setting up the line are very high. Therefore, a large market demand must be available to make this type of assembly viable. Figure 20–3 shows the concept of the continuous assembly line.

The assembly line is like a highway. To keep the product moving, the workers and assembly equipment are set up along the line; and as the product moves along, the various subparts are added by the workers. For complex products such as the automobile, the main highway may have several (see Figure 20–4) side streets which adjoin it and provide subassembled parts.

As the product nears the end of the assembly line, it begins to look more and more like the final product. When it goes through this process, a number of quality inspections are performed to make sure that the assembly is being done correctly. The final part of the assembly line is usually the packaging operation. As the product is packed safely in the container, inspectors give a final check and packers insert the appropriate printed materials for product care and warranties. The product is then shipped to the customer or stored in large warehouses until the items are sold.

AUTOMATION

The use of automatic machines and material-handling devices has increased since World War II for the manufacture and assembly of just about all mass-produced products. Automation refers to machines and equipment that are self-controlled and self-adjusting.

Just as numerical control (NC) is the heartbeat of automated manufacturing processes, the industrial robot is the heartbeat of automated assembly. The robot arm has replaced unskilled and semiskilled workers in areas of assembly such as welding, riveting, screw or bolt fastening, alignment and positioning, painting, inspecting for quality, and packaging. Figure 20–5 illustrates the role of automated robotics systems for the assembly of products.

One of the most difficult aspects of automated assembly is the feeding of parts to automated devices. Typically, the nonautomated factory has made extensive use of forklifts, tote buckets, pallets, and overhead cranes to deliver parts to assembly stations. More modern

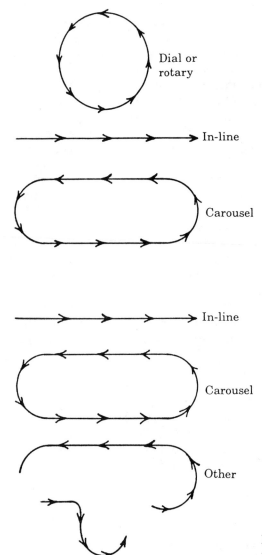

Dial or
rotary

In-line

Carousel

In-line

Carousel

Other

Figure 20–3   Large volumes of products are
produced on the continuous assembly line

plants have made extensive use of conveyor systems and self-contained flexible manufacturing cells (FMCs).

A newer concept in automated product assembly is to move the product to the various
processes by a guided vehicle. Pioneered in the automobile industry, it makes use of manufacturing and assembly cells where operations are performed independent of (but in con-
cert with) the automated line production. When the cell has completed the necessary work,
the guided vehicle returns the product to the main line for continued processing or assembly.

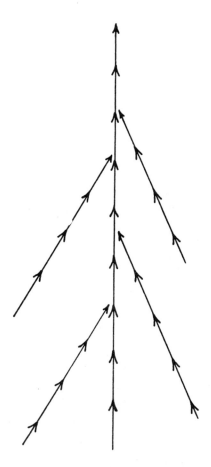

Figure 20–4   Complex assemblies require feeder lines to the main assembly line

The advantage to the automated guided vehicle (AGV) concept is that special options (like an automobile convertible) that would take more time than the standard product can be taken to a special cell, modified, and returned without upsetting the timing of the main line. It allows for greater product flexibility and eliminates the need to combine batch and custom assembly with continuous assembly (see Figure 20–6).

Assembly Systems

A variety of assembly systems are avilable based on the type of product to be assembled and the volume of product expected. The three most common systems are in-line, dial (rotary), and carousel.

In-line assembly systems are conveyors that usually carry a pallet base and deliver parts to stations for assembly. They can be of the circumferential or over-and-under design (see Figure 20–7) and can be controlled by station indexing or allowed to run continuously while

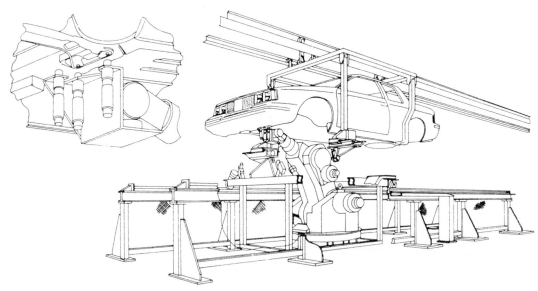

Figure 20–5   Robotics plays a major role in assembly line work *(From Goetsch,* Modern Manufacturing Processes, *Copyright 1991 by Delmar Publishers, Inc. Used with permission)*

parts are removed by special sweepers, drop hoppers, or pallet lifts. If all operations are timed appropriately, the line can be indexed to allow all operations to occur at once and, when completed, indexed to the next station automatically. Usually, the slowest operation will determine the rate of assembly; and line balancing is critical to maximize efficiency.

Figure 20–6   Automatic guided vehicles (mobile robots) are used to organize material-handling problems *(Photo courtesy of Vought Aircraft Company)*

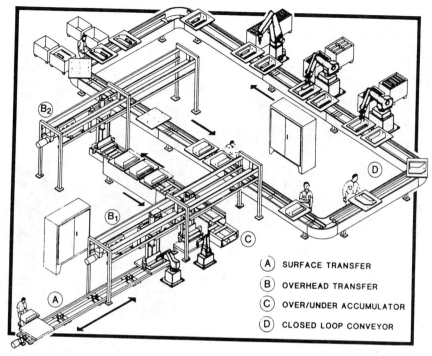

Figure 20–7   Conveyor systems can be of several different configurations

Dial (rotary) machines are used to assemble parts where the operations are located on the outside of a rotary table. As the table moves, the product is positioned for each assembly operation. Figure 20–8 shows how a twenty-station dial index machine would be set up. This "cell" type of assembly technique is best suited for smaller parts and large volume. The food packaging and canning industires have used this technique for assembly successfully for several years.

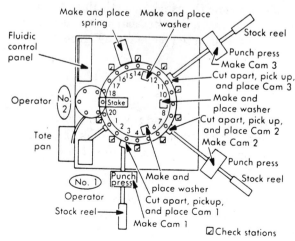

Figure 20–8   Cell assembly is very well suited to small parts *(From Goetsch,* Modern Manufacturing Processes, *Copyright 1991 by Delmar Publishers, Inc. Used with permission)*

Figure 20–9   Automatic conveyors make use of a pallet system *(Courtesy of Amatrol Corp.)*

Carousel assembly conveyors are similiar to in-line systems except that that the pallets are always loaded until removed from the system. Usually set up with an oval design, carousel devices may include multiple-layer configurations that include articulation with automated warehouse storage and retrieval systems. Like the in-line method, the carousel system can be indexed to feed on demand or run continously with raised pallets (see Figure 20–9).

Parts-Feeding Systems

An automatic assembly system is only as good as its ability to get parts into assembly devices on a continuous basis. The basic concept is to be able to load the feeding devices with enough parts to keep up with automated assembly over a period of time. For larger products like airplanes, tractors, or automobiles, parts may be stored in an automated warehouse and delivered to the assembly station on demand by robots, conveyors, and AGVs. Figure 20–10 shows how companies use the automated warehouse to supply parts for assembly.

Smaller parts require more attention to placement orientation. Such devices as parts hoppers, vibrating tables and bowls, centrifugal feeders, centerboard hopper feeders, pin feeders, and pocket hoppers are used to feed parts based on their shape and weight. They all have one

Figure 20–10 The automated warehouse has been used by some companies for many years *(Courtesy of Amatrol Corp.)*

thing in common: parts discrimination and orientation. Their purpose is to continuously feed parts in a consistent manner to assembly devices. Recent breakthroughs in robotic vision systems now allow parts discrimination by robots from random orientation storage bins.

The robot is able to select the part and orient it to proper position for loading the assembly device. Although not perfected entirely, the process has great potential for replacing mechanical parts-feeding systems presently in use.

Automation Assembly Controls

The heartbeat of automated assembly is in the control area. What was once a world of cams and electromagnetic switches has turned into a sophisticated family of microprocesser control devices such as the programmable logic controller (PLC), motion or robotic controls, computers, and unique communications systems.

These controls work on the basis of input and output (I/O) signals that gather data from sensing devices located in specific locations on equipment. The data are fed to the control via analog and digital communication for processing and appropriate response. Figure 20–11 shows the components of a PLC.

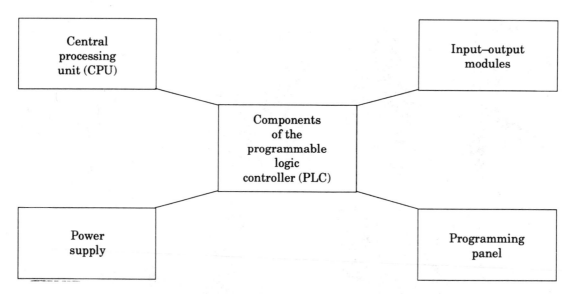

Figure 20–11   PLCs are used as control devices for automation

I/O controls can range from simple start–stop devices to intelligent systems that monitor input status, provide control algorithms and solutions, and directly control outputs based on entry data.

Controllers can be used at the machine level to monitor simple operations, at the station level to control several devices, or at the cell level to control the entire assembly operation (see Figure 20–12).

The major advantage of automated computer control is in quality and flexibility. Quality is increased because of continuous monitoring and adjustment capability. Flexibility is increased because of programmable options that allow reconfiguration of systems based on assembly needs.

Quality Assurance

The assembly of various components for modern products relies on Eli Whitney's interchangeable parts and Henry Ford's assembly line concept. To make these two basic ideas work, a great deal of attention must be given to maintaining standards and quality. All the production planning and sequence timing in the world will not work unless the individual parts are able to be assembled easily and in an efficient manner.

Quality assurance is a method of planning, sampling, inspection, adjustment, and attitude that has become increasingly important with automated systems. When parts are manufactured and assembled in large numbers, it is not possible to inspect each piece. Even if it were possible, inspectors would become fatigued and inspection gauges would become worn out. Therefore, a system of sampling has been developed using statistical analysis (statistical process control [SPC]) to determine (project) the quality of the product. The product may be inspected several times in several different locations to help assure its quality.

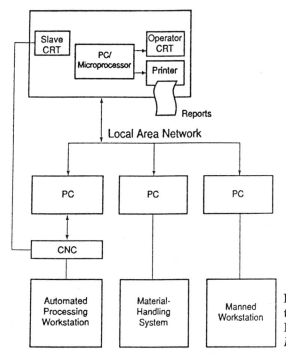

Figure 20–12 Computers control all operations for manufacturing *(From Goetsch,* Modern Manufacturing Processes, *Copyright 1991 by Delmar Publishers, Inc. Used with permission)*

Choosing a sample must be done with impartiality. The random method of inspection can provide the necessary data to calculate the standard deviation (0) of the parts or assemblies being inspected. When the necessary calculations have been made and confirmed, a control chart is constructed, as shown in Figure 20–13. This chart is constructed by plotting the average measurement of a sample against time. Upper and lower control limits are drawn as a distance equal to

$$A, 0 = 30_x$$

above and below the mean line. The value of 30_x is an arbitrary limit that has found acceptability in industry. Thus, control units are set so that only three operations of 1,000 will be found to be defective.

By establishing a control chart, a record of quality performance can be analyzed to make corrections and adjustments to the manufacturing or assembly line.

Making corrections and adjustments can be a very difficult thing to do. The control chart will flag the occurrence of product quality variation but does not in itself tell the manufacturing engineer what is causing the problem. Analysis and isolation of the problem will provide the manufacturing engineer with better direction and information to make changes.

During the adjustment process, statistical process control is continued and used to determine if improvement is made. By maximizing the use of SPC and engineering analysis, product-assembly processes can be fine tuned to meet appropriate standards.

Quality assurance is not just a program of inspection, analysis, and adjustment. It is also a program of attitude. All production systems, regardless of whether they are automated,

For normal distributions, 99.7% of the measurements will lie within 3 standard deviations of the process average. The term *process capability* is applied to ±σ, as this is the spread within which the vast majority of the parts will fall.

Process capability

The process capability is expressed in the units being measured (e.g., inches, volts, pressure). For example, if a milling machine has a capability of 0.005 in, 99.7% of the parts produeced within a ±0.0025 spread.

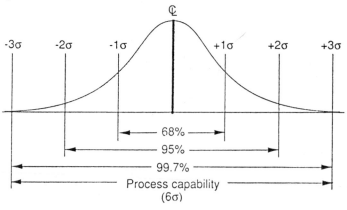

Cp

The Cp value is simply the ratio between the specification and the process capability

$$Cp = \frac{\text{Spread in the specification}}{6\sigma}$$

In the previous example, where the process capability was 0.005, if the tolerance spread was ±0.0025 (0.005), the Cp would be 1.

$$Cp = \frac{\text{Spread in the specification}}{6\sigma} = \frac{0.005}{0.005} = 1.$$

Figure 20–13   Data are plotted on a quality control chart

must employ people to assist in the assembly of parts. Worker attitude about quality can make a significant difference if it is positive and the worker feels as though he or she is making a contribution. Therefore, industry spends a great deal of money on making the worker feel important, appreciated, and needed in an effort to improve overall product quality.

   The process of assembly is an important step for the manufacture of quality goods. Increasingly, as the assembly line becomes more automated and companies depend on subvendors in what is essentially a world market, the need for production control is substantial. Engineering must be able to develop products that have flexible machining and assembling tolerances yet provide high quality and good performance. To do this, better automated systems need to be developed which provide flexible manufacturing assembly and meet the standards of a well-developed quality assurance program.

SUMMARY

When one purchases an outdoor barbecue grill, chances are he or she will have to spend an hour putting all the pieces together. If the directions and drawings are clear and accurate, the job will be easy. If all the holes line up correctly, and there is no hardware missing, the chances of being successful are greatly enhanced.

Imagine now, how long it would take one to assemble an automobile. A truck shows up and drops off 6,000 parts, a barrel of hardware, and a stack of blueprints. Where would a person start and how long would it take to complete the project? How much equipment would one have to purchase, and what technologies would one have to master before he or she begins? The task would be almost impossible.

General Motors, Ford, and Chrysler do it every day and, considering the complexity of assembly, do it very well. In this chapter, the reader has learned about the most common methods of assembly, including the most recent trend on design for manufacturing. Whether it is custom, batch, or continuous assembly, Ford's idea for assembling in a central location (factory) is practiced throughout the world. Cost-effectiveness in assembly can mean as much in profits for a company as the actual manufacture of the product. Assembly is the place where it all comes together and is, in a sense, the "proving grounds" for quality. In control manufacturing and top-quality assembly are the basic ingredients for success.

 REVIEW QUESTIONS _____

1. What is a quality assurance program?
2. Describe the difference between batch assembly and continuous assembly.
3. What role do robots play in assembly processes? List and explain at least six robotic assembly tasks.
4. Why is it important to have interchangeable parts?
5. When is batch assembly appropriate?
6. Describe how "cell" assembly works.
7. Describe the automated guided vehicle concept.
8. What is a quality assurance chart used for? Explain.
9. What role does the worker have in a quality assurance program?
10. Explain the term *automation* as it applies to product assembly.

SUGGESTED ACTIVITIES _____

1. Develop a Quality Assurance Control Chart using random sample numbers.
2. Make up three floor plan drawings to contrast the difference between custom, batch, and continuous assembly.

3. Set up a robotic arm to assemble the parts together using some type of permanent or semipermanent fastening procedure.
4. Visit an assembly plant to see the system in operation.
5. Set up a small assembly line and mass produce a product.

CHAPTER
21

Postprocessing Materials

The final process of manufacturing is called finishing. When the product was processed and assembled, some pre-finishing operations were probably performed. Even so, each product must be postprocessed with an appropriate and enduring finish. Finishes can range from smoothing with coated abrasives to coating materials electrically with chrome or zinc. The choice is usually based on appearance and performance, which enhance the salability of the product.

The appearance of some materials such as wood and metal can be enhanced because of the natural beauty of the material. If left to the elements, their appearance would be destroyed, resulting in rusted metal and bleached or cracked wood. Even plastic and ceramic materials will benefit from appropriate finishing processes. Thus, consideration should be given for finishing during the design and manufacture of all products.

PRE-FINISH CONDITIONING

The preparation of materials to be finished is an important step. During the various processing operations, parts can get marked or coated with oil and grease. If a coating of material were applied without prior preparation, the finish would not look or perform very well. Therefore, most manufacturers use some type of mechanical or chemical method for pre-finish conditioning products.

Mechanical

Mechanical methods of preparing surfaces include a number of coated abrasive or steel wool operations. Metal can be cleaned by using a fine emery cloth followed by steel wool.

Metal castings and some welding operations may also require the use of steel brushes to remove unwanted sand, oil, grease, and welding slag. If excessive rust is a problem, sandblasting may be used to clean and prepare the metal mechanically.

Some metals such as copper and brass oxidize very quickly and need to be mechanically buffed to bring out their luster. Once they have been buffed, a protective coating should be quickly applied to seal the materials from further oxidation.

Wood materials usually have excess glue left over from the manufacturing operation. If the glue is not successfully removed, the stain will not be able to penetrate the wood and leave an unsightly white mark. To remove the glue, a chisel or sanding block should be used with care not to create a new mark or dip in the surface. Certain types of wood also need to have the ends of the grain raised and sanded smooth. By mixing glue (very little) with water and painting the surface (a process called glue sizing), the grain will raise and when it dries, the glue in the water will help keep the grain protruding, allowing you to block sand the surface smooth.

Perhaps the most difficult part of preconditioning wood is removing the milling marks left by the woodworking equipment. Special wood scrapers and block sanding, as shown in Figure 21-1, must be used carefully to create a flat and natural wood surface. Because of the clear coating used for wood materials, defects are magnified.

Figure 21-1   Wood materials need to be scraped and sanded to remove mill (processing) marks

Ceramic materials are usually finished with coated abrasives also. Clay materials may be sanded or filed like wood. They may also be moistened while in the green state and rubbed with special wire or weaved cloth material. Concrete is mechanically conditioned by special block abrasives made of volcanic material. The rubbing action removes excess concrete left over from the molding process or mortar joints left rough from the mason's work.

Glass is polished with special compounds like plastic. The compounds are classified by material and grit size and are used in conjunction with soft-cloth buffing wheels. Sometimes, toothpaste can also be used to precondition and clean plastic materials when the abrasions and scratches are not too deep.

Chemical cleaning of materials is primarily focused on metals. Unlike wood surfaces, polished metal surfaces lack the porosity required for favorable paint adhesion. Therefore, the surface of metal must be mechanically or chemically cleaned and roughened to provide a bond between the base metal and finish coating. Various chemical cleaners are used to accomplish this task:

1. Alkali cleaners with a 5–10% caustic solution. Care must be given to rinse all residue.
2. Acid cleaners with an 8–10% aqueous solution of sulfuric acid or commercial hydrochloric acid is used to remove rust.
3. Phosphoric acid of 5% is used to improve adhesion by providing a very thin crystalline structural coating.
4. Emulsion cleaners of soaps and detergents are used to remove oil and grease.
5. Solvents, such as lacquer thinner, are also effective for removing tars, oil, and grease.

Most materials must be mechanically or chemically cleaned before they can be coated. The residues of manufacturing processes leave unwanted and unsightly markings that would make a product less beautiful and less durable. The next step in the postprocessing of materials is to enhance the product by special decorating techniques.

Decorating

After the material has been pre-finish conditioned, a number of decorating processes can be performed that will add beauty and value to the product. Special designs may be carved into the material or placed over the material as overlays. Both the wood and ceramic industries are heavily involved in decorating as a part of the finishing process.

Perhaps the best-known form of *surface penetration* is wood carving and veneering. Special woods can be added to the surface of a table by making exact grooves (see Figure 21–2) and fitting them with a lighter or darker wood. Even special sunburst designs can be inlaid to enhance the design.

In addition, carved designs can be added to wood and clay materials before they are finished. In fact, carvings are one of humankind's oldest forms of decorating and were used extensively to tell stories on clay pots and wooden plaques.

New to the world of surface penetration is the laser technology of the twentieth century. Beautiful detailed carvings can be burned into wood and metal by the laser as part of a new production technique (see Figure 21–3) for decoration.

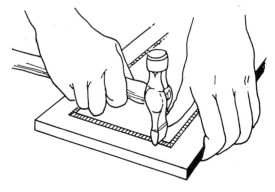

Figure 21–2   Inlays for wood are used to decorate the product *(Courtesy of Stanley Tool Co.)*

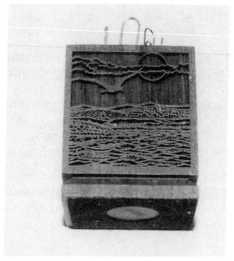

Figure 21–3   Lasers can cut beautiful decorations into the surface of wood

Surface overlays, on the other hand, make use of designs added to the surface as part of product decoration. Ceramic products gain their special colors and designs by adding beads of colored glass, colored glazes, and silk screen–printed patterns. Once fired in the kiln, the patterns become a permanent part of the product (see Figure 21–4).

Cast plastics make use of an inverse form of decorating by embedding the design with clear plastic. Perhaps the best-known overlay of plastic material is the laminated, simulated wood-grain tabletop. The photographed wood pattern combined with the clear plastic material looks exceptionally close to actual wood and provides a more durable finish. Glass also makes use of an overlay process by sandwiching various plastic tinting and nonglare materials.

Mechanical treatment of metals can provide surface decorations that add beauty to the product. Abrasives may be used to create circular designs in nonferrous metals such as aluminum and bronze. It provides the illusion that the material was machined in the circular pattern and offers an alternative to the highly polished look.

Figure 21–4 Color patterns can be screen printed on ceramic dinnerware

Reproductions of antique work make use of distressing techniques to make the product look old. Wood may be dented using chains, metal distressed using ball peen hammers, and clay pots distorted by forming tools. One can even purchase nails with peened heads and modern shanks that simulate original cut nails for decorating cabinets and flooring.

Staining and Coloring

Once the product has been cleaned and decorated, it is appropriate to select the process to enhance the beauty of the product by color. The choice to paint is different from the choice to use a clear coating. If clear coating (popular for wood and ceramics) is to be used, the manufacturer may wish to stain, bleach, or glaze colors and patterns before the final finish is applied.

Stains are solutions which have color matter suspended and are used primarily to change color rather than as a protective coating. As the stain is applied, the vehicle (solution) is wiped away with a cloth and dried via evaporation. Stains can be used on wood and clay products because they are porous. Metals and plastics do not respond well to this type of coloring technique. The most popular use of stain is with wood because it helps highlight the natural beauty of the grain.

Wood stains are classified into four major groups according to their solvent or vehicle. Table 21–1 lists the various types of stains and their characteristics. The four most popular types of stain are water, oil, nongrain-raising (pigment and penetrating oils), and spirit. Water stains are difficult to use because the overlap shows easily, but they give the clearest and most transparent results of all stains. Oil stains are easy to apply and give good appearance with some blocking (hiding of the grain) problems, and they are most widely used stains by the home craftsperson. Nongrain-raising stains are the most recently developed of the group. They have the clarity of water stain without the problems of grain raising associated with all water stains. Last, the spirit type of stains are used primarily for touchup work. Their ability to penetrate old finishes makes them ideal for refinishing antiques and other difficult items to strip and match. However, they have limited production use because they fade easily and are difficult to apply evenly.

Stain	Advantages	Disadvantages	Application
Pigment oil stain	Colorfast Used to imitate maple, mahogany, and walnut Nongrain-raising	Tends to hide grain pattern May affect drying quality and adhesiveness of succeeding coats	Brushing Spraying Dipping
Penetrating oil stain	Nongrain-raising Good color tone and permanence Dries quickly	Tendency to bleed into top coat if not well sealed Retards drying of finish coat	Brushing Spraying Dipping
Preservative oil stain	Resists decay Some types water repelland	Strong odor Dries slowly Not used on furniture	Brushing Spraying Dipping
Water stain	Uniform rich color Excellent color retention Nonbleeding	Raises grain Requires sponging, sizing, and sanding before staining Requires sanding and sealing after staining	Brushing Spraying
Spirit stain	Bright colors Quick drying	Color has tendency to fade with exposure to light Slight grain raising Tendency to bleed	Brushing Spraying

Table 21–1   Wood stains are classified into four major groups

FINISH COATING

Paint

Paint is a liquid mixture of pigment and vehicle (solvent) which adheres to the surface as a colored protective coating. The three most common types of paint are oil-based, oil–resin emulsion, and latex emulsion.

Oil-based paint contains approximately 60% pigment and 40% vehicle by weight. The pigments can be produced chemically or manufactured and are also obtained from three natural sources: (1) mineral, (2) animal, and (3) vegetable. However, the greatest number of pigments are chemically produced or manufactured. White pigment is obtained from white lead (not used today because of its toxic effect), zinc oxide, lithopone, titanium dioxide, and calcium carbonate. Colored pigments are obtained from the earth and some high-temperature manufacturing processes. They are produced as phlalocyanine blue and green, chrome green, chrome yellow, cadmium maroon, chrome orange, zinc yellow, iron yellow, toluidinered, cadmium red, para red, hansa yellow, siennas, umbers, ocre, vandyke brown, ultramarine blue, and iron blue.

The vehicle, which constitutes about 40% of the paint's volume, should contain about 70–90% drying oil and 10–30% solvent dryers. Mineral spirits is the most widely used solvent, and linseed oil is the most important drying oil. Other drying oils used for oil-based paint are castor, soya, fish, tung, oiticica, and perilla.

The oil–resin emulsion paint is an oil-in-resin water-soluble paint which produces hydrophobic coatings similar to oil or alkyd paint finishes. Hydrophobic coatings do not redissolve after removal of the solvent and atmospheric temperature. Oil–resin emulsion paints are easy to apply, dry quickly, do not leave brush marks, and have good holding power. They are easily recognized because of their thick paste-like consistency.

Latex emulsion paints are aqueous emulsions of synthetic resins or rubbers which are also water soluble. The latex acts as the pigment and binder, while the water acts as the carrier. It is particularly suited for masonry materials and plastered walls and ceilings. The three most common types of latex paints are butadiene-styrene, polyvinylacetate (PVA), and acrylic. Table 21–2 provides a chart on the various types of paint used by industry today.

Plastisols and organosols are technically classified as dispersion coatings. Unlike the solvent coatings, they make use of a colloidal system for coating materials. Basically, they are organic coatings consisting of polyvinyl chloride dispersed in a plasticizer; the resultant coating material is called a plastisol. If volatile dilutents are added to the plastisol, the resultant coating material is called an organosol.

As a coating material, plastisols and organosols provide a protective vinyl coating on metal or ceramic materials. The most popular method for applying the material is by dipping (see Figure 21–5) the piece into the liquid. Various thicknesses can be produced by preheating the material to be coated or by dipping the piece several times. Other methods of coating with plastisol and organosol are slush, rotational, injection, and in-place molding.

Glazing is technically a clear coating process but is mentioned in this section because of its ability to solidify or carry colors in ceramic ware. The glaze coating is made up of liquid

Paint

Enamel	Wood Metal	Mineral spirits	12 hrs.	Waterproof Can be brushed or sprayed Tough, hard	
Latex	Wood	Water	4 hrs.	Odorless Water cleanup	

Lacquer	Metal	Lacquer
		thinner
			15 min.	Usually sprayed but can be
				  brushed on or applied by
				  dipping
				Resists water, heat, and
				  a;cohol
					Requires two or more
					  coats
					Toxic fumes

Table 21–2   Various types of paint and their uses

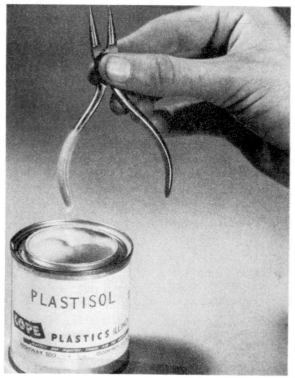

Figure 21–5   Tool handles are coated with plastic

refractory glass. A hand-painted piece of fine china, decayed coffee mug, or screen-printed piece of dinnerware can be sealed with a clear glaze which, when fired in a kiln, becomes a glass coating. Pigment can be added, and plumbing fixtures can be color coated with ease.

Glaze coatings may be applied with a brush, dipped in a tank, or sprayed with a gun. Once fired in a kiln, glaze coatings are very durable.

Clear Coating

Clear coatings are used primarily by the furniture industry to serve as a protection for wood. That is not to say that they cannot be used on other materials because, with the exception of plastic, they are used extensively whenever a clear protective coating is needed.

Clear coatings are technically called resinous varnishes and are made up of a resin and a vehicle. When a pigment is added, the result is enamel paint. Nonpigmented resinous coatings may range in color from water white to dark amber. Most resinous coatings derive their special characteristics from the type of vehicle which is used to carry them. The four most popular resinous vehicles are as follows:

1. Oleoresinous varnish
2. Synthetic varnish
3. Cellulose lacquer varnish
4. Shellac varnish

Oleoresinous varnish is a mixture of oil and resin. Congo resin, kauri resin, pontianak resin, and others are mixed with drying oils such as chinawood oil or linseed oil to gain certain natural characteristics. Oleoresinous varnishes make excellent transparent finishes with great depth of film and good durability.

Varnishes may be classified by their oil content as long, medium, and short. The following oil ranges make up the various oleoresinous varnishes:

Gal. of drying oil	Lbs. of Resin	Class	Use
40 – 100	100	Long	Spar varnish for marine or outdoor applications
12 – 40	100	Medium	Floor varnish for toughness applications
5 – 12	100	Short	Fine furniture with high gloss, smoothness applications

Synthetic varnishes are the new group of "plastic" coatings which have many advantages over the oleoresinous varnishes. The composition of synthetic varnish is very similar to that of oleoresin except that the resin has been manufactured. At first, these resins were of the alkyd type; but research has created a number of new "plastic" coatings which can be applied easily. Most of these synthetic resins will be familiar to the reader because they were discussed in detail earlier in this text. Following is a list of the synthetic resins used for clear coating:

Alkyd resins	Phenolic resins
Amino resins	Epoxy resins
Vinyl resins	Silicone resins
Acrylic resins	Urethane resins

The technician working with protective coatings is very interested in the thermoplastic and thermosetting resins. He or she dissolves the resin material into a proper vehicle and is able to apply it as a protective plastic coating. Perhaps the best known of all the synthetic resins is polyurethane. It is widely used because it is a hard, durable, water-resistant finish that is relatively easy to use.

Cellulose lacquer varnishes are produced from nitrocellulose and have the advantage of drying very quickly by evaporation and forms a film from its nonvolatile constituents. The major disadvantage of nitrocellulose lacquer is in its inability to form a heavy film. Therefore, the product must be coated several times as compared to the oleoresinous and synthetic varnishes.

Recently, progress has been made to increase the thickness of film application by increasing solids from 15% to 35% and heating the material to 160 degrees Fahrenheit. Hot lacquer not only applies more material but does so with increased smoothness and flowability, making the finish more attractive.

Nitrocellulose lacquer is used because of its fast drying, recoatability, and patchability if furniture is damaged in production or in transit to retail stores. It is also an excellent coating for ceramic and metallic material.

Shellac varnish is made from the resinous substance excreted from an insect called tachardia lacca. Most of this resinous substance is gathered and processed in India. The lac resin is dissolved in alcohol to form orange shellac. This resin can also be bleached to form white shellac.

Shellac varnish has the unique ability to seal resinous knots and streaks which would otherwise bleed through the surface of a coating. When diluted, it is an excellent wash coat for fine furniture and will not allow oil stains to bleed like other coatings. It is easy to apply, dries quickly, and produces an extraordinary tough yet elastic finish on wood, glass, and metal.

Metallic Coatings and Finishes

Metallic coatings and finishes include a number of pigment–vehicle coatings and several electrical–chemical finishes. The metallic finishes may be painted on with a brush, electroplated with a thin film of metal, or chemically treated to resist corrosion and wear. Each application has unique advantages, depending on the finish needs of a product.

Metallic pigments are actually tiny flakes of metal suspended and carried by a vehicle such as varnish. Perhaps the best-known metallic pigment is aluminum paint. It is a durable coating which reflects light and heat very well. It is known for its superior holding power and, as a result, is often used as a primer for regular oil-based paints.

In terms of its working properties, aluminum paint is similar to other coatings except that the pigment is made up of aluminum flakes which have an overlapping action (called leafing), which gives it brightness and holding power.

Metallic powders are also available in other colors using brass, bronze, and copper. They are also available in different colors such as blue, red, green, gold, and purple.

Electroplating metallics is a process of electrolysis using chemicals and electrical current to transfer metal coatings from the anode to the cathode. The primary purpose of the electroplating is to coat one metallic (and other materials which are discussed later) with another metallic for protection and beauty.

Electrolysis (see Figure 21–6) is the chemical decomposition or change created by the electric current through the electrolyte. The electrolyte is the conductor (fluid) which allows

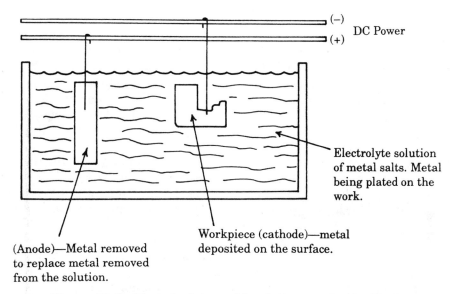

Figure 21–6   Electrolysis is used for plating metal with chrome

the ions to move about and seek the cathode. Essentially, the anode is made up of the material that will be used as the coating, and the cathode is made up of the material to be coated. When a current is passed from the anode to the cathode through the chemical solution (electrolyte), the anode decomposes and the material becomes deposited (attached) to the cathode.

The plating thickness is controlled by the amount of current and the ratio of atomic weight to the valence of the metal in the electrolyte. While a number of metals can be used as a plating material (copper, silver, chrome, etc.), the most popular is chrome. Increasingly, black chrome is being used as a coating because it absorbs light and, in some instances, has the better corrosion resistance than bright chrome.

Electroplating Plastic

The process for plating plastic is different than for metal, requiring a two-phase approach: (1) chemical deposition and (2) electrodeposition.

The chemical deposition process allows the plastic material to be chemically sensitized. An inorganic reducing salt called stannous chloride ionizes, making the stannous ions attach (absorb) to the plastic surface.

Chemical deposition is an autocatalytic process because the action in which metal ions form the aqueous plating solution are deposited on the catalytic surface takes place spontaneously and continuously. The metal deposited must be a catalyst to its own electroless plating bath to effect continuous action on the surface.

The plating of plastic materials has played a significant role in the reduction of weight in the automobile industry. Other manufacturers have increased the use of plastic because of its advantages over metal *and* because of its ability to be plated.

Black oxide has been used extensively in the firearms industry to protect metal from corrosion. Commonly called "blueing," the finish is a black oxide of iron which is produced by the chemical structure of the surface area. The blackening salts (6 pounds per gallon) and soft water are mixed into a solution which is heated to 290 degrees Fahrenheit. The actual immersion time is determined by how deep the "blueing" is desired.

Black oxides are attractive; have good corrosion resistance (if the surface is kept oiled); have good wear resistance; cause minimal dimensional changes; and resist chipping, peeling, blistering, and cracking.

Another method used to provide corrosion resistance is called **phosphralizing**. In this process, iron, steel, or zinc provides a coating that is corrosion resistant by itself or as a base for organic coatings. Insoluble metallic phosphates are formed on the base metal surface in a phosphate bath. The coatings are electrically inert and combine with the surface of the base metal as one.

Iron, zinc, heavy zinc, and manganese are the four basic phosphate coatings used by industry. They can be applied by a spray process or immerged using a bath.

Anodizing the surface of aluminum is an electrolytic oxidation process which takes place in a tank containing an aqueous acid electrolyte. The aluminum part serves as the anode, and the tank (if metal) serves as the cathode.

Essentially, this process provides a protective coating by oxidation that is controlled by the electromechanical process. The quality, range of finish, and color are much improved over the traditional chemical-conversion coating processes. Anodizing provides electrical insulation, corrosion, and abrasion resistance. Because the finish is an integral part of the metal, it will not chop, blister, or peel. Table 21–3 illustrates the anodizing process.

Type of process	Electrolyte composition	DC curent in amps/sq. ft.	DC voltge	Temp. (°F)	Treatment time
Sulfuric acid	H_2SO_4	10–20	10–20	70	10–60 min.
Chromic acid	CrO_3	3–5	35–40	90–120	10–40 min.
Hard coating	H_2SO_4	24–36	20+	20–50	10–60 min.

Table 21–3 How aluminum is anodized

Special Coatings

While most finishes concentrate on protecting the surface of the product, others are used to provide special effects, add beauty, or duplicate the aging process. Such is the case of the special coatings discussed in this section of the chapter. Coatings can provide textures, luminate parts, create smoothness, provide softness, and make new materials look old.

Texture finishes are used to create pleasing designs while, at the same time, covering up defects in the base material. Textured paint is the best known of these finishes and is used on drywall installations to cover up plaster seam imperfections. It is a thick paste-like paint that is applied to the surface for the purpose of texturing or molding. The designs can be done with a paintbrush, scrubbing brush, towel, sponge, or a roller.

Wrinkle finishing is another coating used to cover up defects. It is an excellent one-coat finish for metal, plastic, ceramic, and wood materials. The textured pattern develops because the outer layer of the coating expands during the drying process. Although one coat usually hides surface defects, it will not cover up large pits or holes caused by defective foundry work. Figure 21–7 shows a typical wrinkle finish.

Luminous paint is used widely by the military during wartime when blackouts are in place. The light emitted is bright enough to be used on instrument panels yet does not provide enough illumination to aid the enemy. It also provides the painter and decorator with a material that can produce an almost magical type of decorating.

Phosphorescent paint emits a light in the dark. It works on the principle of irradiation and can last as long as 12 hours. Fluorescent paint reflects segments of visible light and absorbs the rest. When the headlights of an automobile are pointed to a sign with fluorescent paint, it is absorbed and reflected back as if it were electrically alive.

Luminous paints are special-purpose coatings which have a primary function not related to protective coatings. As a postprocessing finish, they play an instrumental role in the safety and advertising industries.

Flocking is a finish that provides a "fuzz-like" covering over the base material. It is available in three different types of fibers: rayon, cotton, and nylon.

Rayon is used to provide a plush velvet look and is the most widely used type of flocking. It has wide acceptance because of its ability to produce a deep pile with rich colors. Cotton flocking looks like suede and is used for applications where a short nap is required. Nylon flock is long wearing and is used in applications where wear is important.

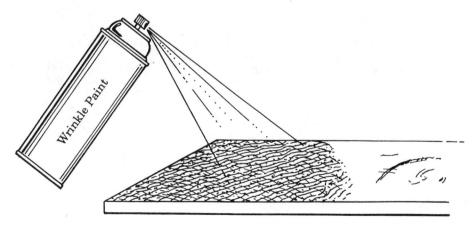

Figure 21–7   The use of a wrinkle-finish paint also can hide some surface defects

Essentially, the base material is covered with an adhesive and the flock material is sprayed on using a flocking gun, mechanically vibrated on, or electrostatically applied. Figure 21–8 shows a flocking application in a jewelry box that helps prevent scratching. The material forms a soft protective coating that adds beauty to the product.

Wax and Liquid Polishes

Wax and polishes are seldom used as the only protective coating. They usually are used to seal and brighten varnish, lacquer, acrylic, shellac finishes or buffed metal, and plastic finishes. As a sealer and brightener, waxes and polishes are made from natural and synthetic materials; and as a postprocessing activity, waxing and polishing are the last steps to complete finish.

The four most common *natural waxes* are beeswax, paraffin, spermacete, and carnauba. Of the four, carnauba is the hardest and must be mixed with softer waxes before it can be used. Synthetic waxes are now available and popular. They are made up of a proper blend of hydrocarbons and fats.

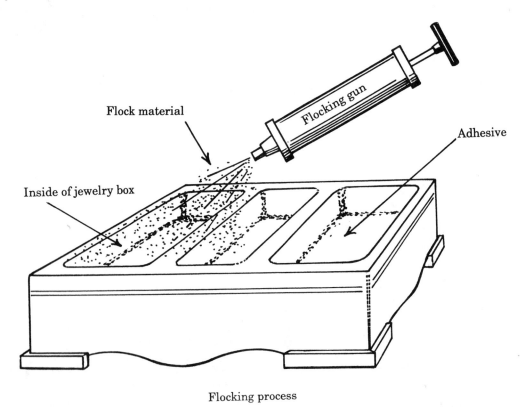

Flock material

Flocking gun

Adhesive

Inside of jewelry box

Flocking process

Figure 21–8   Flocking is another finish that leaves a soft look and touch

Liquid polishes contain silicon because of its ability to repel water. The silicon also has made modern polishes easy to use without the traditional hard rubbing associated with some waxes. As a household item, the spray-on silicon polishes are used to clean and dust fine furniture leaving a protective and lustrous glow. So long-lasting are the silicon polishes that two- and three-year treatments are available for automobile finishes, eliminating the traditional fall and spring wax job.

Wax and liquid polishes not only protect painted and varnished surfaces but also help retain colors, prevent blushing, and help reduce scratches and surface wear. As the last step in the finishing process, they should also be used to maintain the product finish for years of trouble-free service.

 REVIEW QUESTIONS _____

1. What are pre-finish conditions? Explain two methods used to pre-finish condition materials.
2. List and describe five types of chemical cleaners.
3. What types of materials use surface overlays for decorating products? Give examples of products decorated in this manner.
4. What are stains? What materials are they used for?
5. Describe the difference between oil-based paint and latex paint.
6. What role does varnish play with oil-based paint?
7. Is plastisol a paint or plastic coating? Explain.
8. List and describe two ways to provide color for ceramic materials.
9. What is the most popular "plastic" clear coating used for woodworking products? Why?
10. Explain the differences between electroplating metallics and electroplating plastics.

SUGGESTED ACTIVITIES _____

1. Make up samples of wood and apply different finishes. Compare and contrast the differences.
2. Using clay, colored glass, and a glaze finish, have students develop a clay product and fire a hard, clear finish.
3. Set up a tank and conduct black oxidizing or anodizing experiments.
4. Make up test samples of paint, set them up outdoors, and measure their quality over time.
5. Using one of the finishes described in this chapter, conduct the process from preconditioning to postprocessing on a laboratory-produced product. Write up all procedures and explain them to the class step by step.

SECTION

5

RECOVERY

GENESIS

PRE PROCESSING

PROCESSING

POST PROCESSING

Materials Recovery and the Ecological/ Production System

CHAPTER 22

Industrial Waste and Pollution

The system that supports life on earth is unique in this solar system. No other planet in this solar system can support life as it is known on earth. The life-support system for humans and animals is contained in the atmosphere surrounding our planet; in the waters, soil, and vegetation that covers the land; and in the minerals and fossil fuels that are found beneath the surface.

Well-known scientists have compared earth to a spaceship. It is a self-contained system moving through space with finite resources to support life for its passengers. *Finite* means limited, having bounds. The life-supporting resources of the planet earth are not boundless, and many are nonrenewable. The science and technology that has enabled humans to satisfy desires for transportation, recreation, housing, and other material trappings of a higher standard of economic living are also threatening the sensitive balance of the ecological system that provides such basic needs as food, clean air, and water.

Human attitudes and misinformation have helped contribute to the problems associated with the depletion of natural resources and pollution of air, water, and land. These attitudes are based on such faulty assumptions as

1. Air, water, mineral, fuel, and other resources are infinite and will meet ever-increasing demands.
2. Technology will be able to offer solutions to polluted and dwindling resources.

The reality is that neither of these assumptions is accurate. Reserves of certain mineral resources may be depleted by the middle of the twenty-first century, and technology will most likely not be able to reverse pollution of resources devastated by industrial chemicals and agricultural pesticides and herbicides.

POLLUTION

Pollution has become a major international problem. Pollution does not recognize state and natural boundaries. Toxic wastes dumped into waterways or emitted into the air can easily cause devastation to lakes and vegetation thousands of miles away. The international scope of the problem threatens to grow even larger. The developing nations represent 70% of the world's population but currently consume only 20% of the world's energy resources. These countries are becoming more and more industrialized; and in the near future, the 20% could become significantly larger.

The threat to the health of humans and animals is of major proportion. Acid rain and chemical dumping have caused damage to at least 7,000 lakes (Figure 22–1). Some of these lakes have been determined to be lifeless. DDT (dichloro-diphenyl-trichloro-ethane), a toxic chemical pesticide, has been detected in Antarctic penguins and seals. Entire communities have had to leave their homes as a result of life-threatening chemical dumps that were created years before their houses were built. The technology that has given us better food, bigger cities, highways, more efficient transportation, and other conveniences to increase our standard of living has a dark side. Uncontrolled pollution, in a relatively short period of time, could devastate the life-support system humans and animals need for long-term survival. Education and action can prevent this from happening. The following is a brief description of the major environmental concerns.

Carbon Dioxide

Carbon dioxide is an odorless, tasteless gas that is part of the natural process that restores oxygen to the atmosphere. Carbon dioxide is formed during respiration. When a hu-

Figure 22-1   Acid rain has been blamed for devastating this lake *(Courtesy of Photri, Inc.)*

man or animal exhales, they expel carbon dioxide into the air. Carbon dioxide is also formed by the burning of fossil fuels. Nature has a built-in system of using the carbon dioxide to produce oxygen for human consumption. The process, **photosynthesis**, converts carbon dioxide and water into carbohydrates with a simultaneous release of oxygen. Green plants and trees are the vehicle used in this transformation.

This delicate balance in nature that refreshes the air needed for life relies on green plants and vegetation to make the conversion previously described. The problem is that the level of carbon dioxide has increased dramatically as a result of industrialization, increased population, and reduction of green plants and vegetation. Experts believe that carbon dioxide (CO_2) concentrations have increased 15% during the past century and could double by the middle of the twenty-first century if current growth rates continue.

The effects could be devastating. Carbon dioxide in the atmosphere acts much like the glass in a greenhouse. They both pass visible light but trap infrared rays. The increase in carbon dioxide in the earth's atmosphere could cause an alteration in the climate of the world. Temperature increases could cause serious damage to agricultural products and cause the polar ice caps to begin melting. The extent of the impact of increased levels of carbon dioxide is not fully understood by scientists.

Ozone Depletion

Ozone is a gas in the atmosphere that protects the earth from ultraviolet radiation damage. Recently, scientists have gathered data that lead them to believe that the ozone layer in the stratosphere is being depleted by emissions of chlorofluorocarbons (CFCs). CFCs are chemicals used as aerosol propellants, refrigerants, and also produced as by-products of other industrial processes.

The depletion of the ozone layer will allow increased levels of ultraviolet radiation to enter the atmosphere. This can result in increases in the incidence of skin cancer as well as cause damage to agricultural crops, fish, and animals. Although the United States has banned nonessential aerosol uses of CFCs, the worldwide growth in nonaerosol uses has been approximately 7% per year.

Acid Rain

The burning of fossil fuels such as coal, oil, and natural gas produces sulfur and nitrogen oxides that have been emitted into the atmosphere (Figure 22–2). These emissions are converted into sulfates and nitrates and then into acids. The acids return to earth in the form of precipitation (rain, snow, fog, etc.). The acidity of land and water is being affected through this process. The damage to trees, vegetation, crops, and aquatic life is significant. Forests in many parts of the world have felt the effect of acid precipitation. Forests in North America and Europe appear to be affected more than others. Some species of trees in Germany are nearly extinct, and recent reports from national parks in the United States have provided nearly conclusive evidence of the devastating effect of acid precipitation.

Acid precipitation also affects life in the lakes of these same areas. Lakes in the United States and especially Canada have been acidified by acid precipitation. In Ontario, Canada,

Figure 22–2 Factory smokestacks may emit pollutants that cause acid rain *(Courtesy of Photri, Inc.)*

nearly 5,000 lakes have been threatened by acid precipitation. It has become an international problem of major proportions. Acid precipitation does not recognize international borders. Emissions of sulfur and nitrogen oxides in the United States have drifted north into Canada. The problem has become so serious that the topic is among the top agenda items at summits between the United States and Canada.

The effect of continued pollution by acid precipitation on the food supply is not known for sure. What is known, however, is that plants do not do well in acid soil and if the current trend continues, it could threaten human health. It has already cut crop yields in Europe and the United States, causing losses of approximately $1 billion.

Acid pollution can be controlled, and private and public agencies are attempting to reduce it in one or more of the following ways:

1. Use low-pollutant fuels.
2. Prevent the formation of acid causing pollutants during combustion.
3. Screen pollutants from exhaust and fuel gases.
4. Conserve energy, thus reducing level of emissions.

These solutions, however, are not free of problems. In many cases, they are more expensive than what the operator can afford. Burning low-sulfur fossil fuels, for example, could help tremendously. The problem is that, although these fuels are plentiful in the west, much of the fossil fuel is needed along the east coast. Therefore, fuels from the east and midwest are used and the product is higher in sulfur content.

DDT Pesticide Pollution

A discovery by a German chemistry student in 1874 led to the production and use of one of the most controversial and perhaps deadliest chemicals of this century. When Othmar Zeidler discovered DDT at the end of the nineteenth century, he was completing a chemistry assignment. He recorded his findings from his experiments and filed his findings away.

In 1939, a Swiss chemist, Paul Mueller, was attempting to find a chemical that would stop the imported potato beetle. Mueller found the synthetic organic chemical of Zeidler's to be very potent. Shortly after Mueller's discovery, the U.S. Armed Forces began using DDT. They found it to be inexpensive, effective, and very potent.

The military use caused DDT to become a very popular insecticide, and its use soon spread worldwide. DDT became so popular that by 1960 it was a $30-million business in the United States. Thousands of tons were being sprayed and dusted on crops across the world (Figure 22–3).

Figure 22–3 Specially equipped airplanes spray pesticides on crops *(Courtesy of Photri, Inc.)*

As with many technological developments, there was little foresight of the long-range effects of DDT. When DDT was discovered in the milk supply, restrictions were imposed. Unfortunately, it was too late to avoid serious consequences. DDT is in the soil and seeping into water supplies. Even though it has been restricted, the levels in the water supply are expected to rise before they begin to fall.

The length of time it takes for DDT to break down is not known exactly. It has been estimated, however, that it has a half-life of at least 15 years. This means that half of a given application will break down in 15 years, half of the remainder in another 15 years, and so on.

DDT is not selective in killing insects. It kills the desirable insects along with the undesirable. It is also believed to have killed other wildlife, including birds and fish. DDT has been found in Coho salmon and threatens to eliminate 80% of the commercial sale of these fish. The level of DDT found in the Coho salmon was 20 ppm (parts per million). The Food and Drug Administration recommends a maximum level of 5 ppm. Environmentalists estimate that it could take 35 to 40 years before residues fall to the recommended safe levels.

DDT could also have an effect on oxygen production. The green plankton in the ocean is extremely important in producing oxygen. Between 50% and 70% of the earth's oxygen is produced at sea through photosynthesis. Scientists have shown that the levels of DDT in the ocean cause severe reduction in the rates of photosynthesis.

Solid Waste

Approximately 39% of the solid waste is generated from mineral and fossil-fuel mining, milling, and processing industries. According to the Council on Environmental Quality, eighty mineral industries generate solid waste. Eight of the eighty account for 80% of the total. The eight, in order of waste produced, are as follows:

Copper	Iron and steel	Bituminous coal	Phosphate rock
Lead	Zinc	Alumina	Anthracite

The mineral waste accounts for approximately two billion tons per year in the United States. Other sources of solid waste include agricultural, residential, and commercial.

Of all solid-waste categories, agricultural waste is the largest by far. In 1969, there were 2.2 billion tons of agricultural solid waste. This waste represented animal and slaughterhouse waste, residue from crop harvests, orchard and vineyard prunings, and the like. The solid waste accumulating in the feedlots is beginning to present a particular problem to wells and streams. The runoff from the lots is reaching freshwater wells and streams and poisoning the water.

The residential and commercial wastes represent about 110 million tons per year. This waste includes paper and paper products, plastic, rags, and drums of useless by-products. The toxicity of the by-products has been of particular concern lately. One of the most publicized problems with solid waste occurred in 1987. A large barge containing solid commercial waste was refused dumping rights at several locations in the United States and Central America. The barge traveled the east coast for weeks while responsible parties searched for an acceptable solution.

Water Pollution

Many of the pollutants already discussed have threatened the supplies of fresh water available for use by humans. According to a U.S. Geological Survey, ground water provides about 50% of the drinking water and 20% of all water used in the United States. Seventy-five percent of the contaminated sites identified by the EPA targeted for cleanup pose a threat to ground water.

Ground water is increasingly seen as a threatened resource because it is so widely used as drinking water; once contaminated, it is extremely costly to clean. The problem is international in scope and poses a health threat to some of the most highly populated areas of the world.

In the United States, the regions affected mostly by industrial waste are the northeast, the Ohio River basin, the Great Lakes states, and the Gulf Coast states. Industry contributes to 60% of all U.S. water pollution, with the principal offenders being the paper, organic chemical, petroleum, and steel industries.

The pollution of water is of growing international concern. Approximately three-fourths of the freshwater use is for irrigation of agricultural fields. With the population expected to continue to grow, the demand for water is predicted to double by the year 2000. One hundred and forty-eight of the world's major river basins are shared by two or more countries, and agreement between countries over the use of the waters will be essential to the conservation and maintenance of these waterways for agriculture and human consumption.

Parts of the earth, especially the Third World countries, already suffer from severe water shortages and drought. Waterborne diseases are reaching epidemic levels in many areas. In 1975, the World Health Organization (WHO) said that about 60% of the population in developing countries lacked adequate water supplies.

Air Pollution

The pollution of the air by industrial plants and transportation systems, unlike other pollutants, can be seen by even the casual observer. The reports of smog over major metropolitan areas have become commonplace in the United States. Nightly news programs report the air quality for these regions and alert individuals with respiratory illnesses of the dangers of going outside when the pollution index is high.

Automobiles account for a major portion of the air pollution problem in the United States. Road traffic is the source of 65% of all emissions of nitrogen oxide, carbon monoxide, and hydrocarbons. About 90 million tons per year of these emissions can be associated with the burning of fossil fuels in cars and other vehicles. Industrial processes account for about 8% of the aforementioned emissions.

In addition to the emissions of gases such as hydrocarbons and oxides of nitrogen, particulate matter (small particles) are also threatening the air quality. These particles range in size from invisible microscopic particles to those that are visible, such as soot and smoke. Common particulate matter produced through industrial processes includes fluorides, lead, and asbestos. These have been identified as a threat to human health.

The Fight against Pollution

America's interest in the state of the natural resources of the country began to gain momentum during the early 1960s. A best-selling book, *Silent Spring,* by Rachel Carson started Americans thinking seriously about the environment.

Public concern, which was nationwide, initiated government action eventually leading to legislation and policy to protect the environment from devastation. In President Nixon's State of the Union Address of January 22, 1970, a new philosophy and policy on protecting the environment was described. The following is a quote from that address that is representative of the concern held by the public and government agencies.

> The great question of the seventies, is shall we surrender our surroundings, or shall we make our peace with nature and begin to make reparations for the damage we have done to our air, our land, and our water?
>
> Restoring nature to its natural state is a cause beyond party and beyond factions. It has become a common cause of all people of America. It is a cause of particular concern to young Americans—because they more than we will reap the grim consequences of our failure to act on programs which are needed now if we are to prevent disaster later.
>
> Clean air, clean water, open spaces—these should once again be the birthright of every American. If we act now they can be.
>
> We still think of air as free. But clean air is not, and neither is clean water. The price tag on pollution control is high. Through our years of past carelessness we incurred a debt to nature, and now the debt is being called.

During Nixon's administration, he was responsible for funding efforts to clean up the environment that had price tags in the billions of dollars. He proposed programs that covered all of the major environmental areas, including air, water, and solid waste. He established the Environmental Protection Agency in October 1970 to coordinate the environmental interests of the nation and named a Council on Environmental Quality to advise and assist the White House on conservation questions.

Among the most significant pieces of legislation designed to protect the environment was Public Law 90-148, better known as the Clean Air Act of 1971. Its purpose is to control emissions of motor vehicles, regulate the content of fuels, and demonstrate the feasibility of low-emission vehicles.

Public Law 90-148 required that

1. In 1975, automobiles achieve a 90% reduction in emissions of hydrocarbons and carbon monoxide
2. In 1976, automobiles achieved a 90% reduction in the emissions of oxides of nitrogen from average levels measured on 1971 automobiles

This caused domestic and foreign automobile manufacturers to scramble to develop devices that would meet new emission standards. The catalytic converter was one such device that was installed in the exhaust system to reduce emissions of nitrogen oxide, carbon monoxide, and hydrocarbons.

The converter contains either beads or a honeycomb-like structure coated with a precious metal combination of platinum and rhodium. The exhaust gases pass through the converter and are oxidized. Carbon monoxide and the hydrocarbons are converted to carbon dioxide and water, and nitrogen oxide is reduced to nitrogen. Under ideal conditions, the emissions can be reduced by the 90% required by the Clean Air Act. Improperly maintained equipment, however, can greatly reduce the effectiveness of emission control equipment.

Toxic Substance Control

Prior to 1976, the only toxic chemicals regulated by the EPA were pesticides. In 1976, the EPA was ordered by the U.S. District Court to begin regulating a long list of metal and synthetic organic pollutants, many of which had been detected in wastewater discharges from industry and municipal treatment plants.

The Resource Conservation and Recovery Act (RCRA) of 1976 was perhaps a direct result of growing concern over possible links between chemicals and cancer. This act gave the government the authority to regulate the chemical industries and the handling of chemical wastes. It was a comprehensive program for controlling hazardous waste and also providing technical assistance and grants to states to help improve waste management.

Among the statutes set forth by this act is the requirement that firms generating wastes with certain characteristics must identify them as hazardous, obtain EPA identification numbers, comply with transportation requirements, and keep careful records which must be filed with the EPA. The characteristics of hazardous wastes have been identified by the EPA. According to their standards, hazardous wastes are ignitable, corrosive, reactive, or toxic.

Amendments to the 1976 RCRA were added in 1984, placing severe restrictions on treatment, storage, and disposal of hazardous wastes in land management facilities. Land disposal has been the primary disposal method because it is less expensive than incineration or other nonland-based methods. Land disposal, however, has caused contamination of ground and surface water and has resulted in possible health problems. The 1984 amendments are designed to prevent the approximately 264 million metric tons of hazardous waste from threatening the public health.

EPA Superfund

Incidents that occurred during the late 1970s led to the passage of the Comprehensive Environmental Response, Compensation, and Liability Act of 1980. This act created what has become popularly known as the Superfund. Although no single incident caused the creation of this act, events at the Love Canal site in Niagra Falls, New York, provided the impetus.

During the 1940s and 1950s, the Love Canal site was used as a dump site for hundreds of tons of toxic waste which was later covered and developed for residential housing. Health problems among the residents of the site caused the permanent evacuation of hundreds of residents.

The purpose of the Superfund is to provide for liability, compensation, cleanup, and emergency response for hazardous substances released into the environment and the cleaning of inactive hazardous waste disposal sites. The $1.6-billion fund was established to pay for EPA response to hazardous dumping. Table 22–1 lists the substances most commonly removed from waste sites by the EPA.

Type of Waste	Percentage of Removal
PCBs	23.3
Pesticides	13.9
Heavy Metals	13.9
Unspecified Organics	9.3
Toluene	8.5
Cyanide	6.9
Benzene	5.4
Paints	5.4
Caustic Soda	4.7
Acids	4.7
Ethyl Benzene	3.9
Trichloraethylene	3.9
Xylene	3.9
Others	6.2

Table 22–1 Hazardous Substances Removed by the EPA

Employee Rights and Employer Responsibilities

Employees who are exposed to potentially dangerous substances in the workplace are protected by state and federal legislation popularly labeled as "Right to Know Laws." These laws are designed to make sure employees are aware of the effects of toxic substances encountered at their place of employment. These laws also provide for annual training for employees who are routinely exposed to toxic substances.

Containers of toxic substances must be labeled (Figure 22–4) with chemical names and appropriate warnings and Material Safety Data Sheets (MSDS) must be provided by the supplier. Information describing emergency and nonemergency actions which can be taken in the event of exposure to such chemicals must be readily available to employees.

Employees may refuse to work with a substance on the "toxic substance list" if the employer has not supplied the employee with the MSDS after requested in writing. An employee may not be discharged or otherwise disciplined or discriminated against in any manner by an employer for exercising their rights under the law.

According to the Department of Labor–Occupational Safety and Health Administration (OSHA) regulation 29CFR Part 1910, importers, manufacturers, or distributors must provide the following information on containers of hazardous chemicals:

1. Identity of the hazardous chemicals
2. Appropriate hazard warnings
3. Name and address of the chemical manufacturer, importer, or other responsible parties

Figure 22–4   Toxic substances must be labeled with chemical names and appropriate warnings

The regulations also require the following information on the material safety data sheets:

1. The identity used on the label—its chemical and common name
2. Physical characteristics of the the hazardous chemical such as vapor pressure and flash point
3. Physical hazards of the chemical, including potential for fire, explosion, and reactivity
4. The health hazards of the chemicals, including signs and symptoms of exposure and any medical conditions which are generally recognized as being aggravated by the exposure to the chemical
5. The primary route of entry into the body
6. The OSHA permissible exposure limit and any other exposure limit used or recommended by the chemical manufacturer, importer, or employer preparing the MSDS
7. Whether the chemical is listed in the National Toxicology Program (NTP) Annual Report on Carcinogens or has been found to be a potential carcinogen in the International Agency for Research on Cancer
8. Any generally applicable precautions for safe handling and use which are known to the chemical manufacturer, importer, or employer preparing the material safety data sheets, including appropriate hygiene practices, protective measures during repair and maintenance of contaminated equipment, and procedures for cleanup of spills and leaks

9. Any generally applicable measures which are known to the chemical manufacturer, importer, or employer preparing the material safety data sheets, such as appropriate engineering controls, work practices, or personal protective equipment
10. Emergency and first aid procedures
11. The date of preparation of the MSDS or the last change to it
12. The name, address, and telephone number of the chemical manufacturer, importer, or other responsible party preparing or distributing the MSDS who can provide additional information on the hazardous chemical and appropriate emergency procedures, if necessary

SUMMARY

The damage that has occurred to the environment in a relatively short period of time is evidence of the lack of foresight humans have had in the use of technology. The earth's ecological system is a delicately balanced system, and only time will tell whether the damage done is permanent or reversible.

If the will and desire are strong enough, the science and technology that created the problem will produce a solution to the global pollution problems that face current and future generations. Public awareness and education could greatly influence the attitudes of all responsible for keeping the environment safe for humans, plants, and animals. The responsibility belongs to everyone—from the single individual to the largest corporation and to local, state, and federal governments.

 REVIEW QUESTIONS _____

1. What factors may compound the problems of pollution and natural resources depletion in the future?
2. What is carbon dioxide? What threat does it pose?
3. What effect will ozone depletion have on the earth's ecology? What has been done to reverse the effects of chlorofluorocarbons on the ozone layer?
4. What is the major source of acid precipitation? Why has acid precipitation become as big a political problem as a scientific one?
5. What methods are being used to reduce acid precipitation?
6. Describe the negative effects of DDT on the environment.
7. Define _half-life_.
8. How is the oxygen supply affected by DDT?
9. What are the major sources of solid waste?
10. To what extent does transportation contribute to air pollution?
11. What best-selling novel and author brought public attention to the pollution problem?

12. What are the major components of the Clean Air Act of 1971? How effective has this legislation been?
13. What is the purpose of the Superfund established under provisions of the Comprehensive Environmental Response, Compensation, and Liability Act of 1980?

SUGGESTED ACTIVITIES _____

1. Interview city officials to determine if local environmental protection laws exist. Prepare a report describing your findings.
2. Interview local landfill owners to determine what EPA laws apply to their operation and how they are being addressed. Report to the class on your findings.
3. Choose a local manufacturer and write a report describing its pollution-control devices and policies.
4. Many new ideas have been proposed recently to reduce the amount of solid waste. As a class project, design and manufacture a useful product from solid waste items found at home, residence halls, local fast-food restaurants, factories, or landfills.

CHAPTER
23

Recycling and Reclaiming Industrial Materials

Humankind, the producer and consumer, has been concerned about virgin resources and discarded goods for many years. Recycling materials is not a new concept of the twenty-first century but a concept that has always been the concern of the producer and consumer. Early consumers were able to turn corn husks into mattresses and old clothes into quilts. In colonial America, nails were so difficult to obtain that old homes and barns were burned down just to retrieve the nails for new construction. The village blacksmith would reuse wrought iron to make his wares many times over because the metal was so scarce.

Recycling is receiving much more attention today. However, as our materials become much more complex, the volume increases at alarming rates, and our landfills are overwhelmed with the throwaway society. The average American will use and discard total solid materials exceeding 600 times his or her weight during a lifetime. Collection and disposal of these materials has caused major problems for urban areas. Cities now truck or barge more than 160 million tons of solid-waste materials hundreds of miles to incinerators or landfill sites each year. In too many cities, there is a Mount Trashmore next to a major highway that signals the problems of solid-waste disposal in America.

In 1988, solid waste was disposed of primarily by landfill (see Figure 23–1), leaving only 20% for recycling and incineration split equally. These statistics are bound to change because of the very limited space available for future landfills, ecology problems with landfill run-off, and the need to conserve energy and raw materials by industry. The Office of Technology Assessment (OTA) has called for increased management efforts to control municipal solid

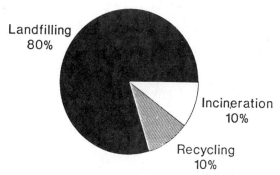

Landfilling
80%

Incineration
10%

Recycling
10%

Figure 23–1   The distribution of disposal methods for solid waste *(Office of Technology Assessment,* Facing America's Trash, *Copyright 1991 by Van Nostrand Reinhold, p. 6)*

waste (MSW) and to create a strategy for managing waste in the future. Their strategy is based on two concepts: one centered on appropriate product design and the other on appropriate disposal. Figure 23–2 shows the comprehensive approach to recycling and disposal that would maximize the use of solid waste.

GREEN REVOLUTION

Product Design

The green revolution is here, and design for recyclability is the way of the future. From automobiles to teapots, companies like GE Plastics, Dow Chemical, and Chrysler Corporation are responding with new engineering designs that can return products for reclaiming after their initial use. The Dow Chemical Co. of Midland, Michigan, and Mobay Corporation of Pittsburgh have developed a process to reuse all scrap from the reaction–injection molding (RIM) process used in automotive parts manufacture. Now engineers can design automobiles with more plastic parts that will be recycled in the future. Such an automobile has been designed by Chrysler called the Dodge Neon (see Figure 23–3), that consists primarily of recyclable materials. Coded body parts identify plastic type to make sorting easier.

At Battelle in Columbus, Ohio, engineers are exploring pyrolysis or depolymerization to recycle mixed plastics. Unlike other processes that yield a lower-quality plastic material, pyrolysis yields monomers, which are the basic unit of virgin plastic. This exciting process, which accepts mixed plastic materials and returns a usable virgin plastic, has great potential for the design engineer who is environmentally responsible.

At Polymer Solutions, a joint venture between GE Plastics and Fitch Richardson Smith, engineers have designed the "Ukettle," a completely recyclable energy-efficient appliance that outperforms stovetop models and microwaves. The new kettle is a DFD (design for dissembly) product that makes recycling inexpensive by using two new plastic materials and snap-fit fasteners. The new engineering thermoplastics are injection molded Noryl, a modified polyphenylene, and molded Lomod, a thermoplastic elastomer—both materials devel-

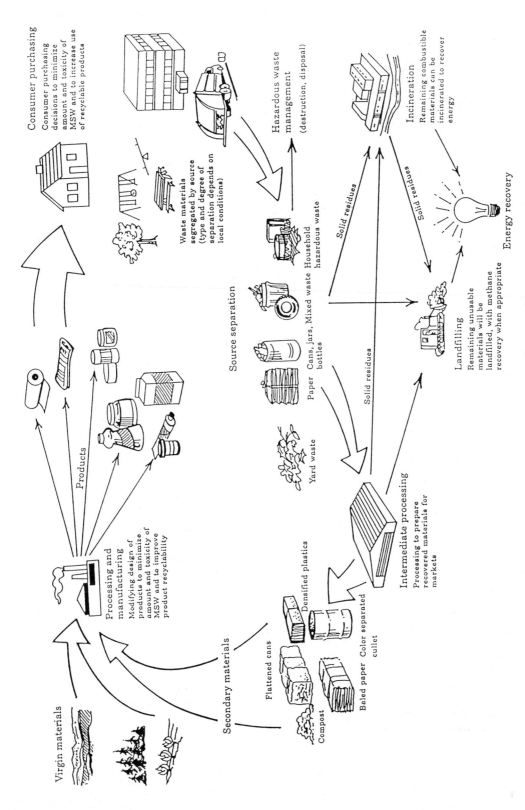

Figure 23-2 The comprehensive approach to recycling and reclamation

Figure 23–3 The Dodge Neon concept car consists primarily of recyclable material. Coded body parts identifying plastic type make sorting easier

oped by GE Plastics. The Ukettle is held together by snap fasteners that allow the parts to be broken apart easily when the appliance is discarded.

The concept of cradle-to-grave design is also taken seriously by the engineers at Design Continuum Inc., a Boston/Milan, Italy, design and development firm. They have developed eight tips (see Figure 23–4) for design engineers that want to be environmentally correct and reduce the pollution caused by throwaway products.

Dematerialization

Doing more with less has not been the flair of advanced industrialized nations in the past. Indeed, industrial affluence has been the precursor to the material society that saw the development of larger homes, multiple automobiles, boats, stereos, and an abundance of toys in a throwaway society. Recent ecological trends and warnings concerning water, air, ozone, and hazardous wastes have changed the attitude of people who recognize the limitations of our ecological system, resulting in regulatory restrictions for manufacturing. In addition, the economics of less natural resources, such as metals, wood, oil, coal, and gas, and worldwide economic competition have also played a role in the idea of doing more with less via a concept called **dematerialization.**

As can be seen in Figure 23–5, the major factors that affect the dematerialization design process are quality, waste generation, ease of manufacture, production cost, replace or repair capability, and size and complexity. Perhaps the best-known example of dematerialization is the American automobile. Motivated by corporate average fuel economy (CAFE) standards for gas mileage, the automobile industry has reduced the weight (and materials) of the average automobile by 400 pounds and increased the efficiency of gasoline engines through engineering to meet the standards and reduce emissions. Other products from toasters to water skis can be designed with dematerialization concepts in mind to reduce the ecological impact of material goods. One must be careful not to confuse cheap throwaway products with those that are made durable through the dematerialization design process because quality and longevity are key to the concept of using our materials more effectively.

Eight tips for 'green' design

Boston—Engineers at Polymer Solutions and Design Continuum believe that a few guidelines go a long way in designing for recyclability. Their tips can make the difference between "reused" and "refused."

1. **Determine feasibility.** Consider recycled products and restraints. Is there a market for recycled material? Can a more "friendly" material be substituted?

2. **Consider current materials.** Check the origin of materials. See if the material or process can be improved for recyclability or biodegradability. Use materials that can be recycled together.

3. **Evaluate manufacturing processes.** Understand side effects of emissions and processes, such as abusive molding, and look for alternatives. Evaluate the environmental impact of materials from extraction through disposal. Consider more energy-effective processes. Reduce energy consumption by eliminating unnecessary manufacturing steps.

4. **Incorporate source reduction.** Use material handling programs to lower the cost of manufacturing and minimize waste in production. Avoid secondary finishes—painting, plating, coatings. Avoid toxic materials and heavy metals that contaminate material. Minimize assembly operations.

5. **Design for long life.** Make the product easier of less expensive to repair and maintain than replace. Design for component replacements. Simplify and standardize connections.

6. **Design for disassembly.** Use two-way snap-fits with break points for ease of assembly and disassembly. Design for easy separation, handling, and cleaning. Provide standardized, easy-to-read bar coding or molded-in logos to streamline separation operations. Apply tight tolerance design principles to reduce the need for fasteners and keep separations simple.

7. **Review packaging and shipping.** Reduce the amount of packaging required. Design reusable shipping vessels.

8. **Recycle.** Identify potential reuses of the product. Maximize the useful value of all materials. Incorporate recycled material. Develop a recycling method and design for recycling opportunitites. □

Figure 23–4 Design plays a very important role for product life and reclamation ("Green Revolution Comes to Engineering," Design News, 9 September 1994, p. 104, Andrea Baker, new products editor. Used by permission)

RECYCLING

Product Disposal

The OTA has indicated its support for materials and energy conservation as stated in the RCRA as national goals but sees local government as the decision makers on how the solid-waste problem is handled. Most local authorities would agree that recycling is the dis-

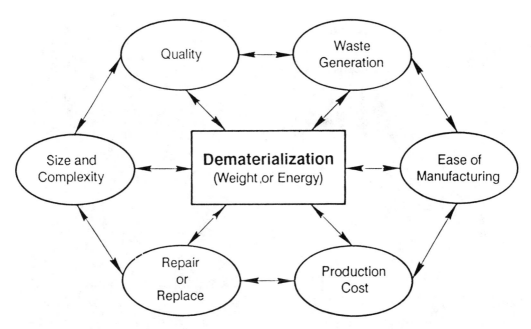

Figure 23–5   Factors affecting, and affected by, the dematerialization process *(Ausubel and Sladovich, Eds.,* Technology and Environment, *Washington, D.C.: National Academy of Engineering, National Academy Press, 1989, p. 56)*

posal method of choice compared to landfilling and incineration, even though incineration does provide new energy resources. During the 1980s, goals for recycling were increased from 10% to 25% in an attempt to cut disposal costs and relieve the overcrowded landfills. The federal government has encouraged education and funded research to increase the awareness (see Figure 23–6) of the public about the importance of recycling our solid waste. With the recent adoption of recycled paper products and the elimination of plastic food containers by McDonald's and other fast-food chains, it would appear that the public is ready for the inconveniences caused by recycling procedures that include separation and special collection procedures. The technology, too, has improved, allowing us to handle more materials more efficiently and quickly. Let us take a look at recycling technology and the procedures that are necessary to return materials back into production.

The level of recycling varies according to material and market price. While the overall level in the 1980s was 10%, some materials, as can be seen in Figure 23–7, have exceeded that rate by several times. As with all supply-and-demand systems, the more that materials are recycled, the lower the market price for discarded materials will become, making recycling and collection efficiency a top priority.

Recycling actually consists of three different activities: collection, preparation, and materials reclamation. Not all recycling activities are aimed at the consumer. Many of our recycling efforts are preconsumer orientated, occurring within the manufacturing plant. Actually, manufacturing wastes are more appropriately divided into three categories:

Please write the Environmental Defense Fund at 257 Park Ave South, NY, NY 10010 for a free brochure.

Figure 23–6   The federal government has encouraged recycling through organizations and advertising *(Office of Technology Assessment,* Facing America's Trash, *Copyright 1991 by Van Nostrand Reinhold, p. 19)*

"Home scrap," produced and reused inside a production facility
"Prompt industrial scrap," produced in an intermediate stage of processing and returned to the basic production facility for reuse
"Old scrap" (postconsumer), generated by the product's final consumer

As one might expect, manufacturers have always practiced preconsumer recycling simply as a matter of economics and a way to keep retail prices competitive. Today's focus is on postconsumer recycling and how that process can be accomplished efficiently. The aluminum can is the most viable and popular recycled item today. Because it has a government-

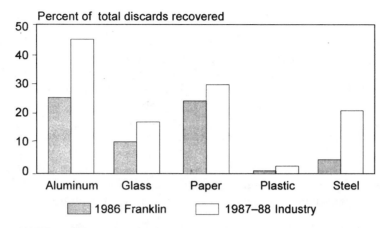

NOTE: Industry estimate for paper includes pre-consumer scrap; industry estimate for steel includes higher total for white goods plus ferrous scrap recovered at incinerators.

SOURCE: American Paper Institute, 1987 *Annual Statistical Summary, Waste Paper Utilization* (New York, NY: June 1988); K. Copperthite, U.S. Department of Commerce, personal communication, 1989; Franklin Associates, Ltd., *Characterization of Municipla Soild Waste in the United States, 1960 to 2000 (Update 1988)*, final report, prepared for the U.S. Environmental Protection Agency (Prairie Village, KS: March 1988); B. Meyer, Aluminum Association , personal communication, 1989; K. Smalberg, Steel Can Recycling Institute, personal communication, 1989; Smalberg Steel Can Recycling Institute, personal communication, 1989; Society of the Plastics Industry, personal communication, 1988.

Figure 23–7   The level of recycling varies according to material *(Office of Technology Assessment,* Facing America's Trash, *Copyright 1991 by Van Nostrand Reinhold, p. 28)*

imposed five-cent deposit fee built into the original price, consumers are willing to take extra time to return cans to the recycling (crushing) machines to retrieve their deposit. Even if they do not, someone else will collect discarded cans and return them for the deposit-fee reimbursement. States that do not have the can deposit-fee requirement do not have a high return rate.

Solid waste is collected in a variety of ways: as mixed wastes, with commingled recyclables, or with separated recyclables. How they are collected determines the way in which they will be handled. If they are separated at curbside, then special collection trucks will be used that will keep them separated until recycling. If they are not separated at curbside, then a centralized materials-recovery facility (MRF) will be used to separate the mixed waste. In 1989, about twelve MRFs were in operation in the United States.

Glass

Recycled glass is called **cullet**, and about 90% of it comes from glass containers. Around 11 million tons of glass containers are produced each year; and of that, about 2.5 million tons of cullet are recycled and used to make the new containers. The drawback to using cullet for making new glass is color and the lack of "fining" agents needed to reduce bubbles. Actually, cullet is desirable for glassmaking because it improves the melting efficiency; and the most common mix of glass contains 25% cullet.

Although cullet itself is 100% recyclable, it lacks special "fining" agents that prevent air bubbles. As a result of potential problems, specifications as shown in Figure 23–8 are used to maintain the quality of glass when mixed with cullet. Because of the different colors of cullet, it is not practical to make glass other than containers through the recycling process.

- Only glass container glass is acceptable

- Permissible color mix levels—
  Flint glass
    95–100% flint; 0–5% amber; 0–1% green; 0–5% other colors
  Amber glass
    90–100% amber; 0–10% flint; 0–10% green; 0–5% other colors
  Green glass
    80–100% green; 0–20% amber; 0–10% flint or Georgia green; 0–5% other colors

- Glass must be free of any refactory materials. Grounds for rejection include:
  —presence of pottery, porcelain, china, dinnerware, brick, tile, clay, and so forth, larger than 1 inch.
  —presence of more than one particle of any of above materials larger than $1/_8$ inch but less than 1 inch in a 200-pound sample.
  —presence of more than two grains of quartzite, sandstone, or sand pebbles larger than U.S. 16 mesh per 10 pounds of sample.
  —any clay particles larger than U.S. 20 mesh or more than 50 particles larger than U.S. 30 mesh per 10 pounds of sample.
  —any alumina silicate refactory heavy minerals larger than U.S. 30 mesh or more than 10 grains larger than U.S. 40 mesh per 10 pounds of sample.
  —presence of zircon, cassiterite, chrome, or similar refactory particles larger than U.S. 60 mesh.

- Glass must be free of metallic fragments and objects, dirt, gravel, limestone chips, asphalt, concrete, and excessive amounts of paper, cardboard, wrap, plastics, etc.

- Large amounts of excessively decorated glass must be kept separate.

SOURCE: Brockway, Inc., "Specifications for Furnace-Ready Cullet," unpublished manuscript (undated).

Figure 23–8 Cullet has restrictions for recycling *(Office of Technology Assessment, Facing America's Trash, Copyright 1991 by Van Nostrand Reinhold, p. 151)*

Aluminum

Aluminum can recycling has increased during the last two decades. Once the steel tabs were removed from aluminum cans, the process for reclaiming the aluminum became profitable. There is a need for aluminum scrap because of the high energy costs required to make aluminum from bauxite ore. In 1988, 77.6 billion aluminum beverage cans were sold; and 42.5 billion cans were recovered for recycling, representing the largest private collection effort in the materials-recovery business.

In secondary aluminum production, scrap aluminum is melted in a furnace, to which alloying elements are added to produce the type of aluminum needed. Scrap aluminum must be identified to maintain alloying levels for different types of aluminum. In the case of factory scrap or beverage cans, the scrap is easily identified and in high demand. For mixed scrap, the process becomes much more difficult. As mentioned earlier in this text, aluminum is an abundant material found in the earth's crust. It is the high energy costs that make aluminum a desirable recycling material. When making aluminum from virgin materials, the smelting costs can range as high as 85%, raising the cost from 19.5 cents per pound (for recycled) to 59.1 cents per pound. As a result of the difference in production costs, the market for scrap aluminum is always high and profitable.

The demand for aluminum beverage cans (see Figure 23–9) is expected to remain strong in the future along with secondary factory scrap. Some challenges from plastic containers can be expected as plastic producers create the technology to make recyclable-plastic products.

Iron and Steel

The major sources for scrap iron and steel are in junked automobiles, food containers, and white appliances. Using scrap steel is commonplace in the steel industry and has helped the industry make up for poor-producing iron ores that need beneficiation before use. Factory (prompt) scrap is also an important source for recycling, especially in the area of cast iron.

In 1986, U.S. steel mills consumed 49.7 million tons of scrap and 44.3 million tons of pig iron to make 81.6 tons of new steel. Of the recycled scrap, about 40% was from old scrap, 20% from prompt industrial scrap, and 40% from home scrap. The process for collecting and delivering all types of steel and iron scrap has been well established for many years because steel recycling is one of our older industries.

The technology of electric arc furnaces (EAFs) has improved in recent years, making scrap recycling more efficient and creating new demand. Over 200 EAFs are in operation today, and they produce in excess of 27 million tons of carbon steel each year using recycled scrap. Also adding to demand are the continuous casting mills that use scrap in the basic oxygen steelmaking to control temperature. These mills are operating at a 40%:60% ratio, with 40% being home or purchased scrap.

Plastics

The number of companies involved in postconsumer recycling of plastics is small but growing. Preconsumer recycling is much higher, with an estimated 95% of all manufactur-

Figure 23–9 Aluminum cans are easily recycled

ing home scrap being recycled. The primary recycling effort for postconsumer scrap is aimed at the beverage container industry. When one goes to a grocery store, the two most common reclaiming machines are for aluminum cans and soft-drink plastic containers known as PET (polyethylene terephthalate) bottles (see Figure 23–10). Additionally, the common milk container made out of high-density polyethylene (HDPE) is successfully recycled at over 17,000 tons per year.

The real issue for plastic recycling is separation of thermosets from thermoforms in a manner that is profitable. Short of success by Battelle's depolymerization program mentioned earlier, which would allow mixing, the public is stuck with the difficult task of separation. Unlike steel and aluminum, different types of plastic are difficult to identify because they can have the same visual appearance. In addition, many plastic products contain more than one type of plastic or other materials such as metal, glass, or wood.

Manufacturers are helping by coding plastic parts (like the aforementioned Ukettle) so that they can be separated at the collection station, or designing total thermoform products with the recycling symbol attached. Recycled plastics have their problems and limitations, however. The Food and Drug Administration (FDA) is concerned with the lack of research

Figure 23–10   Current recycling of plastics from MSW focuses on containers made from polyethylene terephthalate (PET) and high-density polyethylene (HDPE), which are relatively easy to identify and are not degraded significantly by reprocessing. Only about 1% of the plastic in MSW is now recycled *(Office of Technology Assessment,* Facing America's Trash, *Copyright 1991 by Van Nostrand Reinhold, p.169)*

concerning contaminants of recycled plastic and whether recycled plastic should be used in the food industry. Likewise, some manufacturers are concerned with the performance of recycled plastics because, in general, recycled plastics are not as good as virgin-material plastics.

Last, infinite recycling of plastics is technically impossible. Therefore, disposal is the eventual road for all plastic materials with present technology options. Even so, multiple use of plastic materials would save millions of tons of raw materials, energy, and slow down the rate for landfill action of a material that is difficult to decompose. New technology and curbside collection are the keys to increased recycling of plastic materials in the future.

SUMMARY

Humankind has been producing artifacts and products at increasing rates for several hundred years. The depletion of natural resources has, in recent years, caused concern about the recovery of materials from what essentially has been a waste-disposal system. The inevitable fate of these products and goods is to become waste. Using the past approach of disposal, nature is left with a painstakingly slow process of returning waste materials into resources.

The problems associated with recycling and reclaiming materials have always been economic and technological. New environmental protection regulations, increasing energy costs, transportation, social consciousness, and technological advances all play important roles in the decision to dispose or recover materials.

The decision to reclaim or recycle should be made by the engineer during the design stage to create new products. The finite world of resources and our ability to deal with pollution should be the major concern of every citizen in the world. New world standards on water and air quality and regulations on fossil-fuel pollution and nuclear-waste disposal must set the standards for the way we will live in the future. Developing nations will need advanced technology assistance during their industrialization stage in an effort to keep industrial pollution to a minimum. We can no longer rely on after-the-fact solutions to our ecological problems and expect to maintain a quality of life on earth.

In the meantime, engineers must design for the green revolution and dematerialization approaches that reduce the types and amount of materials used in new products. Communities must press forward with recycling programs that reduce the need for landfills and provide important resources for reuse in the industrial sector.

Manufacturing smart is a concept that maximizes technology to work for environmental compatibility. Use of appropriate materials that reduce the need for energy but provide durable service include the metals, plastics, and composites of tomorrow. Designing special fasteners and marking materials are but only a few of the many ways we must utilize in the future so that our products are compatible with our environment.

The paradox of technology is that it can be both the source for environmental damage and the solution for improving the environment. Both problem and solution, technology is the extension of humankind and only as positive or negative as we make it.

 REVIEW QUESTIONS _____

1. How much waste does the average person create in a lifetime?
2. What are two steps of recycling processes?
3. What is the primary reason that prevents recycling on a major scale?
4. Why is material separation a difficult process for recycling?
5. Which is easier to recycle, primary or secondary materials?
6. Why is it not feasible to recycle metal when it is nearly 100% pure steel or aluminum to start with?
7. List and describe two reasons why the reclaiming of secondary materials is difficult.
8. List and describe two trends for the disposal and reclaiming of waste in the future.
9. Why do plastic materials pose special problems for recycling?
10. What types of government regulations can help the reuse of industrial materials?

SUGGESTED ACTIVITIES _____

1. Have students tour a solid-waste landfill site and discuss the types of materials being discarded.
2. Show a film of how a solid-waste incinerator works.
3. Have students design a recycling system and present a report to the class.
4. Have students redesign waste-product materials into a new product design.
5. Design and build a product utilizing all materials in the product (a zero-waste product).

CHAPTER
24

Advanced Material Technology

The resources of the earth, as mentioned in Chapter 22, are not infinite. There is already considerable concern about dwindling reserves of fossil fuels such as oil and coal and mineral deposits for producing primary metals. The demand, however, for products traditionally made of these raw materials has not lessened but has increased. In addition, demands for new products that require special properties not available in these traditional raw materials are being made.

Science and technology is providing the solution to the problems identified above with the introduction of new exotic materials that have the strength of steel and the lightweight qualities of plastics. Plastics that conduct; flexible, woven light-emitting fabric made of glass fibers; and synthetic skin, bones, ligaments, and ear and heart components are only a few examples of the many new materials that are being developed to meet human needs.

Space is the new frontier in materials and materials-processing technology. Scientists must develop materials that will meet the special requirements of extraterrestrial environments of extreme temperature changes, pressures near vacuum, and aerodynamic forces of intensities not normally experienced. Space offers a near gravity-free environment for manufacturing ultrapure materials for supercomputers and new alloys that cannot be produced on earth.

Space also offers new sources of raw materials that could relieve the concern over dwindling supplies on earth. The rich minerals of the moon offer an excellent source of raw materials for manufacturing in space (see Figure 24–1). Retrieving these materials from the moon to use in gravity-free manufacturing in a space station could be easier than trying to transport them from earth.

The future of material science and technology is bright, exciting, and at the edge of a frontier that promises to provide solutions to problems thought to be unsolvable. This chap-

Figure 24–1   Mining raw materials on the moon *(Courtesy of NASA)*

ter provides a brief description of some of the new developments in materials science and technology. Development in this area is occurring so quickly, however, that by the time this book is published, this chapter may need to be revised.

SPACE: THE NEW FRONTIER FOR MATERIALS AND MANUFACTURING PROCESSES

Materials and manufacturing processes promises to be one of the first commercially successful projects of the space station scheduled to be in operation in the 1990s (Figure 24–2). Predictions of its success range from $1 billion by 1990 to a projected 300 billion gross national product (GNP) from space business by the beginning of the next century. Although recent problems, including the tragic accident in 1986, have slowed progress in making manufacturing in space a reality, this new business could generate as many as 10 million jobs in the future.

A major area in the space station design is the Microgravity and Materials Processing Facility (MMPF) (Figure 24–3). A section of this wing of the space station is known as the Manufacturing and Technology Laboratory, which NASA plans to man permanently. The

Figure 24-2   Space station *(Courtesy of Photri, Inc.)*

processes in this facility will be highly automated with robots performing dull, repetitive jobs. The primary objective will be to convert research and development into commercial products as soon as possible. Organic molecular crystals and organic polymers have the greatest commercial potential.

Successful production of these materials requires an environment that eliminates even the smallest defects and impurities. That level of purity is not possible on earth. Applications for these materials include superconductors, thermoelectric elements, radiation detectors, and memory chips for computers.

The most promising products to be manufactured in space are pharmaceuticals, semiconductors, and special glass. Other products that are considered possible include alloys, foam metals, ceramics, and latex spheres. The spheres will be used in calibrating highly sensitive, accurate instruments.

Figure 24–3   A new lunar supply base could be built in the future *(Courtesy of NASA)*

New alloys of metals that have been impossible to produce on earth will be manufactured in the gravity-free environment of space. On earth these alloys of greatly differing densities would quickly separate when solidification began. In space, it may be possible to form more than 500 new alloys.

Mining on the Moon

The payload of the space shuttle is limited. Therefore, scientists are exploring methods of mining crucial raw materials in space and converting them to the primary materials needed for space-station construction and space manufacturing. The moon is rich in critical minerals that are needed for manufacturing and construction. Alumina, titanium, iron, and silicon are only a few of the raw materials available on the moon's surface.

The problem of refining these ores by high temperature will be solved through the development of solar furnaces that will produce the temperatures needed to separate the mineral from other impurities. Temperatures greater than 2,800 degrees Fahrenheit will be required. A unique by-product of the refining process will be oxygen.

Many ores such as iron oxide contain trapped oxygen that is released during the refining process. The scientists and technologists hope to contain this oxygen and use it to manufacture liquid hydrogen H_2O_2 for use as propellant on the space shuttle.

The silicon and other raw materials for production of ceramics is abundantly available on the moon. Structural units for construction on the space station and heat shields are some of the planned applications for these materials. High-strength glass composites are also being proposed for space manufacturing. Moon rock is rich in pure feldspar. Feldspar is a principal source of alumina, which is used in making S glass. S glass is used as the reinforcement in conventional high-strength composites.

The low-gravity conditions should provide high-quality, lower-cost raw materials for space manufacturing. Extracting raw materials from the moon and refining them in space will eliminate any need to transport materials from earth—a difficult and expensive operation.

Materials Revolution on Earth

Development in materials used in the aerospace industry over the last few decades is beginning to be considered for more traditional products. Advanced-composite technology is now being planned for use in the automotive industry. The physical properties that these materials offered to aerospace are equally attractive to automakers. They are lightweight, easily formed, and corrosion resistant.

Perhaps the biggest advantage is that these materials can be formed into one large part which could replace several assembled parts. For example, Ford Motor Company has been able to replace ten parts of a floor panel with a single composite molded panel. Another composite front-end panel replaced forty-two metal parts. Researchers estimate the eventual weight reduction from using composites will be as much as 30%. This translates to better fuel efficiency.

The disadvantage is that the process used for manufacturing composite moldings is not fast enough for high-production demands. New methods of forming advanced composites will need to be developed in order for them to be used in high-volume production.

The market for advanced composites is predicted to reach $12 billion by the beginning of the twenty-first century. Aerospace will be the fastest growing sector with more and more graphite epoxies and special reinforced fiberglass materials being used in new commercial and military aircraft. Even the space shuttle is using more of these materials to reduce weight without losing strength. The fairings on the external tanks of the shuttle are now being made of graphite-reinforced epoxy rather than the aluminum previously used. The weight savings has been 40%–50%.

The Remote Manipulator System (RMS) on the shuttle can handle payloads of up to 6,500 pounds. This 50-foot electromechanical arm is the device that sets satellites into orbit and will one day assist in building the space station (Figure 24–4). The material used for the shoulder, wrist, and elbow joints is a composite of themoplastic reinforced with glass fibers (PTFE).

Advanced Polymers and Ceramics

A characteristic of plastics identified in Chapter 6 was that they are insulators rather than conductors. A new breed of conductive polymers, however, is being developed. These new lightweight, noncorrosive, and easily formable materials have many advantages over

Figure 24–4 Remote manipulation system of space shuttle *(Courtesy of NASA)*

the more traditional metal conductors. The addition of conductive materials to a polymer matrix is the technique being used. Stainless steel and pure carbon—as well as the more traditional aluminum, nickel, brass, copper, silver, and gold—are being imbedded in the polymer and formed into the desired shape.

The conductivity can easily be controlled to produce the desired effect. Electronic components including switch contacts, resistors, and electrodes are easily molded from these new conductive polymers.

New advanced-composite thermoplastics (ACTPs) have been developed to combine the high strength of thermoset composites with the toughness of thermoplastics. These new composites are made with strong fiber reinforcements, including silicon carbide, alumina, boron glass, carbon, and polyethylene fibers.

These new ACTPs match the strength/weight and stiffness/weight ratios of metals and are tough, corrosion resistant, and wear resistant. In addition to the excellent physical properties is the fact that these materials are considerably lower in cost.

The new ACTPs have many uses. Their applications include printed circuit boards, fiber-optic connectors, bearings, glass, and other uses requiring lightweight, wear- and abrasion-resistant, dimensionally stable materials.

A new family of ceramics has been developed and will join the more commonly known silicon carbide and silicon nitride. SIALON, a structural ceramic, is a low-density; high-strength; and wear-, abrasion-, and high temperature–resistant synthetic ceramic.

SIALON derives its name from its chemical formula, $Si_3Al_3O_3N_5$. It is synthesized by reacting silicon nitride, silica, alumina, and aluminum nitride together. The properties of SIALON make it ideal for engine and turbine parts. Its low density and low coefficient of friction translates to reduced weight and reduced frictional loss. SIALON also offers better insulation than metallics.

New Materials in Medicine

The revolution in materials technology is providing the medical community with raw materials that are fully compatible with human tissue. These materials are being used to construct artificial limbs, to replace missing or diseased areas, and to manufacture vital organs such as artificial hearts to meet the increasing demand for replacements of human hearts (Figure 24–5). Other applications of these materials include artificial skin for burn victims, ligaments, and naturally colored fillings for teeth.

Polymer bones made of polymethyl methacrylate (PMMA) reinforced with glass fibers offer the patient a replacement bone that is fully compatible with human tissue and stronger than the original. The bending strength of PMMA (28,000 to 200,000 psi) exceeds that of natural bone (25,000 psi). Finger bones, wrists, elbows, and other artificial replacements are being made of silicone elastomers that are also compatible with human tissue and durable.

Polymer-based plates and rods are being used to stabilize broken bones. These new materials eventually dissolve in the body, eliminating the need for additional operations to remove the commonly used metal stabilizers. Some human joints are being made from a fluorocarbon epoxy known as fluoroepoxy. This material is especially suitable for joints because of its self-lubricating quality. Carbon particles are released from the fluoroepoxy that act as a lubricant for the joint.

Silicone is being used to fabricate bones for fingers, toes, ears, and noses. The silicone material is compatible with human tissue and is extremely durable. A silicone elastomer is being used as artificial skin that is used temporarily until the body can generate new skin to repair damaged areas. The artificial skin gives strength to the area and prevents bacteria from entering the wound until the new skin forms.

Polyurethane reinforced with a flexible mesh fabric is also being used as an artificial skin. The advantage of this material is that it is strong, durable, and can be stretched like natural skin. It is also **permeable** (allows air and moisture to pass through).

Polyurethane is also being used in the manufacture of artificial heart units. It is one of only a few materials that meet the physical property requirements of long-term flexure endurance and biocompatibility. During a ten-year design life, the artificial heart must flex over 400 million times without missing a beat. Every year, approximately 2,000 biological hearts are available for transplants; however, 100,000 patients need them. The need for artificial hearts is obvious. Unlike many transplanted natural hearts, the body does not reject polyurethane.

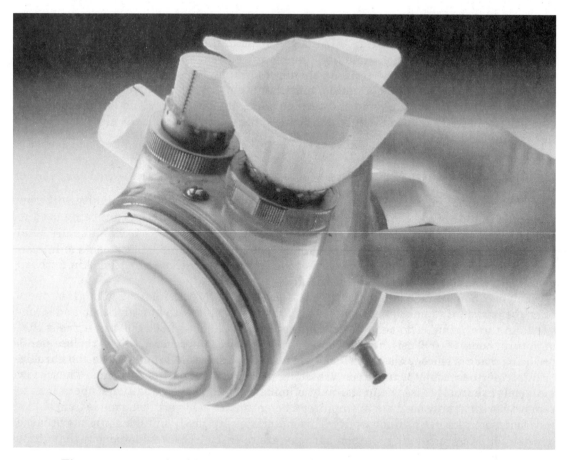

Figure 24–5   Artificial heart made of advanced plastics *(Courtesy of AMOCO Co.)*

Other applications of new materials include silicone elastomers for replacement of damaged ear parts and ceramic composites to replace the traditional silver used to fill teeth. These and other applications in medicine have been made available through the research conducted by material scientists and technologists in the public and private sectors.

SUMMARY

The revolution in material technology has just begun. The future promises to bring even better and more improved materials for use in electronics, aerospace, automotives, and medicine. The majority of the applications for these materials will be very positive. They will improve the human condition and expand the frontiers of outer space.

More important, these materials and the access to resources in outer space will help preserve the reserves of nonrenewable raw materials that are available on earth. This preservation of natural resources should help ensure a standard of living for future generations that would equal or surpass that of the present.

 REVIEW QUESTIONS _____

1. How is science and technology addressing the problem of dwindling natural resources and demand for new materials?
2. How have developments in space technology advanced the more traditional industries like automotives?
3. What raw materials are available on the moon?
4. Why will it be easier to mine raw materials on the moon for space manufacturing than transport them from earth?
5. List several medical applications of new materials technology.

SUGGESTED ACTIVITIES _____

1. Select one of the topics listed below and prepare a five- to eight-page report detailing the history, development, applications, benefits, and potential for technology transfer to other commercial markets.

   Composites and advanced engineered materials in sports
   Materials used in the construction of the space station
   Advanced engineered materials for the medical and bioengineering fields
   Materials used in advanced aircraft design, especially stealth technology

2. Gather research and write a three- to four-page report describing the type and composition of materials that could be used in residential construction ten to fifteen years from now.

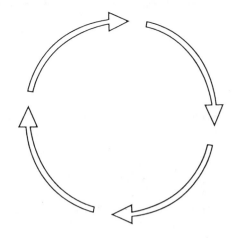

Glossary

A

Abrasive: A substance that is used to remove material by friction

Allotropic: A material that can exist in two or more crystalline forms

Alloy steels: Steels formed by the addition of metal or nonmetal alloying elements to pure molten steel

Anisotropic: A material having properties that differ, depending on the direction in which they are measured

Annealing: A metal softening process using heat treatment; relieves internal stresses created during the hardening process

Anodizing: A conditioning process that applies a protective oxide coating to aluminum; it is done in a tank containing an acid solution and is very similar to electroplating

Applied research: Research that is done using accumulated knowledge and facts to solve a specific problem

Aspect ratio: The relationship of the length of the reinforcing material in a composite to its diameter

Assembly: The process of joining together various parts and subassemblies

Assembly drawing: A drawing that shows both the completed product and how its parts fit together

Atom: The smallest particle of an element; the building block of all materials

Atomic weight: The sum of the weights of the protons and neutrons in the nucleus of an atom

B

Ball clay: A very plastic sedimentary clay originally dug from the ground in blocks or balls in England; it is composed of kaolinite with varying amounts of impurities and organic matter

Bayer process: The commercial refining of bauxite

Bending strength: The measure of a material's ability to resist bending stress, which is a combination of compression and tensile stresses

Beneficiation: The process of removing impurities from iron ore

Bentonite: A very plastic and highly alkaline type of clay that is mixed with other clays to increase plasticity and strength when dried

Brazing: An adhesion process for metals, similar to soldering, which requires a heat range above 800 degrees Fahrenheit and below the melting point of the base metal and in which the filler metal is distributed between the base metal surfaces to be joined by capillary attraction

Break-even point: The point in a product's life cycle where the cost of production is equal to the revenue gained

Breaking strength: The measure of the force that is necessary to cause a given specimen to fracture

Brick clay: A type of sedimentary clay deposit that has been compacted and consolidated

C

Calendaring: A forming process used in the plastics industry; used to form film and sheet stock

Carbon steels: Steels whose primary elements are carbon and iron; they generally have no alloying elements added

Cartesian coordinate system: A coordinate system that allows any point in space to be described using three coordinates (X, Y, and Z); used to establish the geometry that defines the part or tool path on a CNC machine

Casting: A manufacturing process in which a material in molten, liquid form is poured into a mold and allowed to harden

Cement: A slurry or mixture of limestone, clays containing silica, alumina, and ferric oxide, and a small amount of magnesia; it is used, together with water and sand, to make concrete

Ceramic matrix composites: Composite consisting of two or more ceramic constituents whose matrix (binder) is ceramic

Cermets: Composite materials that consist of a mixture of ceramic and metallic elements

China clay: A kaolin that is used to fire white areas and fine china; it is usually mixed with other clays for plasticity and color

Cold finishing: A technique of further processing steel It consists of cold rolling, cold reduction, and cold drawing of semifinished hot-rolled shapes

Composite: A material formed from the combining of two or more materials or material phases; usually has properties superior to those of its individual constituents

Compressive strength: The measure of a material's ability to resist a squeezing or pressing force or load

Concrete: A combination of cement, water, and an aggregate such as sand, gravel, or stones

Condensation polymerization: A process of polymerization in which a chemical reaction causes a by-product (often water) to be produced in addition to the polymer

Creep: The gradual deformation of a given specimen that is under a constant strain for a given period of time

Cullet: Scraps of waste glass that can be remelted

D

Dendrites: Tree-like crystalline structures

Density: The measure of mass per unit volume of a substance under specified conditions of pressure and temperature

Design engineering process: The phase of product development where possible manufacturing or construction problems are solved

Detail drawing: An accurate, scaled representation of a part or assembly with all necessary information about size, shape, and location dimensions to produce the part or assembly

Drafting: A form of drawing that is used to develop product plans It uses symbols that are universally recognized

Drawface: The area of a pattern that is drawn parallel to the sides of a mold

E

Elastic behavior: The behavior exhibited by a material if it returns to its original shape when the load on it is removed

Electron cloud: Sea of electrons flowing among atoms in metallic materials

Elongation: The total plastic strain before fracture; measured as a percentage of strain along the axis of the specimen

Engineered: Designed for a specific application: engineered materials are designed for use in specific applications where conventional materials such as steel, iron, and wood cannot be used

Eutectic point: The melting point of metal

Exploratory research: Creating or discovering new technologies, gaining new information about materials and processes, and applying this knowledge toward the development of a new product; research designed to come up with a functioning product that can be marketed for a profit

Extrusion: A manufacturing process used with plastics, metals, and ceramics in which the material is forced to flow through one or more die orifices to produce products of a specific cross-sectional design

F

Fatigue strength: The measure of a material's ability to resist failure during repeated loads or stresses

Fiber-strengthened composites: Composite materials with reinforcing materials whose length is greater than their width

Fire clay: A type of clay containing low amounts of alkalis, calcium, magnesium, and iron oxide It is used in the manufacture of brick, saggers, glass-melting pots, crucibles, and various types of refractory mortars and cements

First-angle orthographic projection: Type of orthographic projection in which the object is placed between the observer and the image planes

Fixture: A device fixed to a machine tool that is designed and constructed to support and locate the part while the tool's operation is being completed

Flint clays: Hard clays, almost devoid of plasticity

Following principles: Procedure and choice for selecting the correct combinations of elements of design

Forging: Forming method that uses hammer blows or machine pressure to deform metals into a desired shape

H

Hardness: The measure of a material's ability to resist penetration

Hot rolling: A steel-shaping process that can be used to produce finished mill products; often used to produce semifinished products that require further processing

I

Impact resistance: The ability of plastics to resist an impact force

Impact strength: The measure of a material's ability to withstand the shock of a sudden applied force

Injection molding: A forming process in which plastics are first melted and then forced into a cool mold cavity and allowed to harden

Inorganic materials: Substances that are typically silicon based; these include a family of materials known as ceramics

Interface (or interphase):   The boundary between the reinforcing element and the matrix of a composite

Ionization: The process whereby an atom with few electrons in its outermost, or valence, shell tends to give up those electrons to atoms that need them to fill their valence shells

Isotropy: The quality of exhibiting properties that are the same regardless of the direction they are measured in

J

Jig: A device used in machining that is designed and constructed to hold and locate the work or to guide the tool while an operation is being performed

K

Kaolin: A type of clay containing a small amount of iron particles that is used extensively in the manufacture of whitewares and china

Kerf: The groove or cut made by a saw

L

Laminar composites: Composites with two or more layers of reinforcing material The reinforcing material has two dimensions that are greater than the third

M

Machinability: The measure of how easily a material may be shaped or machined

Matrix: The polymer material that is used to bind together the reinforcing layers in a composite

Mechanical fasteners: Special devices or joint designs that allow almost all materials to be joined together quickly in permanent, semipermanent, or temporary arrangements

Mechanical properties: Properties that describe how a material behaves when static and dynamic loads (forces) are applied to it

Mers: Molecules formed from atoms of different elements; the building blocks of organic polymers

Metal: A general term used to identify a broad classification of materials that are hard, strong, and able to conduct heat and electricity

Metal matrix composites (MMCs): Composite consisting of two or more metallic constituents whose matrix phase (binder) is metallic

Metric system: Standard system of measurement used by the vast majority of the world today; a simple and very logical system of measure which is based on the number 10

Modulus of elasticity (MOE): The ratio of stress to elastic strain

Modulus of rupture (MOR): Measure of bending force required to cause fracture

Molding processes: The forming of a desired part or product in a mold

Molecules: Groups or units of different atoms that are joined together either naturally or artificially

Multiview drawing: The use of several two-dimensional views to represent a three-dimensional object

N

Normalizing: The process of relieving internal stresses often caused by work hardening

Numerical control (NC): A control system in which actions are controlled by numbers, letters, or symbols

O

Organic materials: Substances that are based on carbon chemistry

Orthographic projection (straight or right-angle projection): System of drawing composed of images perpendicular to one or more desired planes of projection

P

Paint: A liquid mixture of pigment and vehicle that adheres to a surface as a colored protective coating

Parison: An inflatable heated hollow tube used in blow molding

Particle-strengthened composites: Composites that are reinforced by spheres, flakes, and other reinforcements of approximately equal dimension on all axes

Permeable: Allowing air and moisture to pass through

Phosphoralizing: A method of providing corrosion resistance where iron, steel, or zinc provides a coating that is either corrosion resistant by itself or that can be used as a base for organic coatings

Photosynthesis: The process that converts carbon dioxide and water into carbohydrates and simultaneously releases oxygen as a by-product

Physical vapor deposition (PVD):   A process in which metal is vaporized then deposited as a coating on a base metal

Plastic behavior: The behavior exhibited by a material if the load on it is great enough to stretch the material beyond its elastic limits, causing permanent deformation

Plasticity: *See* **Plastic behavior**

Polymer: A material that is composed of many small repeating parts, or mers

Pottery clay: A type of clay that is less pure, has good plasticity, is free of course material, has a low iron content, and will vitrify at less than 1,200 degrees Celsius

Powder metallurgy (P/M):   A process used to form parts from metal powder exactly or close to final dimension and finish, with little or no machining required

Precision block: A special layout table used for a point of reference with an angle plate to locate special points for cutting, boring, or fastening procedures

Preform: Primary material that has been preshaped to facilitate molding into final form

Primary processing: The process of getting basic materials ready for use

Product analysis: The identification of the most economical and appropriate method of manufacturing a new product to meet required specifications and regulations

Product design: The process of creating goods that (1) have a pleasant appearance and (2) function effectively

Product profile: A collection of known facts about the requirements of the product based on the problem statement and the research that is used to clarify any unknowns

Proportional limit: The point beyond which stress is no longer proportional to strain The point at which elastic strain becomes plastic strain

Pure research: Research that is done purely to accumulate knowledge and facts; it has no other intended purpose

R

Reduction of area: The total plastic strain on a specimen before fracture; measured as the percentage of decrease in the specimen's cross-sectional area

Refractories: A group of oxides or oxide-aluminum compounds with very high melting points

Reinforcement: Constituent of a composite that provides the strength to the composite

Roll forming: A process used to form thin metal into various cross-sectional shapes

S

Seasoning: The process of reducing the moisture content in lumber

Secondary processing: The process of developing primary materials into goods and items used by the consumer

Semiconductors: A family of materials whose molecular structure lies between that of a ceramic and that of a polymer and exhibits some of the characteristics of each

Set: The idea of bending the teeth of a multiple-tooth blade; the offsetting of the teeth

Sewer pipe clay: A type of clay similar in content to pottery clays, having high dry strength, high density, and low warpage

Shear strength: The measure of wood's ability to resist forces that cause the fibers to tear or slip past one another

Shrinkage: The amount of contraction that takes place as metal cools and solidifies in a mold

Sintering: The process of mixing fine iron ores with powdered coal

Slip clay: A type of clay that is very impure and has a high alkali content; it is commonly used in glazing stoneware and electrical porcelain

Soldering: An adhesion process where metal surfaces are bonded by a filler metal that flows between the surfaces; uses temperatures of less than 800 degrees Fahrenheit

Specialty steels: Term used to identify a group of alloy steels that have excellent corrosion-resistance and heat-resistance properties

Specific gravity: The ratio of the weight of a given volume of a material to an equal volume of water

Spheroidizing: A form of annealing that reduces hardness and increases the machinability of a metal

Stains: Solutions with color matter suspended in them; they are used primarily to change color rather than as a protective coating

Standardization: The process of deciding which types of materials, parts, hardware, and equipment will become the accepted norms for use

Stiffness: The measure of a material's resistance in the elastic range of the stress–strain diagram

Stress: The relationship between tension and compression

Surface mining: A process of ore recovery that requires the removal of surface rock and soil to expose the raw material It is often referred to as strip or open-pit mining

T

Tempering: A heat-treating process done to reduce hardness and increase toughness

Tensile strength: The measure of a material's ability to resist a pulling force

Thermal conductivity: The measure of a material's ability to transmit heat

Thermal properties: Properties that govern a material's behavior when it is exposed to changes in temperature

Thermoplastics: A group of plastics that can be heated and shaped or formed repeatedly

Thermosets: A group of plastics that is cured by heat; once set, they cannot be remelted and reshaped

Third-angle orthographic projection: A type of orthographic projection in which the view planes are between the object and the observer

Thumbnail sketching: A technique of making doodle-type drawings to help the designer refine his or her ideas in trying to solve functional, aesthetic, material, and production problems

Traditional machining: A manufacturing process that uses mechanical force provided by a power-driven device to remove material in chip form from standard and cast forms

U

Ultimate tensile strength: The maximum load force per unit of original cross-sectional area

Underground mining: A method of ore recovery using vertical or inclined shafts where the raw material is too deep to be surface mined safely and effectively

V

Veneering: A procedure by which a sheet of more expensive wood is added to plywood to give it a better appearance

Viscoelasticity: Fluid-like time-dependent deformation of normally solid materials, usually thermoplastics

W

Working assembly drawing: A type of drawing that combines the detail drawings and the assembly drawing; it shows the relationship of the parts to the whole assembly and provides the detailed size and location dimensions in one drawing

Y

Yield point: The point at which a material begins to change shape

Young's modulus: *See* **Modulus of elasticity**

Z

Zero inventory: a material-management term used in just-in-time to refer to material flow directly from supplier to manufacturer as needed in order to eliminate material inventories at the manufacturer's facility

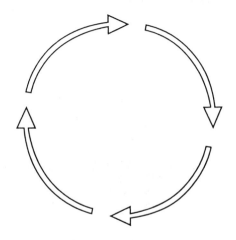

Index

Page numbers in *italic* indicate figures.